# Cellular Quiescence and Dormancy

# Cellular Quiescence and Dormancy

Guest Editor

**Jürg Bähler**

MDPI

Basel • Beijing • Wuhan • Barcelona • Belgrade • Novi Sad • Cluj • Manchester

*Guest Editor*
Jürg Bähler
Department of Genetics
University College London
London
UK

*Editorial Office*
MDPI AG
Grosspeteranlage 5
4052 Basel, Switzerland

This is a reprint of the Special Issue, published open access by the journal *Biomolecules* (ISSN 2218-273X), freely accessible at: https://www.mdpi.com/journal/biomolecules/special_issues/CWM3DE486M.

For citation purposes, cite each article independently as indicated on the article page online and as indicated below:

| Lastname, A.A.; Lastname, B.B. Article Title. *Journal Name* **Year**, *Volume Number*, Page Range. |
|---|

**ISBN 978-3-7258-7939-7 (Hbk)**
**ISBN 978-3-7258-7940-3 (PDF)**
**https://doi.org/10.3390/books978-3-7258-7940-3**

Cover image courtesy of Jürg Bähler

# Contents

# About the Editor

**Jürg Bähler**

Jürg Bähler is a Professor of Systems Biology at University College London (UCL) and the Director of UCL's Institute of Healthy Ageing, and a Group Leader at the Francis Crick Institute (London, UK). The Bähler laboratory studies cellular quiescence, ageing and gene function using fission yeast and turquoise killifish as model systems. Their research involves multi-dimensional genetic/genomic, cellular, computational and molecular approaches to examine how the regulation and function of genes affect the maintenance and traits of quiescent cells. This work addresses an important blind spot in biomedical research by dissecting the roles of previously unknown genes, with the potential to reveal new biology relevant to ageing and associated diseases.

biomolecules

MDPI

Article

# Ecological History Shapes Transcriptome Variation in Quiescent *Saccharomyces cerevisiae*

**Agnieszka Marek [1], Katarzyna Tomala [2] and Dominika Wloch-Salamon [2,*]**

1 Institute of Botany, Faculty of Biology, Jagiellonian University, 30-387 Kraków, Poland; agnieszka.marek@uj.edu.pl

2 Institute of Environmental Sciences, Faculty of Biology, Jagiellonian University, 30-387 Kraków, Poland; katarzyna.tomala@uj.edu.pl

* Correspondence: dominika.wloch-salamon@uj.edu.pl

**Abstract:** Quiescence is a pivotal state for all living organisms and cells. However, recent research indicates a lack of uniformity among quiescent cells. That is, even if the primary feature of quiescence—the ability to restart divisions—is maintained, quiescent cells within populations exhibit variation in their cellular architecture and characteristics. While it is known that the process of entry into quiescence is influenced by a combination of nutrient starvation and temporal factors, the underlying mechanisms remain to be fully elucidated. In this study, we compare the transcriptomes of known homogenous fractions of dense quiescent yeast isolated from populations of different ecological histories. These populations have undergone experimental enrichment of certain types of quiescent cells during cycles of growth and starvation for 300 generations. Transcriptome analysis revealed discrepancies in terms of characteristics associated mainly with energy turnover processes, biosynthesis, and cell wall maintenance. The results of this study suggest that quiescent cells possess the capacity to adapt their transcriptome activity in accordance with their evolutionary history.

**Keywords:** RNA-seq; quiescent cells; life history

## 1. Introduction

Quiescence is a conserved form of the G0 state across various types of eukaryotic cells. Its hallmark is the cell's ability to reenter the cell cycle after prolonged arrest. Understanding this omnipresent process of how cells enter, maintain, and exit quiescence and dormancy has significant implications not only for cancer biology and aging but also for all microorganism survival [1,2].

In single-celled organisms, quiescence could serve as an adaptive strategy enabling populations to withstand periods of starvation, regulate population heterogeneity, and engage in cooperative behaviors like resource recycling. Quiescence has fundamentally shaped microbial ecology by providing a mechanism for cells to predict and prepare for environmental changes, coordinate population responses through chemical signaling, and maintain community structure during periods of resource limitation, thereby serving as a cornerstone of microbial adaptation and survival strategies [3].

The biology of quiescent cells has important medical significance, notably in understanding human aging, stem cell function, and the pathogenesis of diseases, including cancer. In humans, the alternation between quiescence and proliferation not only governs tissue renewal, immune responses, and stem cell maintenance but also impacts tumor biology and persistent infections [2,4,5]. Quiescent cancer cells can evade anti-proliferative

 https://doi.org/10.3390/biom15111588

treatments and remain dormant in tissue niches, only to reactivate and drive metastasis—a major challenge in oncology. Moreover, quiescent microbial cells exhibit increased resistance to antibiotics and antifungals, complicating the treatment of chronic infections. Thus, elucidating the mechanisms and ecological context of quiescent cell states advances both fundamental biology and the development of more effective therapeutic treatments.

Recent research highlights that quiescence is not a static shutdown but rather a dynamic process marked by changes in cellular structure, metabolism, and phenotypic diversity, all of which contribute to evolutionary adaptation and population fitness. Quiescent cells exhibit distinct physiological and molecular characteristics, including reduced transcriptional activity, altered chromatin structure, and enhanced stress resistance, compared to growing cells [6]. The transition into and out of quiescence is tightly regulated by complex signaling networks, including the target of rapamycin (TOR), retinoblastoma (Rb), and cyclin-dependent kinase (CDK) pathways, which control cell cycle progression and metabolic adaptations in response to nutrient availability and stress signals [2,7]. Quiescence is also characterized by specific metabolic reprogramming, such as increased reliance on mitochondrial respiration and autophagy, which helps maintain cellular homeostasis during periods of inactivity [8]. Understanding these regulatory mechanisms is essential for biomedical applications, as dysregulation of quiescence has been implicated in pathological conditions, including cancer, where dormant tumor cells can evade chemotherapy and later contribute to relapse [7,9].

Although studied in various cells and organisms, budding yeast has emerged as a powerful model for studying quiescence in eukaryotic cells due to two key reasons: an experimental toolbox that makes it easy to isolate huge numbers of Q cells and study genes involved in quiescence regulation [10], and evolutionary conservation of many signaling pathways that regulate quiescence in yeast (e.g., TOR, PKA, and AMPK pathways) in higher eukaryotes, including humans [6]. Yeast cells transition into quiescence upon nutrient depletion (e.g., glucose or nitrogen starvation), and exit from the quiescence state is highly reproducible and easy to control in laboratory settings [11]. Yeast quiescent cells exhibit longevity and stress resistance, which provides insights into aging and dormancy mechanisms [12,13]. However, as in the other eukaryotes, yeast quiescence is not a uniform state—cells within a population exhibit different quiescent states, providing a model to study cellular heterogeneity and differentiation in quiescent populations. Observed heterogeneity at the level of populations and individual cells can result from the media used or the length of time in starvation [14,15].

There are currently no unified methods or protocols for studying observed quiescent heterogeneity [16]. However, the Werner–Washburne laboratory has demonstrated that, upon carbon exhaustion, a stationary-phase, haploid, planktonic yeast population can be separated into two subfractions: upper (U) and lower (L). Using a Percoll density gradient [11], these fractions exhibit distinct buoyant densities (approximate U = 1.10 g/mL and L = 1.14 g/mL), revealing clear morphological and physiological differences. The denser L fraction isolated by this method is homogeneous, mostly composed of smaller cells that show synchronous cell cycle reentry, and has high thermotolerance, enhanced viability over extended periods (>87% viability over 28 days), and a high reproductive capacity. The method allows for comprehensive transcriptomic and proteomic analyses, revealing differentially expressed mRNAs between L and the less-dense U fraction [17]. While we are aware that the fractionation method dichotomy fails to capture the whole complex of quiescent cell diversity that "displays various properties, shedding light on a plethora of individual behavior" [15], the dense fraction remains a valuable tool for comparative population-level studies [16].

We previously employed a fractionation method to elucidate the adaptive significance of phenotypic heterogeneity in starved *Saccharomyces cerevisiae* populations. To this end, we compared the survival and regrowth dynamics of monocultures composed exclusively of L or U quiescent cells versus naturally occurring mixed (U/L) quiescent populations across variable ecological conditions, including starvation duration and environmental complexity. L quiescent cells exhibit hallmark features of deep quiescence, accumulate high levels of storage carbohydrates (e.g., trehalose), display thickened cell walls and enlarged vacuoles, and maintain >80% viability after prolonged starvation (>4 weeks). U fraction cells show rapid loss of viability under the same conditions. The experimental data and mathematical modeling both confirm that L monocultures exhibit superior survival and shorter lag phases after long-term starvation, whereas U monocultures outperform L and mixed cultures following very short starvation and nutrient-recycling conditions [14].

Standard laboratory populations of S288c strains maintain rather stable proportions of U/L quiescent cells (15–20%/75–80%). However, when populations are experimentally enriched for one fraction by serially selecting only the U or L fraction over successive feast–famine cycles, the U/L balance evolves: in L selected lines progressively increase the denser fraction, while in U selected lines accumulate lighter cells, demonstrating that the U/L ratio is a heritable, evolvable trait shaped by ecological context [18]. Therefore, the evidence supports the hypothesis for U and L coexistence as a bet-hedging strategy that maximizes population fitness under fluctuating environmental pressures.

Several studies profiling the transcriptome of quiescent yeast cells as whole starved populations, as well as fractionated U and L fractions, confirm the presence of quiescent-specific expression patterns [19,20], global transcriptional shutdown, and the emergence of over 1000 noncoding RNAs specifically enriched in quiescence [21]. This finding indicates the potential existence of a distinct functional transcriptome "core program" involving quiescence. Additionally, if such a "core program" is present, it is expected to be manifested by a uniform L fraction. To date, no studies have directly compared the L quiescent fraction across populations that experienced distinct evolutionary histories during starvation conditions. This has led to the open question of whether and how the quiescent transcriptome adapts to varying environmental contexts.

In this study, we focused on a homogenous fraction of quiescent cells within lines differing by about 300 generations of varying selection-enrichment experimental procedures derived from our previous study [18]. The absence of discernible variation in the transcriptomics of these quiescent cells would support the notion of a "core quiescent" functional program maintained in the quiescence state. Nonetheless, the findings of this study suggest a more dynamic model of quiescence that exhibits a higher degree of adaptability to environmental stresses experienced.

## 2. Materials and Methods

### 2.1. Experimental Strains

This study utilized *Saccharomyces cerevisiae* strains obtained and characterized in our previous studies [18,22]. In short, the ancestral laboratory haploid strain S288c (*MATα, ura3::KanMX4*) [23] was used to establish six experimental lines. These populations underwent 300 generations of selection for enrichment for only one of two specific fractions of quiescent cells: U or L. The experimental regime consisted of alternating environmental conditions. During the *feast phase*, the cells grew on the rich nutrient plate as a colony, with free access to nutrients. The *starvation phase* began when the nutrients were gradually depleted, leading to a transition to quiescence due to nutrient deprivation. After 3.5 days, when the stationary phase was reached, populations were fractionated using a Percoll density gradient [11]. U and L fractions were separated, and one cell type was subjected to subsequent cycles of growth and starvation of Lines_U and Lines_L, respectively. From the final populations, we collected clones that differ in the proportions of U or L cells. For this study, we selected samples from this collection and grouped them based on the enriched quiescent cell type. The QU_group consisted of five clones. Clones were enriched for U cell proportions: U_3I = 50.24%, U_3III = 74.19%, U_4III = 53.01%, U_8III = 51.23%, and U_10III = 49.80%. The average proportion of U in this group of strains was UAv = 55.70% ± 9.31%. The QL_group included three clones enriched for L cells: L_2I = 91.27%, L_5I = 77.64%, and L_3II = 57.67%, with an average L cell proportion of LAv = 75.53% ± 16.90%. The average U fraction proportion from the ancestral population was 14% ± 3.3%. Whole-genome sequencing of these clones was performed in prior studies [18] and showed some SNVs acquired during the experiment (Supplementary Table S1). Genomic data were deposited in the SRA (NCBI BioProject: PRJNA383491, accession number SRP104610). Possible roles of the mutations are discussed in our previous papers [18,22].

### 2.2. Fractionation Assay

Selected clones were incubated overnight in rich medium at 30 °C with shaking. The following day, 20 µL of liquid culture (of each clone) was transferred to 20 mL of YPD medium and incubated with shaking at 30 °C for 7 days. Yeast cells were then separated into U and L fractions using Percoll centrifugation (as described previously [18]), and only the L fraction of quiescent cells was used. Post-centrifugation, quiescent L cells were collected from the bottom of the tubes and subjected to two sequential washes with 8 mL of Tris buffer. After each wash, the supernatant was carefully removed.

### 2.3. RNA Extraction and Purification from Quiescent Cells

The optical density (OD) of each sample was measured to ensure an extraction efficiency yielding approximately ≥20 ng/µL, corresponding to ~3 × $10^8$ cells. Total RNA was extracted from the isolated Q cells using the RiboPure™-Yeast Kit (Thermo Fisher Scientific, Waltham, MA, USA) following the manufacturer's protocol. Each sample had two technical repetitions (1/2) that were then stored at −70 °C.

### 2.4. Transcriptome Analysis

Library construction, sequencing, and the subsequent bioinformatic analysis were carried out by Novogene. Lists of differentially expressed genes (DEGs) were created using the DESeq2 R package (version 1.36.0) [24]. They were then used to perform Gene Ontology (GO) enrichment analysis with the GOrilla tool (accessed on February 2025) [25]. The differentially expressed gene sets (with FDR < 0.05 and positive and negative fold change) were compared with background lists of all ORFs from the RNA-seq analysis. GO terms with padj < 0.05 in the Cellular Component, Molecular Function, and Biological Process categories were combined, and the mean values from two replicates were used to create heat maps. Gene expression heatmaps were generated based on FPKM values. For visualization, expression was scaled per gene by computing row-wise z-scores (mean = 0, SD = 1). Genes were grouped into predefined functional categories GO categories. Heatmaps were generated using the pheatmap package (version 1.0.13) in R, with clustering disabled. [26]. Heat maps were scaled by rows using the average linkage clustering method and the Pearson distance measurement method. Transcriptomic data were deposited in the GEO (Gene Expression Omnibus) Series GSE309454.

## 3. Results

### 3.1. Data Quality

The data quality summary table obtained from Novogene shows a Q20 (%) score ≥97.70 for all samples and a total mapping rate ≥96.64%. A detailed data quality table for each sample is in the Supplementary Files (Supplementary Table S2). A comprehensive correlation analysis was performed among RNA-seq samples using FPKM-normalized gene expression data. The resulting heatmap and hierarchical clustering reveal that biological replicates cluster together and exhibit high pairwise correlation coefficients (≥0.97) (Figure 1), indicating high reproducibility and consistency across sequencing runs. Minor differences observed between distinct sample groups may reflect true biological variation. PCA of log2 FPKM values demonstrates robust separation between the two experimental groups (L_Lines and U_Lines), as depicted by the distinct clustering of samples on PC1 (Figure 2). Replicates within each group cluster closely, underscoring the consistency and reproducibility of gene expression profiles within experimental conditions. The substantial variance captured by PC1 and PC2 suggests that group identity is a key driver of gene expression variation, supporting a strong biological effect and validating the experiment's design. Overall, the data quality is robust, supporting downstream comparative and functional analyses.

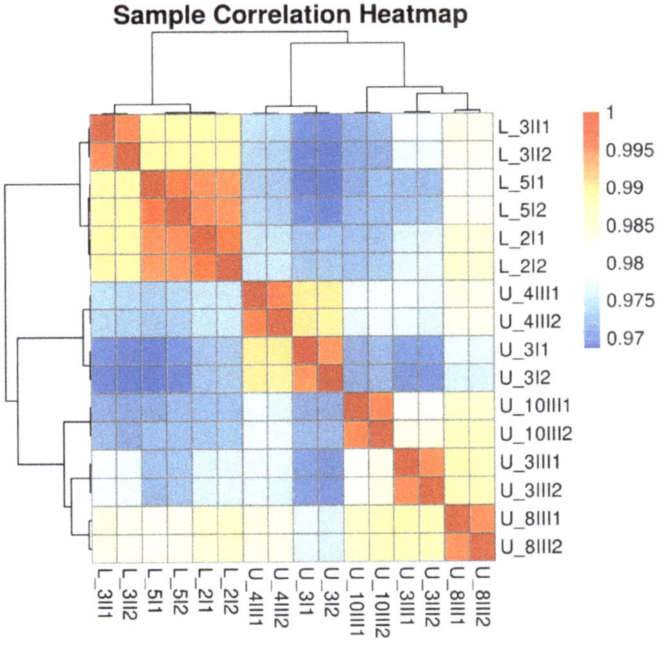


**Figure 1.** Sample correlation heatmap depicting pairwise Pearson correlation coefficients between RNA-seq samples based on FPKM-normalized gene expression profiles. Technical replicates are indicated by 1 or 2 in the sample name; biological replicates are indicated by U or L in the sample name. Each square represents the correlation of a specific pair of samples, with the color scale ranging from blue (lower correlation, 0.97) to red (higher correlation, 1.0).

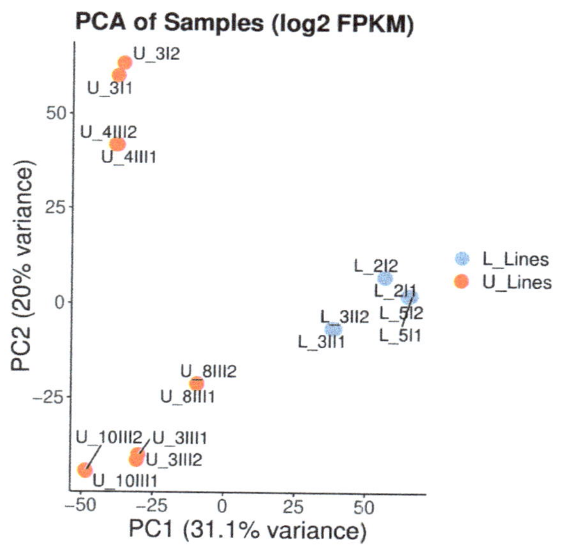


**Figure 2.** Principal component analysis (PCA) plot illustrating sample distribution based on log2-transformed FPKM gene expression profiles. Each point represents an individual sample, colored by biological replication group (blue for L_lines, orange for U_lines).

### *3.2. Differentially Expressed Genes*

Differential expression analysis of a total of 6462 genes compared shows 2837 significant genes (1476 upregulated and 1361 downregulated) with adjusted *p*-values below 0.05 (the histogram of raw *p*-value distribution of all genes is in Supplementary Figure S1). Lists of differentially expressed genes (DEGs) are provided in Supplementary Materials (Supplementary Table S3). In the comparison of expressions of U and L lines, positive FC denotes higher expression in U_lines. Among the significantly expressed genes (FDR < 0.05), 1050 show large effect sizes ($|\log_2 FC| > 0.5$), with 698 upregulated in U lines and 352 in L lines. All DEGs are listed in Supplementary File, Table S3.

Focusing specifically on genes that exceed both thresholds ($|\log_2 FC| \geq 1$ and $-\log_{10}$ *p*-value $\geq 20$, corresponding roughly to $p \leq 1 \times 10^{-20}$), there are nine genes most significantly downregulated in U_lines versus L_lines (Figure 3). Downregulation of these genes suggests a coordinated suppression of nucleotide metabolism, stress response factors, and protein modification machinery. YMR175W (SIP18) is a phospholipid-binding hydrophilin essential for osmotic stress tolerance, YKL198C (PTK1) is a putative serine/threonine kinase that regulates polyamine (spermine) uptake, YLR128W (DCN1) is a scaffold-type E3 ligase component required for cullin neddylation, YOR352W (TFB6) is a nonessential TFIIH subunit that facilitates dissociation of the Ssl2 helicase during transcription initiation, and YLR323C (CWC24) is a general splicing factor required for stable U2 snRNP binding to primary transcripts. Three ORFs (YGR201C, YLR297W, and YLR297W) encode uncharacterized proteins of unknown function; YBR242W is a 5′-deoxynucleotidase involved in deoxyribonucleoside monophosphate degradation, while YGR146C (ECL1) is a gene whose overexpression is known to extend chronological lifespan [27].

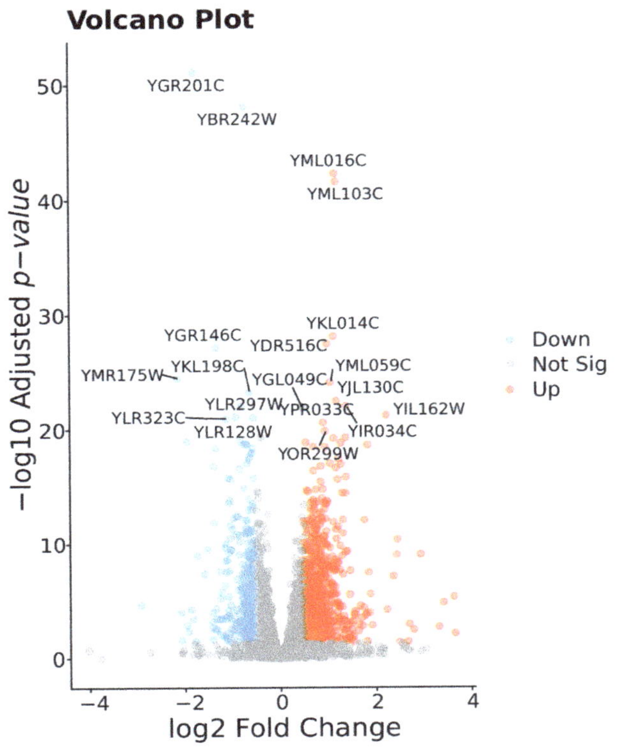


**Figure 3.** Volcano plot comparing gene expression between U_line and L_line groups of clones. Colored dots represent 1050 genes with increased expression in L (orange—698 genes) and U (blue—352 genes) groups (FDR < 0.05 and $|\log_2 FC| \geq 0.5$). Gray dots indicate a non-significant gene (YGL049C). X-axis ($\log_2$ fold--change): positive values indicate genes upregulated in U_lines versus L_lines; negative values indicate downregulation in U_lines versus L_lines. A total of 20 genes with the highest significance are annotated by name.

The eleven genes most significantly upregulated in U versus L lines of cells encompass diverse functions critical for metabolic maintenance, stress resistance, and cellular integrity. YML016C (PPZ1), a Ser/Thr phosphatase, regulates potassium homeostasis, pH balance, and cell cycle progression under stress. YKL014C (URB1) is essential for 60S ribosomal RNA processing and large-subunit biogenesis. YJL130C (URA2) and YIR034C (LYS1) sustain de novo pyrimidine and lysine biosynthesis, respectively, supporting nucleotide and amino acid pools during nutrient limitation. YML059C (*NTE1*) mediates phosphatidylcholine turnover, modulating Opi1p localization to repress phospholipid biosynthesis. YIL162W (*SUC2*) enables extracellular sucrose hydrolysis for alternative carbon uptake. YOR299W (BUD7) participates in exomer-dependent cargo sorting and bud-site selection, reinforcing cell wall assembly in the stationary phase. YPR033C (*HTS1*) encodes the histidyl-tRNA synthetase responsible for charging tRNA^His in both mitochondria and cytoplasm. YDR516C (EMI2), although sporulation-specific, is upregulated in U-enriched lines, suggesting a potential link between meiotic factors and quiescence. Finally, YML103C (NUP188) forms part of the nuclear pore complex scaffold, modulating nucleocytoplasmic transport and nuclear integrity during $G_0$. Together, these upregulated genes define a program that integrates ion and pH homeostasis, macromolecular synthesis, membrane remodeling, and stress-adaptive structures to optimize survival and frequent exit from quiescence.

### *3.3. Ontology of the RNA-Seq Upregulated in L_ or U_line Genes Show Differences Between Quiescence Cells from Populations of Different Enrichment Histories*

Sets of all genes with different expression for groups of clones (U_lines and L_lines) were compared with background lists of all ORFs from the RNA-seq analysis. Following GOrilla analysis, the groups with increased transcribed genes (of FDR less than 0.01) were grouped within several GO categories (process, component, and function) (data are provided in Supplementary Table S4. Heat_Map_RNAseq_GO). Six groups of Gene Ontology terms are upregulated in the U-line group of clones and encompass all three categories: biological process (BP), molecular function (MF), and cellular component (CC). These categories are as follows: amino acid-related (BP and MF), biosynthesis (translation, biosynthesis, RNA processing, and ribosomes) (BP, MF, and CC), cell wall and actin filaments (BP and CC), chaperones (BP and MF), energy-related processes, and nitrogen-related processes (Figure 4A).

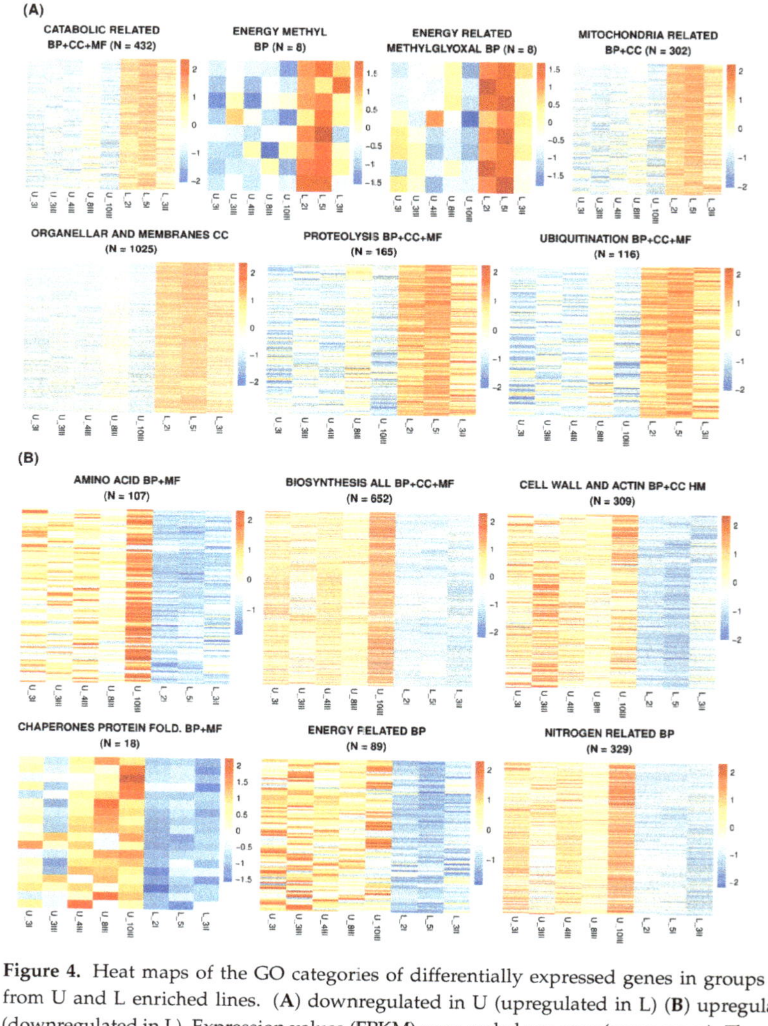


**Figure 4.** Heat maps of the GO categories of differentially expressed genes in groups of clones from U and L enriched lines. (**A**) downregulated in U (upregulated in L) (**B**) upregulated in U (downregulated in L). Expression values (FPKM) were scaled per gene (row z-score). The color scale represents z-scores. Number of genes per panel was indicated in the title.

A total of seven groups were identified as being overexpressed within the L_line group of clones. These groups are as follows: catabolic-related (BP, MF, and CC), CoA-related (BP), methylglyoxal (BP), mitochondria-related (BP and CC), organellar- and membrane-connected (CC), and proteolysis and ubiquitination (BP, MF, and CC) (Figure 4B).

## 4. Discussion

In the last decade, major progress has been made to increase understanding of yeast quiescence entry, maintenance, and exit, as described in comprehensive reviews [13,28,29]. However, there are still many important questions left: *Is quiescence a single state, or could there be many different types of quiescence? How different is quiescence when it is triggered by different signals? What is the degree of cell-to-cell variation in signaling, chromatin dynamics, and gene expression during quiescence entry, maintenance, and exit?*

Here, a comparison is made between the transcriptomes of the lower fraction (L) quiescent cells within the population that experienced different environmental stress conditions. It was determined that the transcriptome reflects the ~300 generations of evolutionary history of the population, thereby rendering it more prepared for environmental conditions to which it has been previously exposed. The downregulation of genes observed in quiescent L cells from U _lines reflects a history of elimination of all lower fraction QL cells before regrowth. This has resulted in an elevated prevalence of U types of Q cells in comparison to their ancestral counterparts, as well as L-line enrichment.

A total of seven GO groups were identified as being downregulated in U_line groups (Figure 4). A subset of the downregulated genes within these categories appears to be of possible significance. The downregulation of *DCN1*, a crucial component in the orchestration of cullin neddylation, emerges as a salient feature. A reduction in *DCN1* expression has been shown to attenuate ubiquitin–proteasome pathway activity. *DCN1* is one of the 116 genes involved in the ubiquitination process (PP, CM, and MF, as shown in Figure 4), thereby decelerating protein turnover and preserving ATP. This energy-saving strategy has also been observed in quiescent mammalian cells, where proteostasis adjustments are implicated in supporting long-term survival [30,31] which might explain why it is more expressed in the L_lines, which are prepared for longer quiescence periods [18].

Concurrent repression of *TFB6*, which encodes a previously unidentified, dissociable subunit of the yeast general transcription factor TFIIH complex, might be meaningful. The TFIIH subunit is critical for controlling the release of the Ssl2 helicase during transcription initiation [32], suggesting an active remodeling of the transcriptional machinery. By shifting TFIIH away from its holoenzyme form, cells more effectively enforce the global transcriptional repression characteristic of $G_0$ entry and maintenance. Although Tfb6 itself is nonessential under standard growth conditions, its function may hypothetically be analogous to other nonessential regulators (e.g., Xbp1 and Msa1/2) that fine-tune the chromatin and transcriptional reprogramming underlying quiescence [33].

Loss of Cwc24p following decreased *CWC24* expression might mirror the general downshift in spliceosomal components noted in quiescent yeast, reducing energy-intensive mRNA maturation processes when transcription of intron-containing genes nearly ceases. Such coordinated repression of splicing factors has been proposed to prevent the accumulation of unprocessed transcripts and to reallocate resources to stress resistance [34]. Stress response pathways are also modulated as *SIP18* and *PTK1* encode proteins essential for osmotic stress tolerance and polyamine uptake, respectively. Their downregulation suggests that quiescent cells deprioritize adaptation to external environmental fluctuations in favor of basal maintenance, reflecting a narrowed physiological focus [19,35].

In the course of quiescence research, uncharacterized genes and open reading frames (ORFs) have been frequently identified. This phenomenon was also observed in our study.

particularly with YGR201C, YLR297W, and *ECL1*, likely representing quiescence-specific regulators. It is worth noting that *ECL1* overexpression extends chronological lifespan, hinting that its downregulation in U_lines might fine-tune survival pathways or sensitize cells to reactivation cues [27].

We identified six groups of Gene Ontology terms that were upregulated in the U-line group of clones (Figure 2). In the U-line enrichment set, a few upregulated genes seem to have a role in quiescence. During quiescence, cells must maintain membrane integrity over extended periods while adapting to changing metabolic conditions. Proper lipid homeostasis is crucial for membrane stability, and the cellular adaptations required for long-term survival are conserved from bacteria to humans [36]. Thus, *NUP188*, involved in membrane lipid homeostasis upregulation, might represent a strategic reprogramming of lipid metabolism to stabilize membranes and suppress unnecessary biosynthesis and RNA processing. It may also prepare for efficient exit from $G_0$, which is expected to happen more often for U_line cells.

Another potentially interesting gene is *SUC2*, which encodes the secreted enzyme invertase, which hydrolyzes extracellular sucrose into glucose and fructose. Its transcription is tightly repressed in high-glucose conditions by the Mig1 repressor and derepressed when glucose levels fall via activation of the Snf1 kinase pathway and associated factors (e.g., Cat8 and Sip4), which also regulate gluconeogenic and stress-protective genes during entry into quiescence. Thus, SUC2 induction reflects broader remodeling of transcriptional networks that balance energy conservation, stress tolerance, and readiness for reactivation. By maintaining a low-flux steady-state of sugar uptake, invertase activity may keep key signaling pathways—such as Ras/PKA or TORC1—at threshold levels that facilitate rapid sensing of improved nutrient conditions, shortening the lag phase upon return to growth [37]. *SUC2* upregulation in U_lines might represent a strategic adaptation: it enables scavenging of alternative carbon sources, feeds storage polysaccharide synthesis, integrates with global stress and nutrient-sensing networks, and primes cells for swift reentry into the cell cycle when nutrients become available.

During quiescence, maintaining proper ion homeostasis becomes critical for long-term survival and stress tolerance. *PPZ1* encodes a serine/threonine phosphatase that regulates K+ transport and pH homeostasis. Research demonstrates that Ppz1 controls the activity of Trk K+ transporters, affecting membrane potential, intracellular pH, and cell cycle progression. In addition, overexpression of *PPZ1* causes a pronounced slow-growth defect [38]), which might be positive in U enriched lines.

While overall ribosome biogenesis decreases during quiescence, specific rRNA processing pathways may be maintained or even enhanced to support the production of specialized ribosomes adapted for the quiescent state [39]. This might explain the upregulation of *URB1*, which is involved in ribosomal RNA processing. Such regulation allows for the tailored production of ribosomes capable of handling the unique cellular environment and requirements of quiescent cells [40]. The de novo pyrimidine pathway is regulated at the enzymatic level by feedback inhibition of the initial enzyme, Ura2, via the final product, UTP.

Additionally, several pathway genes are upregulated at the transcriptional level in response to pyrimidine starvation. *URA2* encodes carbamoylphosphate synthetase for pyrimidine biosynthesis, and *LYS1* encodes saccharopine dehydrogenase for lysine biosynthesis. The upregulation of these biosynthetic enzymes during quiescence suggests that, despite overall metabolic downregulation, cells maintain the capacity to produce essential building blocks, ensuring readiness for cell cycle reentry when nutrients become available [41].

*SED1* encodes a major structural cell wall protein specifically expressed in the stationary phase. Research demonstrates that Sed1p becomes the most abundant cell wall protein in stationary-phase cells, accounting for 30% of extractable cell wall proteins. *SED1* expression increases 2.7-fold in the stationary phase and is required for lytic enzyme resistance. This protein exemplifies the specialized cell wall adaptations necessary for quiescent cell survival [6,42]. While specific mechanistic details for *TBS1*—General Stress Tolerance are limited, its identification as enhancing general stress tolerance mechanisms aligns with the comprehensive stress resistance programs activated during quiescence entry.

## 5. Conclusions

Our study confirms the assumption that quiescence in *Saccharomyces cerevisiae* is not a uniform physiological state but rather a dynamic condition shaped by present and past conditions. Our comparative transcriptomic analysis of dense (L) quiescent cells from populations with distinct enrichment histories reveals several genes that show differential expression patterns reflecting approximately 300 generations of selection pressure. The observed transcriptional differences encompass critical cellular processes, including energy metabolism, stress response, protein homeostasis, and cell wall maintenance, indicating that quiescent cells retain remarkable adaptive plasticity even within morphologically homogeneous fractions.

The contrasting gene expression profiles between L cells within U_ and L_ lines support a model of quiescence as an evolvable trait rather than a fixed cellular program. U_lines upregulated genes associated with biosynthesis, translation machinery, and rapid reactivation pathways, reflecting adaptation to environments favoring quick exit from quiescence. Conversely, L_lines showed enhanced expression of catabolic processes, mitochondrial functions, and proteolysis machinery, consistent with optimization for prolonged survival during extended starvation. These findings challenge the notion of a universal "core quiescent program" and instead reveal that quiescent cells can fine-tune their molecular machinery based on ecological context and selection history. Understanding this transcriptional plasticity has important implications for microbial ecology, biotechnology applications, and our broader comprehension of cellular dormancy mechanisms across diverse biological systems.

**Supplementary Materials:** The following supporting information can be downloaded at https://www.mdpi.com/article/10.3390/biom15111588/s1. Table S1. Single-nucleotide variation present in the clones used in this study. Table S2. Data quality summary table obtained from Novogene. Figure S1. Histogram of the raw *p*-value distribution of all genes. Table S3. Lists of differentially expressed genes (DEGs). Table S4. Heat_Map_RNAseq_GO.

**Author Contributions:** Conceptualization, D.W.-S.; methodology, A.M., K.T. and D.W.-S.; formal analysis, K.T. and D.W.-S.; resources, D.W.-S.; data curation, D.W.-S. and K.T.; writing—original draft preparation, D.W.-S.; writing—review and editing, A.M., K.T. and D.W.-S.; visualization, supervision, D.W.-S.; funding acquisition, D.W.-S. All authors have read and agreed to the published version of the manuscript.

**Funding:** The work was funded by the National Science Center, Poland, via OPUS grants to D.W.-S. (grant numbers: 2017/25/B/NZ8/01035 and 2024/53/B/NZ8/02259), and OPUS grant to K.T. (grant number: 2022/47/B/NZ8/00537).

**Institutional Review Board Statement:** Not applicable.

**Informed Consent Statement:** Not applicable.

**Data Availability Statement:** Transcriptomic data were deposited in GEO (Gene Expression Omnibus) Series GSE309454.

**Acknowledgments:** We would like to express our gratitude to Hanna Tutaj. This paper would not have been possible without her knowledge and assistance. During the preparation of this manuscript, the authors used DeepL and Perplexity for the purposes of language editing. The authors have reviewed and edited the output and take full responsibility for the content of this publication.

**Conflicts of Interest:** The authors declare no conflicts of interest.

## References

1. Daignan-Fornier, B.; Laporte, D.; Sagot, I. Quiescence through the prism of evolution. *Front. Cell Dev. Biol.* **2021**, *9*, 745069. [CrossRef]
2. Valcourt, J.R.; Lemons, J.M.; Haley, E.M.; Kojima, M.; Demuren, O.O.; Coller, H.A. Staying alive: Metabolic adaptations to quiescence. *Cell Cycle* **2012**, *11*, 1680–1696. [CrossRef] [PubMed]
3. Bradley, J.A. Microbial dormancy as an ecological and biogeochemical regulator on Earth. *Nat. Commun.* **2025**, *16*, 3909. [CrossRef]
4. Cheung, T.H.; Rando, T.A. Molecular regulation of stem cell quiescence. *Nat. Rev. Mol. Cell Biol.* **2013**, *14*, 329–340. [CrossRef]
5. Coller, H.A.; Sang, L.; Roberts, J.M. A new description of cellular quiescence. *PLoS Biol.* **2006**, *4*, e83. [CrossRef] [PubMed]
6. De Virgilio, C. The essence of yeast quiescence. *FEMS Microbiol. Rev.* **2012**, *36*, 306–339. [CrossRef] [PubMed]
7. Mitra, M.; Batista, S.L.; Coller, H.A. Transcription factor networks in cellular quiescence. *Nat. Cell Biol.* **2025**, *27*, 14–27. [CrossRef] [PubMed]
8. Laporte, D.; Massoni-Laporte, A.; LEfranc, C.; Dompierre, J.; Mauboules, D.; Nsamba, E.T.; Royou, A.; Gal, L.; Schuldiner, M.; Gupta, M.L., Jr. A stable microtubule bundle formed through an orchestrated multistep process controls quiescence exit. *eLife* **2024**, *12*, RP89958. [CrossRef]
9. Balayan, V.; Guddati, A.K. Tumor dormancy: Biologic and therapeutic implications. *World J. Oncol.* **2022**, *13*, 8. [CrossRef]
10. Gray, J.V.; Petsko, G.A.; Johnston, G.C.; Ringe, D.; Singer, R.A.; Werner-Washburne, M. "Sleeping beauty": Quiescence in *Saccharomyces cerevisiae*. *Microbiol. Mol. Biol. Rev.* **2004**, *68*, 187–206. [CrossRef]
11. Allen, C.; Buttner, S.; Aragon, A.D.; Thomas, J.A.; Meirelles, O.; Jaetao, J.E.; Benn, D.; Ruby, S.W.; Veenhuis, M.; Madeo, F.; et al. Isolation of quiescent and nonquiescent cells from yeast stationary-phase cultures. *J. Cell Biol.* **2006**, *174*, 89–100. [CrossRef]
12. Sun, S.; Gresham, D. Cellular quiescence in budding yeast. *Yeast* **2021**, *38*, 12–29. [CrossRef]
13. Breeden, L.L.; Tsukiyama, T. Quiescence in *Saccharomyces cerevisiae*. *Annu. Rev. Genet.* **2022**, *56*, 253–278. [CrossRef]
14. Opalek, M.; Smug, B.; Doebeli, M.; Wloch-Salamon, D. On the Ecological Significance of Phenotypic Heterogeneity in Microbial Populations Undergoing Starvation. *Microbiol. Spectr.* **2022**, *10*, e00450-21. [CrossRef] [PubMed]
15. Sagot, I.; Laporte, D. The cell biology of quiescent yeast–a diversity of individual scenarios. *J. Cell Sci.* **2019**, *132*, jcs213025. [CrossRef] [PubMed]
16. Opalek, M.; Tutaj, H.; Pirog, A.; Smug, B.J.; Rutkowska, J.; Wloch-Salamon, D. A Systematic Review on Quiescent State Research Approaches in *S. cerevisiae*. *Cells* **2023**, *12*, 1608. [CrossRef] [PubMed]
17. Aragon, A.D.; Rodriguez, A.L.; Meirelles, O.; Roy, S.; Davidson, G.S.; Tapia, P.H.; Allen, C.; Joe, R.; Benn, D.; Werner-Washburne, M. Characterization of differentiated quiescent and nonquiescent cells in yeast stationary-phase cultures. *Mol. Biol. Cell* **2008**, *19*, 1271–1280. [CrossRef]
18. Wloch-Salamon, D.M.; Tomala, K.; Aggeli, D.; Dunn, B. Adaptive roles of SSY1 and SIR3 during cycles of growth and starvation in *Saccharomyces cerevisiae* populations enriched for quiescent or nonquiescent cells. *G3 Genes Genomes Genet.* **2017**, *7*, 1899–1911. [CrossRef]
19. Klosinska, M.M.; Crutchfield, C.A.; Bradley, P.H.; Rabinowitz, J.D.; Broach, J.R. Yeast cells can access distinct quiescent states. *Genes Dev.* **2011**, *25*, 336–349. [CrossRef]
20. Cap, M.; Stepanek, L.; Harant, K.; Vachova, L.; Palkova, Z. Cell differentiation within a yeast colony: Metabolic and regulatory parallels with a tumor-affected organism. *Mol. Cell* **2012**, *46*, 436–448. [CrossRef]
21. Greenlaw, A.C.; Alavattam, K.G.; Tsukiyama, T. Post-transcriptional regulation shapes the transcriptome of quiescent budding yeast. *Nucleic Acids Res.* **2024**, *52*, 1043–1063. [CrossRef]
22. Marek, A.; Opalek, M.; Kałdon, A.; Mickowska, B.; Wloch-Salamon, D. Hypersensitive SSY1 mutations negatively influence transition to quiescence in yeast *Saccharomyces cerevisiae*. *Yeast* **2021**, *38*, 102–116. [CrossRef]
23. Cubillos, F.A.; Louis, E.J.; Liti, G. Generation of a large set of genetically tractable haploid and diploid Saccharomyces strains. *Fems Yeast Res.* **2009**, *9*, 1217–1225. [CrossRef] [PubMed]
24. Anders, S.; Huber, W. Differential expression analysis for sequence count data. *Genome Biol.* **2010**, *11*, R106. [CrossRef] [PubMed]
25. Eden, E.; Navon, R.; Steinfeld, I.; Lipson, D.; Yakhini, Z. GOrilla: A tool for discovery and visualization of enriched GO terms in ranked gene lists. *BMC Bioinform.* **2009**, *10*, 48. [CrossRef] [PubMed]
26. Babicki, S.; Arndt, D.; Marcu, A.; Liang, Y.; Grant, J.R.; Maciejewski, A.; Wishart, D.S. Heatmapper: Web-enabled heat mapping for all. *Nucleic Acids Res.* **2016**, *44*, W147–W153. [CrossRef]
27. Azuma, K.; Ohtsuka, H.; Murakami, H.; Aiba, H. Extension of chronological lifespan by ScEcl1 depends on mitochondria in *Saccharomyces cerevisiae*. *Biosci. Biotechnol. Biochem.* **2012**, *76*, 1938–1942. [CrossRef]

28. Caligaris, M.; Nicastro, R.; Hu, Z.; Tripodi, F.; Hummel, J.E.; Pillet, B.; Deprez, M.-A.; Winderickx, J.; Rospert, S.; Coccetti, P.; et al. Snf1/AMPK fine-tunes TORC1 signaling in response to glucose starvation. *eLife* **2023**, *12*, e84319. [CrossRef]
29. Smets, B.; Ghillebert, R.; De Snijder, P.; Binda, M.; Swinnen, E.; De Virgilio, C.; Winderickx, J. Life in the midst of scarcity: Adaptations to nutrient availability in *Saccharomyces cerevisiae*. *Curr. Genet.* **2010**, *56*, 1–32. [CrossRef]
30. Kurz, T.; Ozlü, N.; Rudolf, F.; O'Rourke, S.M.; Luke, B.; Hofmann, K.; Hyman, A.A.; Bowerman, B.; Peter, M. The conserved protein DCN-1/Dcn1p is required for cullin neddylation in *C. elegans* and *S. cerevisiae*. *Nature* **2005**, *435*, 1257–1261. [CrossRef]
31. Zhang, S.; Yu, Q.; Li, Z.; Zhao, Y.; Sun, Y. Protein neddylation and its role in health and diseases. *Signal Transduct. Target. Ther.* **2024**, *9*, 85. [CrossRef]
32. Murakami, K.; Gibbons, B.J.; Davis, R.E.; Nagai, S.; Liu, X.; Robinson, P.J.; Wu, T.; Kaplan, C.D.; Kornberg, R.D. Tfb6, a previously unidentified subunit of the general transcription factor TFIIH, facilitates dissociation of Ssl2 helicase after transcription initiation. *Proc. Natl. Acad. Sci. USA* **2012**, *109*, 4816–4821. [CrossRef] [PubMed]
33. Li, L.; Miles, S.; Melville, Z.; Prasad, A.; Bradley, G.; Breeden, L.L. Key events during the transition from rapid growth to quiescence in budding yeast require posttranscriptional regulators. *Mol. Biol. Cell* **2013**, *24*, 3697–3709. [CrossRef] [PubMed]
34. Coltri, P.P.; Oliveira, C.C. Cwc24p is a general *Saccharomyces cerevisiae* splicing factor required for the stable U2 snRNP binding to primary transcripts. *PLoS ONE* **2012**, *7*, e45678. [CrossRef]
35. Shen, W.; Gao, Z.; Chen, K.; Zhao, A.; Ouyang, Q.; Luo, C. The regulatory mechanism of the yeast osmoresponse under different glucose concentrations. *iScience* **2023**, *26*, 105809. [CrossRef]
36. de Mendoza, D.; Pilon, M. Control of membrane lipid homeostasis by lipid-bilayer associated sensors: A mechanism conserved from bacteria to humans. *Prog. Lipid. Res.* **2019**, *76*, 100996. [CrossRef]
37. Persson, S.; Welkenhuysen, N.; Shashkova, S.; Cvijovic, M. Fine-Tuning of Energy Levels Regulates SUC2 via a SNF1-Dependent Feedback Loop. *Front. Physiol.* **2020**, *11*, 954. [CrossRef]
38. Yenush, L.; Mulet, J.M.; Ariño, J.; Serrano, R. The Ppz protein phosphatases are key regulators of $K^+$ and pH homeostasis: Implications for salt tolerance, cell wall integrity and cell cycle progression. *Embo J.* **2002**, *21*, 920–929. [CrossRef]
39. Yang, G.; Schmid-Siegel, M.; Heissenberger, C.; Kos-Braun, I.C.; Prechtl, M.; Meca-Laguna, G.; Rocha, M.; Wagner-Schrittwieser, A.; Pils, V.; Meixner, B.; et al. 2′-O-ribose methylation levels of ribosomal RNA distinguish different types of growth arrest in human dermal fibroblasts. *J. Cell Sci.* **2024**, *137*, jcs261930. [CrossRef] [PubMed]
40. Xue, S.; Barna, M. Specialized ribosomes: A new frontier in gene regulation and organismal biology. *Nat. Rev. Mol. Cell Biol.* **2012**, *13*, 355–369. [CrossRef]
41. Ljungdahl, P.O.; Daignan-Fornier, B. Regulation of amino acid, nucleotide, and phosphate metabolism in *Saccharomyces cerevisiae*. *Genetics* **2012**, *190*, 885–929. [CrossRef] [PubMed]
42. Shimoi, H.; Kitagaki, H.; Ohmori, H.; Iimura, Y.; Ito, K. Sed1p is a major cell wall protein of *Saccharomyces cerevisiae* in the stationary phase and is involved in lytic enzyme resistance. *J. Bacteriol.* **1998**, *180*, 3381–3387. [CrossRef] [PubMed]

biomolecules

MDPI

*Article*

# Mitochondrial Translation Inhibition Triggers an Rst2-Controlled Transcriptional Reprogramming of Carbon Metabolism in Stationary-Phase Cells of Fission Yeast

**Ying Luo [1,2], Shaimaa Hassan [1], Saniya Raut [1] and Jürg Bähler [1,*]**

1 Institute of Healthy Ageing, Department of Genetics, Evolution & Environment, University College London, London WC1E 6BT, UK; ying.luo@ucl.ac.uk (Y.L.); shaimaa.hassan.18@ucl.ac.uk (S.H.); saniya.raut.23@ucl.ac.uk (S.R.)
2 Jiangsu Key Laboratory for Microbes and Genomics, School of Life Sciences, Nanjing Normal University, Nanjing 210023, China
* Correspondence: j.bahler@ucl.ac.uk

**Abstract:** Mitochondria possess their own genome, which encodes subunits of the electron transport chain, rendering mitochondrial protein translation essential for cellular energy metabolism. Mitochondrial dysfunction affects nuclear transcription through the retrograde response. We applied RNA-seq to investigate whether and how the inhibition of mitochondrial translation by chloramphenicol (CAP) affects transcriptome regulation in proliferating or stationary-phase cells of *Schizosaccharomyces pombe* growing in fermentative or respiratory media. Stationary-phase cells in glucose medium exhibited the strongest transcriptome response to CAP, characterized by expression signatures similar to those observed under other stresses, including the retrograde response. The induced genes were also significantly enriched in cytoplasmic carbon metabolism pathways, reflecting a transcriptional reprogramming from respiration to fermentation. The transcription factors Scr1 and Rst2, regulators of carbon catabolite repression (CCR), controlled a common set of carbon metabolism genes in CAP-treated stationary-phase cells, and they showed opposing effects on the lifespan of these cells. Rst2 was required for the induction of carbon metabolism genes and maintained nuclear localization in CAP-treated stationary-phase cells. A systematic genetic interaction screen revealed functional relationships of Rst2 with processes related to stress and starvation responses. These findings uncover a complex transcriptional program in stationary-phase cells that adapt to inhibited mitochondrial translation, including stress- and retrograde-like responses, contributions of the CCR factors Scr1 and Rst2, and adjustment of carbon metabolism to deal with mitochondrial dysfunction.

**Keywords:** *S. pombe*; retrograde response; carbon catabolite repression; Scr1; transcription factor; transcriptome; mitochondrial metabolism; stress response; RNA-seq; genetic interactions

## 1. Introduction

Mitochondria are double-membrane-bound organelles in most eukaryotic cells that are primarily responsible for energy production through the oxidative phosphorylation (OXPHOS) system. In addition to their role in cellular bioenergetics, mitochondria are involved in a wide range of important biological processes, including the regulation of intermediary metabolism, calcium homeostasis, ferroptosis, apoptosis, and aging [1–3]. Mitochondria possess their own genome (mtDNA), which encodes several essential


https://doi.org/10.3390/biom15101354

proteins required for the assembly and function of the electron transport chain (ETC). In the fission yeast *Schizosaccharomyces pombe* (*S. pombe*), the mtDNA is ~19 kb in size and encodes seven key subunits of the ETC (*cytb* of complex III; *cox1*, *cox2*, and *cox3* of complex IV; *atp6*, *atp8*, and *atp9* of ATP synthase), two rRNAs (*rns* and *rnl*), one small subunit of ribosomal protein (*rps3*), and a full set of tRNAs [4].

The expression of mtDNA-encoded proteins relies on a dedicated mitochondrial translation machinery, which is more closely related to the bacterial counterpart than to the cytosolic one, reflecting the endosymbiotic origin of mitochondria [5,6]. Unlike bacterial translation initiation, which requires three essential initiation factors (IF1, IF2, and IF3), mitochondrial translation initiation typically involves only two initiation factors, mtIF2 and mtIF3 [7,8]. Mammalian mtIF2 proteins contain a unique 37-amino-acid insertion domain that exhibits functional similarities to bacterial IF1 and is required for efficient mitochondrial translation in vitro [9]. *S. pombe* contains two mitochondrial translation initiation factors, Mti2 (mtIF2) and Mti3 (mtIF3), with Mti2 playing a more critical role. Deletion of *mti2* disrupts mitochondrial protein synthesis and impairs mitoribosome assembly [10]. A recent study has further demonstrated that the insertion domain of Mti2 is critical for its proper folding and for maintaining mitochondrial translation efficiency in *S. pombe* [11]. Another distinct feature of mitochondrial translation is the requirement for translational activators, which function in conjunction with the initiation factors. In the budding yeast *Saccharomyces cerevisiae*, activators such as Sov1, Cbp1, Pet309, and Mss51 regulate the translation of specific mitochondrial mRNAs [6]. Similarly, Ppr10 and Mpa1 are key activators required for efficient translation in *S. pombe* [12,13]. Due to the essential roles of mitochondria in cellular homeostasis, mitochondrial dysfunction is associated with various human diseases, including neurodegenerative disorders such as Parkinson's and Alzheimer's [1,14].

Mitochondrial dysfunction can impact nuclear transcription through retrograde (mitonuclear) signaling pathways, enabling mitochondria to communicate with the nucleus and trigger adaptive gene expression programs, a process known as the retrograde response [15–17]. The retrograde response enables cells to maintain cellular homeostasis by compensating for mitochondrial problems, such as impaired oxidative phosphorylation or loss of mitochondrial membrane potential. The retrograde response is conserved across yeasts, mammals, and other eukaryotes [18–20]. For example, in *S. cerevisiae*, chemical inhibition of ETC complexes induces a retrograde response [21], and in *S. pombe*, mutations in respiratory genes lead to similar coordinated changes in nuclear transcription [16,22].

Cells can also adapt their carbon metabolism in response to available sugars in the environment using nutrient-sensing regulatory networks. Glucose is the preferred carbon source for most cells, and many microorganisms have evolved regulatory mechanisms to suppress the utilization of alternative carbon sources when glucose is available. This process is referred to as carbon catabolite repression (CCR), enabling cells to respond to changing environmental carbon sources by adjusting the expression of genes involved in sugar transport, carbon metabolism, and cell growth [23–25]. In *S. pombe*, the CCR is mainly mediated by two conserved $C_2H_2$ zinc finger transcription factors, Scr1 and Rst2 [26–28]. Scr1 functions primarily as a transcriptional repressor under glucose-sufficient conditions, and its activity requires the corepressors Tup11 and Tup12 to establish full repression. Scr1 represses the expression of *fbp1* in glucose-sufficient conditions. Upon glucose depletion, Scr1 is hyper-phosphorylated by the AMP-activated protein kinase (AMPK) Ssp2, leading to its export from the nucleus. This enables the transcriptional activator Rst2 to bind to the promoter of *fbp1*, thereby activating its expression [26–28]. ChIP-seq analysis of Scr1, Rst2, Tup11, and Tup12 suggests a model by which the transcriptional network governing carbon metabolism involves a dynamic balance between Scr1-mediated repression and Rst2-mediated activation at shared target genes [25].

Here, we investigate whether and how the inhibition of mitochondrial translation affects the nuclear transcriptome of *S. pombe* in different carbon sources and growth phases. Proliferating *S. pombe* cells will grow mainly by fermentation in standard glucose medium, and once glucose becomes limiting, they will enter a non-dividing state (stationary phase) when mitochondrial respiration increases. Stationary phase and other non-dividing cell states have been much less studied than cell proliferation. We find that only during the stationary phase in glucose medium does inhibition of mitochondrial translation trigger an extensive transcriptional reprogramming of carbon metabolism, along with a general stress-like transcriptional response. We show that Scr1 and Rst2 contribute to this adaptive reprogramming process by regulating the expression of a shared set of genes, many of which are involved in carbon metabolism. In stationary-phase cells, Rst2 functions as a primary transcriptional activator and remains in the nucleus only when mitochondrial translation is inhibited. The retrograde response and CCR have been studied in isolation in *S. pombe*. Our findings reveal new aspects of the transcriptional response to mitochondrial defects: upon inhibition of mitochondrial translation, stationary-phase cells launch a gene expression program that combines a retrograde-like response with a response controlled by CCR regulators.

## 2. Materials and Methods

### 2.1. Fission Yeast Strains and Growth Media

*S. pombe* strains used in this study are listed in Supplementary Table S1. The deletion strains Δ*rst2* and Δ*scr1* were generated by homologous recombination using pFA6a-kanMX6 [29] in the wild-type strain *972 h*$^-$ background. The *S. pombe* Bioneer haploid deletion library v5.0, containing 3420 deletion mutants, was obtained from the Bähler lab stock [30,31]. The query strains (*h*$^+$ *rst2*:: *natMX6*) for SGA with a nourseothricin cassette were constructed in a *975 h*$^+$ background by overlapping PCR [32]. As a control query strain for SGA, we used a wild-type strain (*h*$^+$ *natMX6@(NC_003421.2:1959559_1959560ins)*), with genotype named according to [33], in which a nourseothricin cassette was inserted, hereafter referred to as Nat$^R$ wild type. This insertion does not affect the fitness of the wild-type strain (Supplementary Figure S5). All deletion strains were verified by colony PCR using check primers, with all primers used in this study listed in Supplementary Table S2. Strains expressing Rst2 C-terminal-tagged GFP from its endogenous promoter were generated by overlapping PCR, as described [32]. The GFP-tagged strain does not affect cell growth and mitochondrial function (Supplementary Figure S6).

*S. pombe* cells were cultured in rich yeast extract and supplements medium (YES: 0.5% yeast extract, 225 mg/L adenine, histidine, leucine, uracil, and lysine hydrochloride) supplemented with 3% glucose for fermentative growth or 3% glycerol and 0.1% glucose for respiratory growth. Malt extract agar (MEA) media was obtained from Formedium (PCM 0810).

### 2.2. RNA-Sequencing Experiments

The schematic overview of samples for total RNA extraction following CAP treatment is shown in Figure 1a. Briefly, cells were precultured overnight in YES medium containing 3% glucose. The cultures were subsequently harvested, washed with dd$H_2O$, and diluted with fresh YES containing 3% glucose (glucose medium, fermentative condition) or 3% glycerol (glycerol medium, respiratory condition) and 0.1% glucose to an initial $OD_{600}$ of 0.2 (Time 0). Chloramphenicol (CAP) was then added to a final concentration of 2 mg/mL, where indicated. To control for any solvent effects, 2% ethanol (EtOH) was added to untreated samples, as CAP was dissolved in EtOH. Cell growth was monitored by measuring $OD_{600}$ every 2 h up to 18 h and every 6 h thereafter (Supplementary Figure S1b). For the preparation of the Δ*rst2* and wild-type (wt) samples, cells were precultured overnight in YES medium, diluted to an initial $OD_{600}$ of 0.2 (Time 0), and then harvested for RNA extraction at 6 h (exponential phase) and 30 h (stationary phase). Three independent biological repeats were prepared for each condition.

Total RNA was extracted using the PureLink RNA Mini Kit (Thermo Fisher Scientific, Waltham, MA, USA) following the manufacturer's instructions. Cells were homogenized with 0.5 mm acid-washed beads (Sigma-Aldrich, St. Louis, MO, USA) using a FastPrep instrument (MP Biomedicals, Santa Ana, CA, USA). RNA was purified with RNA Mini Columns (Qiagen, Hilden, Germany) and digested with DNase I for 15 min at room temperature to remove genomic contamination. RNA quality was assessed using the High-Sensitivity D5000 ScreenTape Assay on the TapeStation System (Agilent Technologies, Santa Clara, CA, USA). Strand-specific RNA-seq libraries were prepared using poly(A) mRNA enrichment from total RNA samples and sequenced on an Illumina platform with 150 bp paired-end reads in two lanes. Raw FASTQ files were provided for downstream analysis. Sequencing quality was assessed by FastQC (version 0.12.1, https://www.bioinformatics.babraham.ac.uk/projects/fastqc/, accessed on 29 July 2024). Transcript abundance was estimated by quasi-mapping to *S. pombe* reference genome sequences and quantified with Salmon (https://combine-lab.github.io/salmon/, accessed on 29 July 2024) [34]. Differential expression analysis was performed by DESeq2 (version 1.44.0). A linear model was constructed incorporating different carbon sources (glucose vs. glycerol), mitochondrial translation status (untreated vs. CAP treated), and growth phases (exponential vs. stationary) as independent factors. The designed model formula used for DESeq2 was specified as "~ carbon source * mitochondrial translation status * growth phase". The *S. pombe* genome sequence and gene annotation files were obtained from PomBase [35]. Gene ontology (GO) terms and KEGG pathway enrichment analysis were performed by DAVID [36,37] with default parameters and visualized by R (version 4.4.1) [38]. Overlap analysis used in this study was performed using two approaches. Pairwise overlaps between two gene sets were assessed by the GeneOverlap R package (version 1.40.0) [39]. For overlaps among three gene sets, we used the SuperExactTest R package (version 1.1.0) [40].

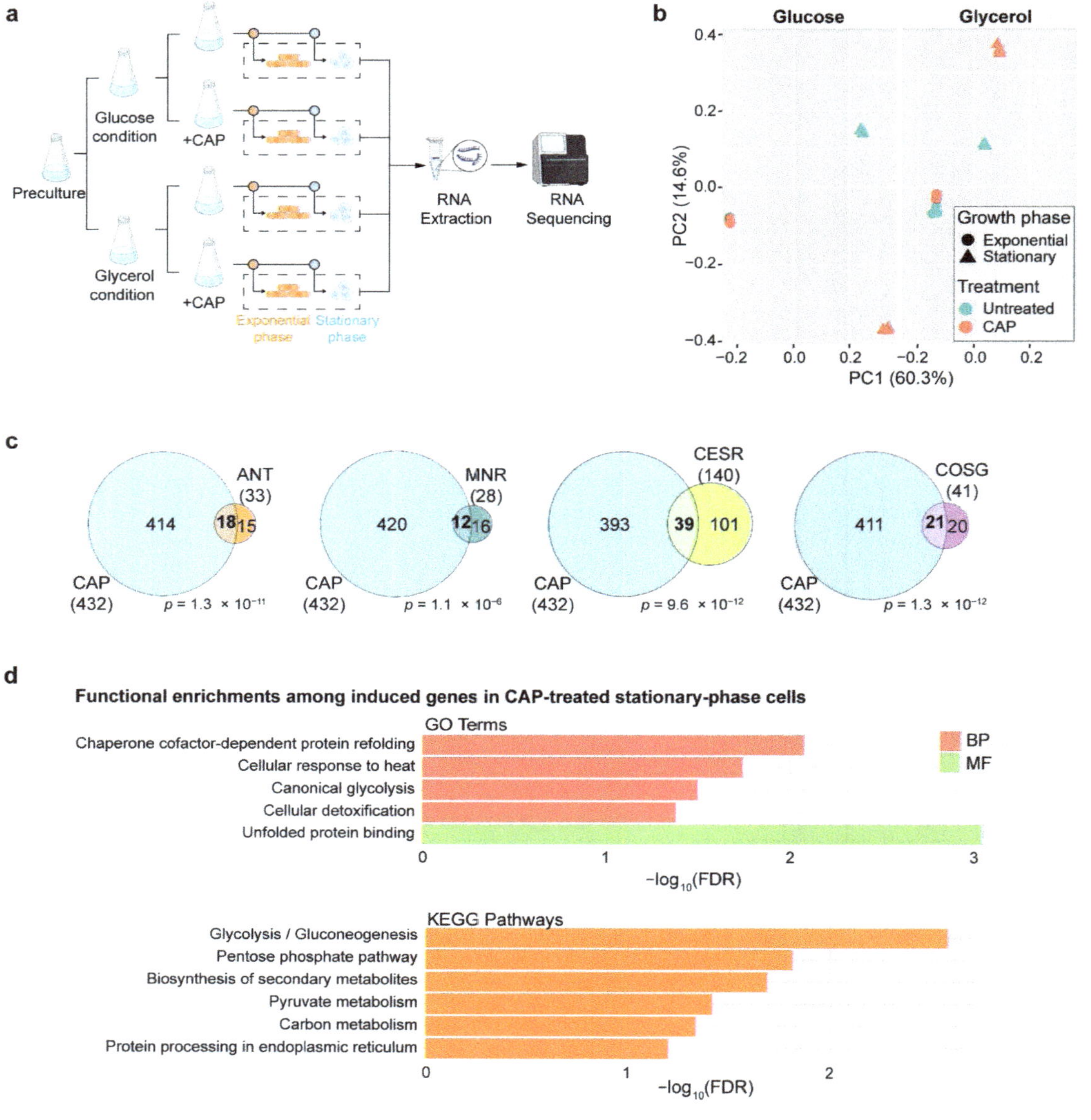


**Figure 1.** Mitochondrial translation inhibition primarily affects the transcriptome during the stationary phase. (**a**) Experimental design for RNA-seq samples used in this study. For experimental details, see the Materials and Methods Section. Three independent biological replicates were prepared for each condition. (**b**) Principal component analysis (PCA) based on rlog-transformed normalized expression values of transcriptome responses to CAP treatment under different carbon sources and growth phases. Each dot represents one biological replicate. Dot shapes indicate the growth phases (circle: exponential; triangle: stationary), and dot colors indicate CAP treatment (turquoise: untreated; pink: treated). The separate panels for glucose and glycerol conditions are derived from

the same PCA using all samples and thus share the same coordinate axes. PC1 and PC2 account for 60.3% and 14.6% of the total variance, respectively. (**c**) Overlap analysis of CAP-induced genes with gene sets induced in response to other stress conditions. Statistical significance of the overlaps was calculated using Fisher's exact test with the GeneOverlap R package (version 1.40.0) [39], with the resulting *p*-values of the overlaps indicated. The background gene set used for the analysis corresponds to all genes detected in the RNA-seq dataset in this study. (**d**) GO term and KEGG pathway enrichment analysis of 432 induced genes following CAP treatment during the stationary phase in glucose medium. GO terms are grouped by category: BP (biological process) and MF (molecular function), and were selected for non-redundancy, specificity, and significance.

### 2.3. *Quantitative Cell Growth Assays in Liquid Culture*

Quantitative growth assays were performed using 48-well FlowerPlates in a BioLector microbioreactor (m2p-labs GmbH, Baesweiler, Germany), as previously described [31]. Briefly, wild-type, Δ*rst2*, and Δ*scr1* strains were precultured overnight in YES medium and subsequently diluted to an initial $OD_{600}$ of 0.2 in fresh YES. After ~4 h of cultivation to mid-exponential phase, the cultures were further diluted with 1.5 mL of fresh YES media supplemented with either 3% glucose or 3% glycerol and 0.1% glucose to an initial $OD_{600}$ of 0.02 (Time 0). Mitochondrial translation was inhibited with the treatment of CAP at a final concentration of 2 mg/mL. Cultures were incubated in three replicates at 32 °C with shaking at 1000 rpm and 85% humidity. Growth was monitored in real time, and growth data were normalized to the initial time point (Time 0). The mean growth curves were fitted, and growth rates and lag period were calculated by grofit (version 1.1.1.1) [41]. All pairwise comparisons were analyzed using one-way ANOVA, followed by Tukey's honest significant difference test. Data visualization was performed in R (version 4.4.1) [38].

### 2.4. *Spot Assays*

The wild-type, Δ*rst2*, and Δ*scr1* strains were precultured overnight in YES medium at 32 °C, and cultures were subsequently diluted to an initial $OD_{600}$ of 0.2. CAP (dissolved in EtOH) was added to a final concentration of 2 mg/mL to inhibit mitochondrial translation, and 2% EtOH was added as a control for solvent effects. A total of 3 $OD_{600}$ of cells were collected at 6, 12, 24, 36, and 48 h, and 10-fold serial dilutions were prepared and spotted onto YES plates. Plates were incubated at 32 °C for 3 days before being photographed.

The spot assays for wild-type strains in the SGA analysis were performed as follows. The wild-type, Δ*ade6* (see Section 2.6), and $Nat^R$ wild-type strains were precultured overnight in YES medium at 32 °C. Cultures were diluted to an initial $OD_{600}$ of 0.1 and grown for 4 h to ensure collection during the logarithmic growth phase. The $OD_{600}$ was then adjusted to 0.2, and serial dilutions were prepared and spotted onto YES plates, either without additives (control) or supplemented with the indicated stress agents: oxidative stress (0.075% MMS), heat stress (38 °C), low glucose (0.1%), non-fermentable carbon source (3% glycerol + 0.05% glucose), ER stress (75 fg/mL tunicamycin), TOR inhibition (5 μM Torin), and rapamycin + caffeine (100 μg/mL + 10 mM). Plates were incubated at 32 °C for 3 days before scanning (Supplementary Figure S5).

### 2.5. *Chronological Lifespan (CLS) Assays*

CLS assays assess the viability of non-dividing cells during aging by measuring colony-forming units (CFUs) at different time points. CLS assays were performed as previously described [42]. Briefly, cells were precultured overnight in YES medium at 32 °C, diluted to an initial $OD_{600}$ of 0.02, and cultured to stationary phase (Day 0). CFUs were measured daily and normalized to the CFUs at Day 0 (100% cell viability). At each time point, aliquots of aging cultures were collected, 3-fold serially diluted, and spotted onto YES plates using a RoToR HDA pinning robot (Singer Instruments, Somerset, UK). Three independent biological replicates were prepared for each condition. Plates were incubated at 32 °C for 3 days and subsequently imaged. The scanned images were analyzed using the R package DeadOrAlive (version 1.0.0, https://github.com/JohnTownsend92/DeadOrAlive, accessed on 20 January 2025) [42].

### 2.6. Synthetic Genetic Array (SGA) Analysis

A genome-wide genetic interaction analysis was performed to identify the genes that genetically interact with *rst2*. The SGAs were carried out as previously described [43]. The deletion strain Δ*rst2* ($h^+$ *rst2*:: *natMX6*) was used as a query strain to mate with the *S. pombe* Bioneer haploid deletion library v5.0, consisting of 3420 deletion mutants marked with *kanMX6*. Traditionally, the Δ*ade6* strain was used as a control strain, but we observed that this strain features some traits different from wild type, likely altering the growth and chronological lifespan (Supplementary Figure S5a,b). We, therefore, generated a new prototrophic control strain, $Nat^R$ wild type (as described in Section 2.1), which shows the same fitness as the regular wild-type strain (Supplementary Figure S5c,d). The query strains were mated with the library mutant strains to create double-mutant strains on MEA plates. After incubation for 3 days at 25 °C for mating and sporulation and 3 days at 42 °C to eliminate the parental cells, the resulting colonies were selected on YES agar plates containing G418 and nourseothricin. Once colonies were sufficiently grown, they were printed onto either YES or YES containing 2 mg/mL CAP agar plates, followed by incubation at 32 °C for 3 days. The plates were imaged by an EPSON V800 scanner. Each condition was performed in three biological repeats.

The SGA image analysis was conducted in R (version 4.4.1) [38]. The colony sizes were quantified by the R package gitter (version 1.1.4) [44]. Systematic gene IDs were assigned to colony data according to the Bioneer library plate number and row/column position. Small (<50 pixels) and absent colonies were excluded to minimize false positives potentially caused by technical issues, such as an incomplete library or poorly pinned mutants. The colony sizes were subsequently normalized to reduce spatial and plate-to-plate variation effects by median smoothing and row–column median normalization [45]. Genes located within 250 kb of the query loci were excluded to avoid effects from linked loci. Genetic interaction scores (GISs) were calculated as the $\log_2$-transformed ratio between the normalized colony sizes of the mutant strain and the wild-type control. Genetic interaction hits were defined as genes with a GIS beyond the threshold of ±0.15 and a $p$-value less than 0.1.

### 2.7. Fluorescence Microscopy

Overnight cultures of cells expressing Rst2-GFP were diluted to an initial $OD_{600}$ of 0.2 and grown for the indicated hours (6, 12, 18, 24, 30, and 36 h) before analyzing by fluorescence microscopy. The green fluorescence signals were captured using a Zeiss Axio Imager 2 microscope (Carl Zeiss, Jena, Germany) with excitation wavelengths of 488 nm.

## 3. Results

### 3.1. Mitochondrial Translation Inhibition Affects Genome Regulation Mainly During Stationary Phase

To investigate the effect of mitochondrial translation inhibition on the transcriptome in *S. pombe*, we performed RNA-sequencing before and after treatment inhibiting mitochondrial translation with chloramphenicol (CAP) [46,47]. We sequenced the transcriptomes of cells grown in glucose and glycerol media, both in the exponential and stationary phases. Exponentially growing cells require low mitochondrial respiratory activity in glucose media (fermentative condition), whereas they require high mitochondrial activity in glycerol media (respiratory condition). Stationary-phase cells have stopped proliferating due to carbon source depletion and require higher mitochondrial respiratory activity than exponentially growing cells in glucose media. The experimental design for the eight RNA-seq conditions is illustrated in Figure 1a. We carried out three independent biological repeats for each condition. The numbers of differentially expressed genes following CAP treatment under each condition are summarized in Table 1. The stationary-phase cells in glucose medium exhibited the strongest transcriptome response to CAP. The annotated gene lists and transcriptome-wide expression data are provided in Supplementary Dataset S1. Principal component analysis (PCA) based on regularized log (rlog)-transformed normalized expression values of the transcriptome signatures in all conditions revealed that the largest variance in the data is attributed to differences between exponentially growing and stationary-phase cells, irrespective of CAP treatment, while the second-largest variance is based on differences between CAP-treated and untreated cells in stationary-phase cells, particularly in glucose medium (Figure 1b and Table 1). In exponentially growing cells, on the other hand, CAP treatment had only a subtle effect on the transcriptome, especially in glucose medium (Figure 1b and Table 1). These results are consistent with the much higher requirement for mitochondrial respiration in stationary-phase cells, particularly in glucose medium. However, with glycerol as the carbon source, high mitochondrial respiration is required even in exponentially growing cells to sustain energy production [48]. Thus, the altered expression of only 23 genes in response to CAP treatment in this condition is surprising. Nevertheless, cell growth was almost completely abolished in CAP-treated cells in this condition (Supplementary Figure S1a), indicating that mitochondrial translation is essential for proliferation in glycerol medium, as expected.

**Table 1.** Numbers of differentially expressed protein-coding transcripts after CAP treatment in the different experimental conditions. Induced and repressed genes are defined as transcripts that show expression changes in CAP-treated cells relative to untreated control cells in the same condition, with a significance threshold of $p < 0.05$ and $|\log_2(\text{Fold Change})| > 1$. A detailed list of these annotated genes is provided in Supplementary Dataset S1.

| Condition | Induced | Repressed |
|---|---|---|
| Glucose_Exponential | 9 | 2 |
| Glucose_Stationary | 432 | 456 |
| Glycerol_Exponential | 15 | 8 |
| Glycerol_Stationary | 141 | 250 |

Unlike in exponentially growing cells in glycerol medium, CAP treatment triggered a strong response in stationary-phase cells, with 391 genes showing altered expression (Table 1). Gene ontology (GO) and KEGG pathway analysis revealed that the induced genes were enriched in glycolysis/gluconeogenesis as well as genome maintenance processes, including DNA integration, recombination, and polymerase activity, while the repressed genes were enriched in nucleotide metabolism-related pathways and chromatin functions (Supplementary Figure S2). These results suggest a potential trade-off between energy metabolism and genome maintenance under mitochondrial translation stress.

CAP-treated stationary-phase cells grown in glucose medium showed the strongest transcriptome response, with 888 genes showing altered expression levels (Table 1). Therefore, the remainder of the results focus on this condition.

### *3.2. Mitochondrial Translation Inhibition During Stationary Phase Induces Genes Involved in the General Stress and Retrograde Responses*

We compared the 432 genes that were induced in CAP-treated stationary-phase cells grown in glucose medium with the following gene sets induced under other stressful conditions: (1) response to treatment with antimycin A (ANT), an inhibitor of the mitochondrial ETC complex III [16]; (2) mitonuclear retrograde (MNR) response, comprising genes altered in response to ANT as well as in two respiratory-deficient mutants [16]; (3) the core environmental stress response (CESR), which includes genes responsive to oxidative stress, heavy metal stress, heat shock, osmotic stress, and DNA damage [49]; (4) the core oxidative stress genes (COSGs), which are induced in three different oxidants [50]. We observed significant overlaps between our set of induced 432 genes and all 4 stress-related gene sets induced in different conditions (Figure 1c). Various numbers of CAP-induced genes overlapped with different combinations of the other stress-related lists (Supplementary Figure S3). In contrast, the genes induced in the other CAP-treated conditions did not show significant overlaps with genes induced in any of the four stress conditions, and no significant overlaps were observed among the genes repressed in the CAP-treated conditions.

To further elucidate the biological processes associated with the genes induced in CAP-treated stationary-phase cells in glucose medium, we performed gene ontology (GO) and KEGG pathway enrichment analyses. The 432 genes were significantly enriched in terms related to stress response and protein homeostasis processes (Figure 1d). These enrichments are consistent with the significant overlaps with the responses to other stressful conditions (Figure 1c). Together, these findings indicate that mitochondrial translation inhibition triggers a transcriptional response similar to that observed in other stressful conditions, but only in stationary-phase cells grown in glucose medium.

### *3.3. Mitochondrial Translation Inhibition During Stationary Phase Induces Carbon Metabolism Genes Functioning in Non-Mitochondrial Pathways*

When glucose is used as a carbon source (fermentative condition), exponentially growing cells primarily rely on cytosolic glycolysis to convert glucose to pyruvate, which is then fermented to ethanol [51]. During the stationary phase, however, cells shift to mitochondrial respiration, whereby pyruvate is oxidized through the tricarboxylic acid (TCA) cycle and oxidative phosphorylation to generate ATP [52]. Figure 1d shows that the genes induced in CAP-treated stationary-phase cells grown in glucose medium were also significantly enriched in several pathways related to carbon metabolism, including glycolysis, the pentose phosphate pathway, and pyruvate metabolism (Figure 1d). In contrast, no significant enrichments were detected among the genes repressed in response to CAP treatment. To further characterize this response, we examined all genes associated with major carbon metabolic pathways in PomBase [53], including glycolysis, the TCA cycle, the pentose phosphate pathway, fermentation, and glycerol metabolism. Figure 2a shows the expression changes of all these genes in CAP-treated stationary-phase cells grown in glucose medium, grouped by the different metabolic pathways. Notably, the pathways that include most of the induced genes all function in the cytoplasm, outside of mitochondria and upstream of the TCA cycle, most notably genes functioning in glycolysis and the pentose phosphate shunt (Figure 2b). Together, these findings indicate that the inhibition of mitochondrial translation activates a substantial metabolic reprogramming that enhances the cytosolic carbon flux, likely as a compensatory mechanism shifting energy metabolism from respiration to fermentation to maintain cellular homeostasis when mitochondrial oxidative phosphorylation is impaired.

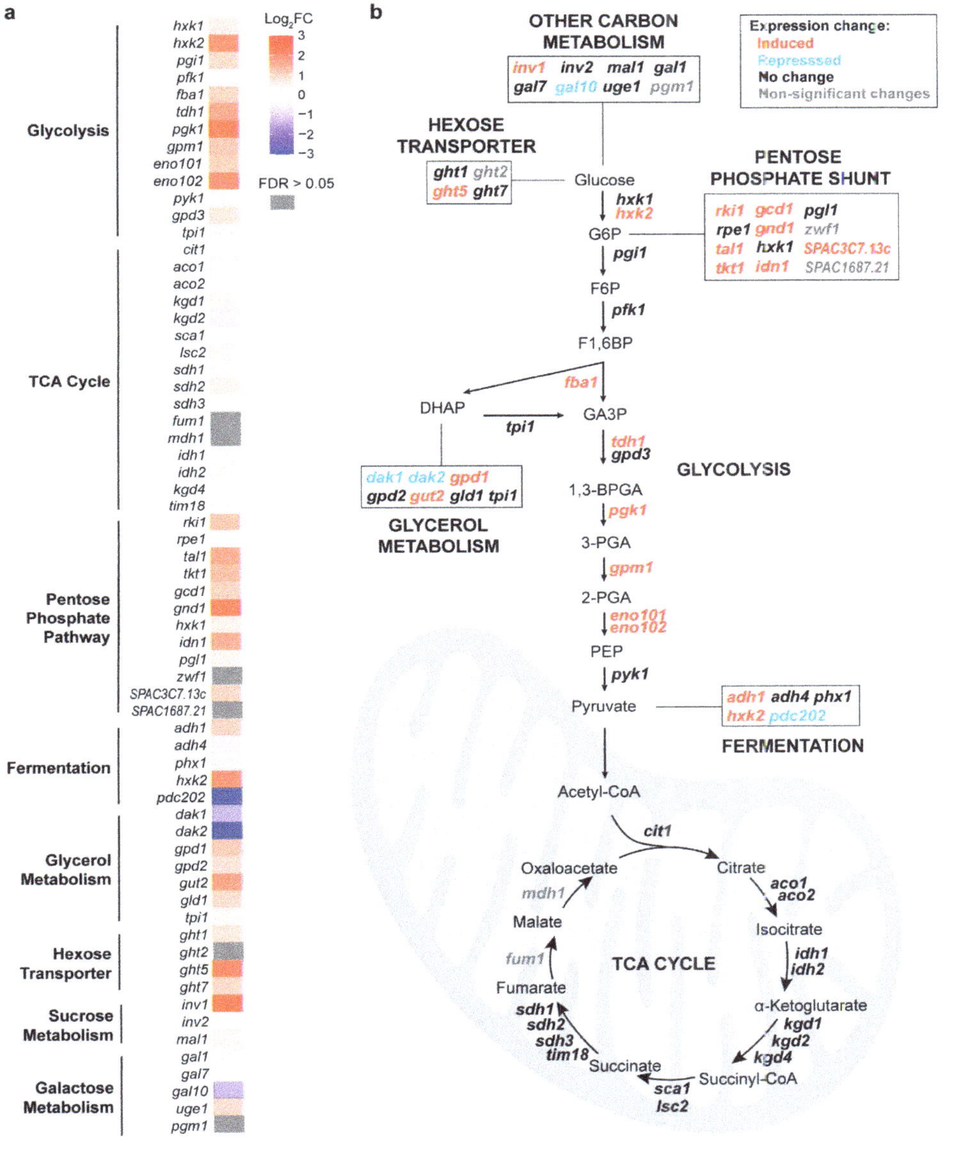


**Figure 2.** Expression changes of carbon metabolism genes in response to CAP treatment of stationary-phase cells in glucose medium. (**a**) Heatmap of expression fold changes (FCs) for genes involved in major carbon metabolism pathways. The $\log_2$FC values are color-coded as indicated, with grey boxes indicating non-significant expression changes (FDR > 0.05). (**b**) Carbon metabolism network scheme to map gene expression changes in response to CAP treatment in stationary-phase cells grown in glucose medium. Genes are colored according to their $\log_2$FC shown in (a): red indicates significantly induced genes ($\log_2$FC > 1), blue indicates significantly repressed genes ($\log_2$FC < −1), black indicates small fold changes (−1 < $\log_2$FC < 1), and grey indicates non-significant changes (FDR > 0.05).

### 3.4. The Scr1 and Rst2 Transcription Factors Regulate Common Metabolic Genes upon Inhibition of Mitochondrial Translation in Stationary-Phase Cells

The CAP-induced gene expression response appears to be a complex program that likely involves multiple regulatory factors. The Scr1 and Rst2 transcription factors are key regulators of the CCR (see the Introduction Section). ChIP-seq experiments have identified 140 Scr1 and 413 Rst2 target genes in glucose and glycerol media, respectively, where these factors are most active [25]. As expected, these two gene sets share many common target genes (Figure 3a). To investigate the involvement of Scr1 and Rst2 in the CAP-induced gene expression response, we performed an overlap analysis between their published target genes and our 432 genes induced in CAP-treated stationary-phase cells grown in glucose medium. These genes showed significant overlaps with both the Scr1 and Rst2 target genes (Figure 3b). Moreover, many of the carbon metabolism genes analyzed above (Figure 2) were included among the Scr1 and Rst2 target genes (Figure 3c). In contrast, the 456 repressed genes in CAP-treated stationary-phase cells showed no significant overlaps with Scr1 or Rst2 target genes (Figure 3d).

We then tested whether the Scr1 and Rst2 target genes showed altered expression in response to CAP treatment in exponentially growing or stationary-phase cells in glucose medium. On average, both the Scr1 and Rst2 target genes were significantly induced in stationary-phase cells, while they showed only minor expression differences in exponentially growing cells (Figure 3e). These findings further support the conclusion that Scr1 and Rst2 play an important role in regulating the strong transcriptional response to mitochondrial translation inhibition in stationary-phase cells.

To dissect the regulatory relationship between Scr1 and Rst2, we looked for their target genes that were significantly induced or repressed in our RNA-seq data of CAP-treated stationary-phase cells. Notably, the target genes common to Scr1 and Rst2 included significant numbers of both induced and repressed genes, although fewer of the latter (Figure 3f). In contrast, no overlap was observed between induced Scr1 target genes and repressed Rst2 target genes, or vice versa. These results indicate that Scr1 and Rst2 cooperatively regulate a set of common genes, and most of these genes are induced in response to CAP treatment. KEGG pathway analysis of the genes that are positively regulated by both Scr1 and Rst2 revealed significant enrichments in several processes related to carbon metabolism (Figure 3g). Collectively, these findings support a model of functional convergence, in which Scr1 and Rst2 jointly promote the reprogramming of carbon metabolism in response to mitochondrial translation inhibition.

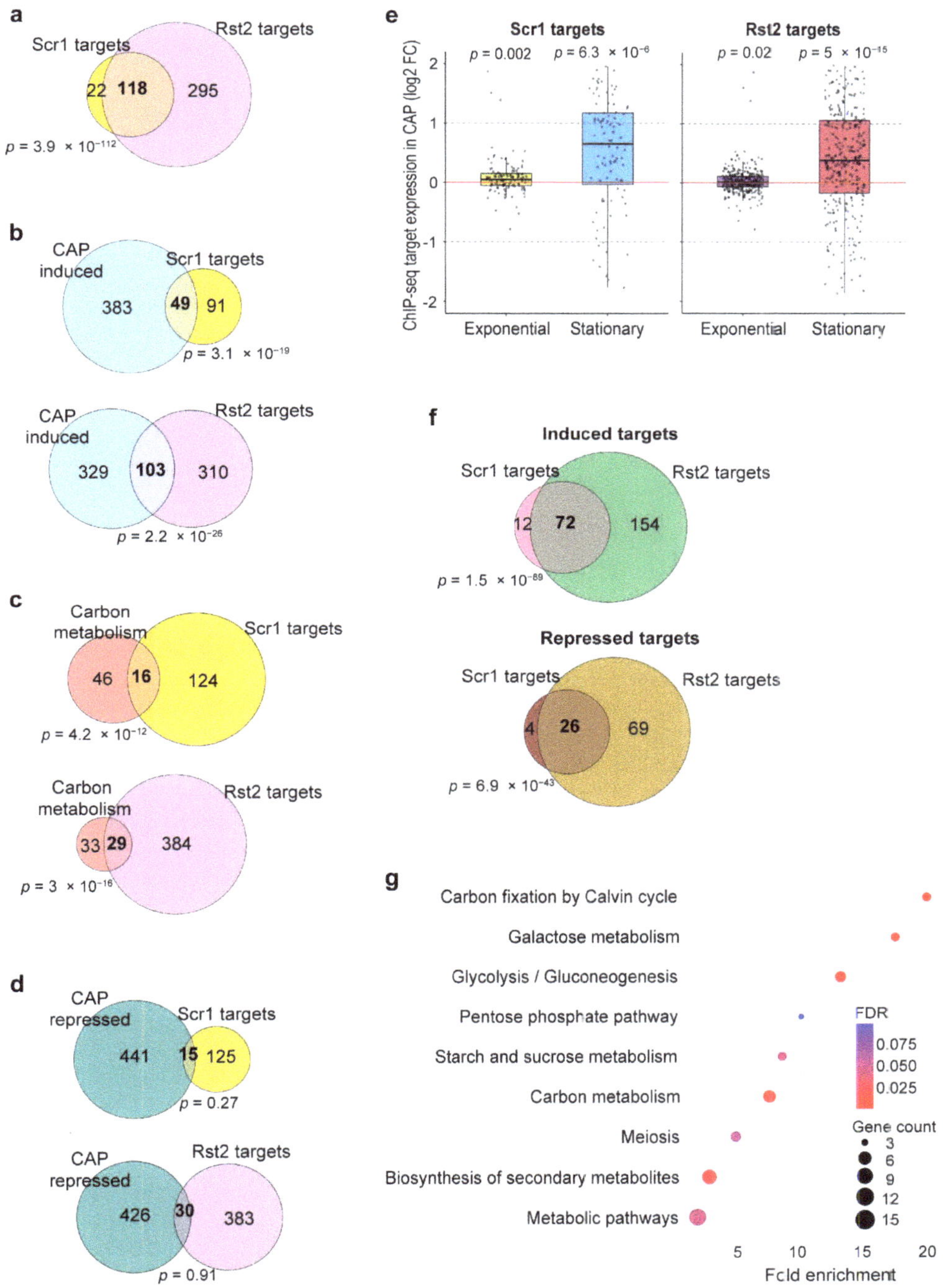


**Figure 3.** Scr1 and Rst2 regulate shared metabolic genes in response to CAP treatment. (**a**) Venn diagram to show the overlap between published Scr1 and Rst2 target genes [25]. Statistical significance of overlaps was calculated using Fisher's exact test with the GeneOverlap R package (version 1.40.0) [39], with the resulting $p$-values of the overlaps indicated. The background gene set used for

the analysis corresponds to all genes detected in the RNA-seq dataset in this study. (**b–d**) Overlap analysis of CAP-induced genes (**b**), carbon metabolism genes in Figure 2 (**c**), and CAP-repressed genes (**d**) with Scr1 and Rst2 target genes, with the resulting *p*-values of the overlaps indicated. Statistical significance of overlaps was calculated using Fisher's exact test with the GeneOverlap R package (version 1.40.0) [39], with the resulting *p*-values of the overlaps indicated. The background gene set used for the analysis encompasses all genes detected in the RNA-seq dataset in this study. (**e**) Transcriptional response of Scr1 and Rst2 target genes to CAP treatment. Box plots show $\log_2$ fold changes in gene expression of Scr1 and Rst2 targets in CAP-treated versus untreated cells during exponential and stationary phases, as indicated. Substantial changes in gene expression (mainly induction) of Scr1 and Rst2 targets were observed predominantly during the stationary phase. Red horizontal lines mark $\log_2$ fold change = 0 (no change), while dashed lines mark $\log_2$ fold-change thresholds of $\pm 1$. The *p*-values were calculated using the Wilcoxon signed-rank test [54]. (**f**) Overlap analysis of Scr1 and Rst2 targets that were induced (top) or repressed (bottom) based on the RNA-seq data shown in (e), with the *p*-values of the overlap indicated. (**g**) KEGG pathway enrichment analysis of Scr1 and Rst2 target genes that are induced in our data. The most enriched pathways are related to carbon metabolism. Dot size represents gene count per pathway, and dot color indicates FDR values, as shown on the right. The analysis was performed using DAVID [36,37] with default parameters and visualized by R (version 4.4.1, R Foundation for Statistical Computing, Vienna) [38].

### *3.5. Scr1 and Rst2 Antagonistically Affect Cellular Growth and Viability upon Inhibition of Mitochondrial Translation*

To further investigate the roles of Rst2 and Scr1 in CAP-treated cells, we constructed Δ*rst2* and Δ*scr1* deletion strains. We then assessed various traits of the cells lacking Rst2 or Scr1 as a function of CAP treatment. Without CAP treatment, Δ*rst2* cells showed a slightly faster growth rate, shorter lag phase, and higher final cell mass compared to wild-type cells, while Δ*scr1* cells showed a slower growth rate, prolonged lag phase, and lower final cell mass (Figure 4a). With CAP treatment, all three strains exhibited reduced overall growth patterns, but with the same relative differences of increased growth patterns in Δ*rst2* and decreased growth patterns in Δ*scr1* compared to wild-type cells (Figure 4a). To assess whether CAP affects cell viability over time, we performed serial-dilution spot assays at multiple time points following CAP treatment. In untreated cells, deletion of either *rst2* or *scr1* did not significantly affect cell viability at any time points compared to wild type (Figure 4b). CAP-treated cells of all three strains, on the other hand, showed markedly reduced viability at later times of culturing (36 h and 48 h), with Δ*rst2* showing slightly higher viability and Δ*scr1* slightly lower viability compared to wild-type cells (Figure 4b). This result suggests differences in the chronological lifespan (CLS) of stationary-phase cells between the three strains. To test this prediction, we conducted CLS assays to measure the longer-term viability of stationary-phase cells. Without CAP treatment, all strains showed similar CLS profiles over time. With CAP treatment, however, Δ*rst2* cells showed a substantially extended CLS compared to wild-type cells, maintaining some viability up to Day 9, whereas Δ*scr1* cells showed a dramatically shortened CLS, with a sharp decrease of viability already by Day 1 (Figures 4c and S4).

To explore whether the opposing effects of Scr1 and Rst2 on the CLS could be attributed to distinct target genes, we compared the transcriptional profiles of Δ*scr1* [25] and Δ*rst2* (this study) with previously identified long- and short-lived mutants, in which gene deletions extend or shorten the CLS, respectively [42]. Among the genes repressed in Δ*scr1* but not in Δ*rst2* cells, two deletion mutants (*abp2* and *cnp3*) have been reported to shorten the CLS [42]. Notably, *abp2* showed a 2.7-fold reduction after CAP treatment. Conversely, among the genes repressed in Δ*rst2* cells but not repressed (or not detected) in Δ*scr1* cells, three deletion mutants (*cyp4*, *mug27*, and *gal1*) have been reported to extend the CLS [42]. These distinct gene expression patterns might explain the opposite effects of Scr1 and Rst2 on the CLS under conditions of mitochondrial translation inhibition. Collectively, these findings indicate that Scr1 and Rst2 play antagonistic roles in the cellular adaptation to mitochondrial translation inhibition during the stationary phase.

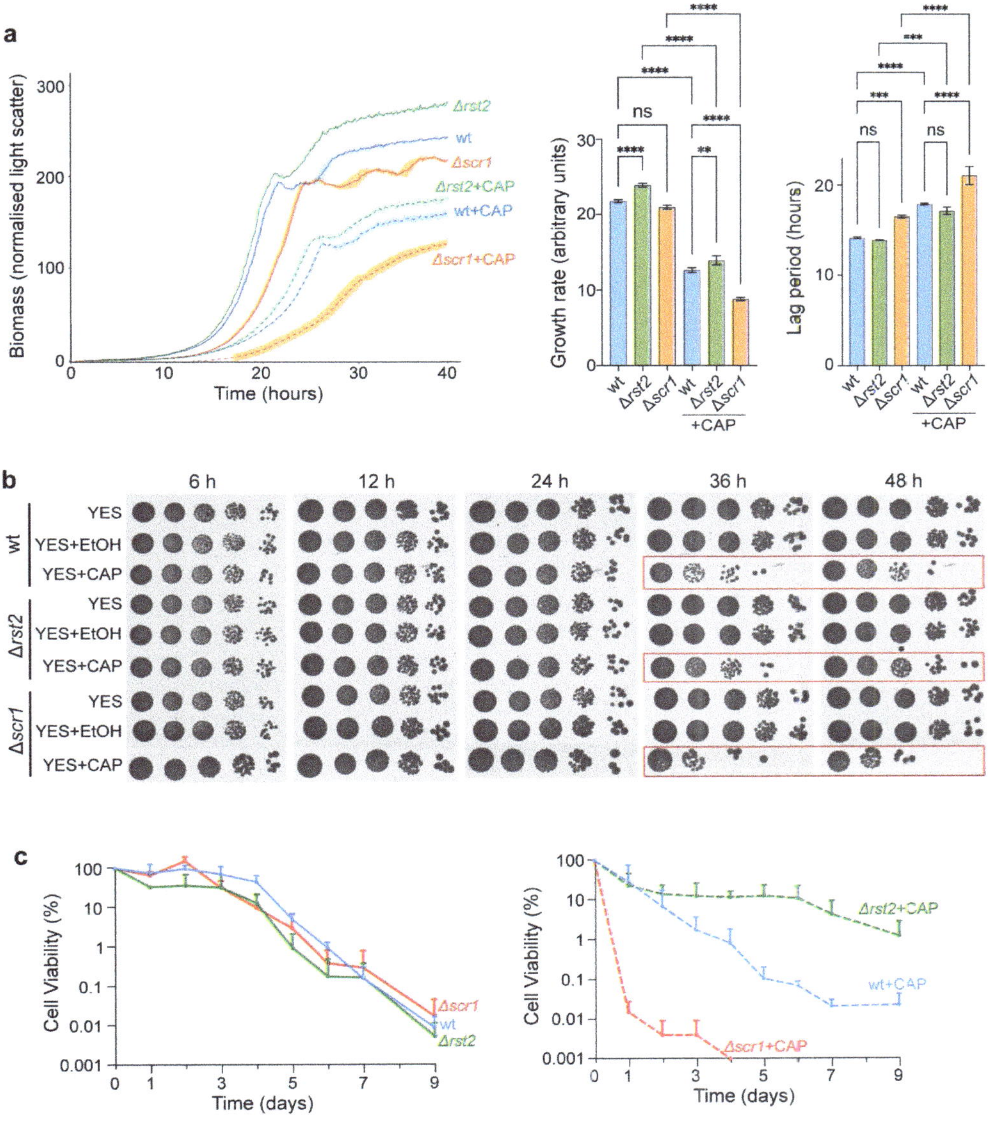


**Figure 4.** Scr1 and Rst2 affect growth and viability of CAP-treated cells. (a) Growth measurement of wild-type (wt), Δ*rst2*, and Δ*scr1* strains cultured in YES medium with or without CAP treatment. Left graph: growth curves were measured using a microbioreactor, as described [31]. Mean growth curves (biomass accumulation) are shown, along with the standard deviation (shaded regions) for three independent replicates. Growth rates and lag phases (middle and right panels) were quantified from the left graph using the R package grofit (version 1.1.1.1) [41]. The statistical analysis was conducted

by one-way ANOVA, followed by Tukey's honest significance test (ns, not significant; **, $p < 0.01$; ***, $p < 0.001$; ****, $p < 0.0001$). (**b**) Viability assays of wild-type, Δ*rst2*, and Δ*scr1* strains with and without CAP treatment, with 2% EtOH serving as a solvent control. Cells grown to an initial $OD_{600}$ in the media indicated on the left were collected after the indicated time points, and 10-fold serial dilutions were spotted onto YES solid plates. The plates were photographed after 3 days. The red boxes highlight markedly reduced viability at 36 h and 48 h. (**c**) Chronological lifespan assays of wt, Δ*rst2*, and Δ*scr1* strains without (left graph) or with (right graph) CAP treatment. Cells were grown in YES medium to stationary phase (Day 0). A proxy for cell viability was determined over time by a robotics-based colony-forming unit (CFU) assay [42], with values normalized to Day 0 (100% viability). CFUs were measured daily from Day 0 to Day 7 and at Day 9. At each time point, aliquots of aging cultures were collected, 3-fold serially diluted, and spotted onto YES plates, which were imaged after three days. Three independent biological replicates were prepared for each condition. Error bars represent standard deviation (SD) and are shown only in the upward direction for clarity on a log-scale *Y* axis.

### *3.6. Rst2 Activates Carbon Metabolism Genes in Stationary-Phase Cells upon Inhibition of Mitochondrial Translation*

The gene expression changes in response to CAP treatment are most pronounced during the stationary phase, a state triggered by glucose limitation. Thus, Rst2 is expected to play a dominant regulatory role in this condition based on its known function in the absence of glucose [25]. We, therefore, analyzed the specific contribution of Rst2 to this adaptive response. To this end, we performed RNA-seq of Δ*rst2* and wild-type strains for both exponential and stationary phases (Figure 5a). PCA based on rlog-transformed normalized expression values of the transcriptome signatures in these conditions revealed that the biggest transcriptional differences were caused by differences in the growth phases (Figure 5b). In addition, deletion of *rst2* had a major impact on the transcriptome during the stationary phase, but not during the exponential phase (Figure 5b). This phase-specific transcriptome response resembles the pattern observed for CAP treatment (Figure 1b). A total of 137 genes were differentially expressed in Δ*rst2* compared to wild-type cells during the stationary phase, with 30 genes induced and 107 genes repressed. These results indicate that Rst2 primarily functions as a transcriptional activator, consistent with the analyses of Rst2 target genes described above (Figure 3e,f) and the known function of Rst2 in the CCR response [25].

We then checked whether the genes whose expression is lowered in stationary-phase Δ*rst2* cells are part of the CAP-induced response. To this end, we compared the repressed genes in Δ*rst2* stationary-phase cells with both the induced Rst2 target genes (Figure 3f) and the genes induced in CAP-treated stationary-phase cells grown in glucose medium. All three lists shared substantial numbers of genes (Figure 5c). These results indicate that Rst2 is required for the regulation of a substantial proportion of the genes induced in CAP-treated stationary-phase cells. Accordingly, KEGG pathway analysis of the repressed genes in Δ*rst2* stationary-phase cells revealed significant enrichments of carbon metabolism pathways (Figure 5d). Genes involved in meiosis were also significantly enriched (see also Figure 3g), consistent with the established role of Rst2 in activating *ste11* expression, which encodes a key transcription factor regulating sexual differentiation [55,56].

To assess whether CAP treatment affects the subcellular localization of Rst2, we tagged the genomic *rst2* gene with GFP to express a Rst2-GFP fusion protein. We then monitored the dynamics of Rst2-GFP in cells grown in glucose medium with or without CAP (Figure 5e). In untreated cells, Rst2 was predominantly localized to the nucleus during the exponential phase but gradually shifted to the cytoplasm during the stationary phase. In contrast, in CAP-treated cells, Rst2 remained localized in the nucleus during the stationary phase (Figure 5e). This nuclear retention further points to Rst2 contributing to the transcriptional response to CAP treatment on stationary-phase cells. Together, these findings establish Rst2 as an important regulator for the transcriptional reprogramming of carbon metabolism in stationary-phase cells adapting to the inhibition of mitochondrial translation.

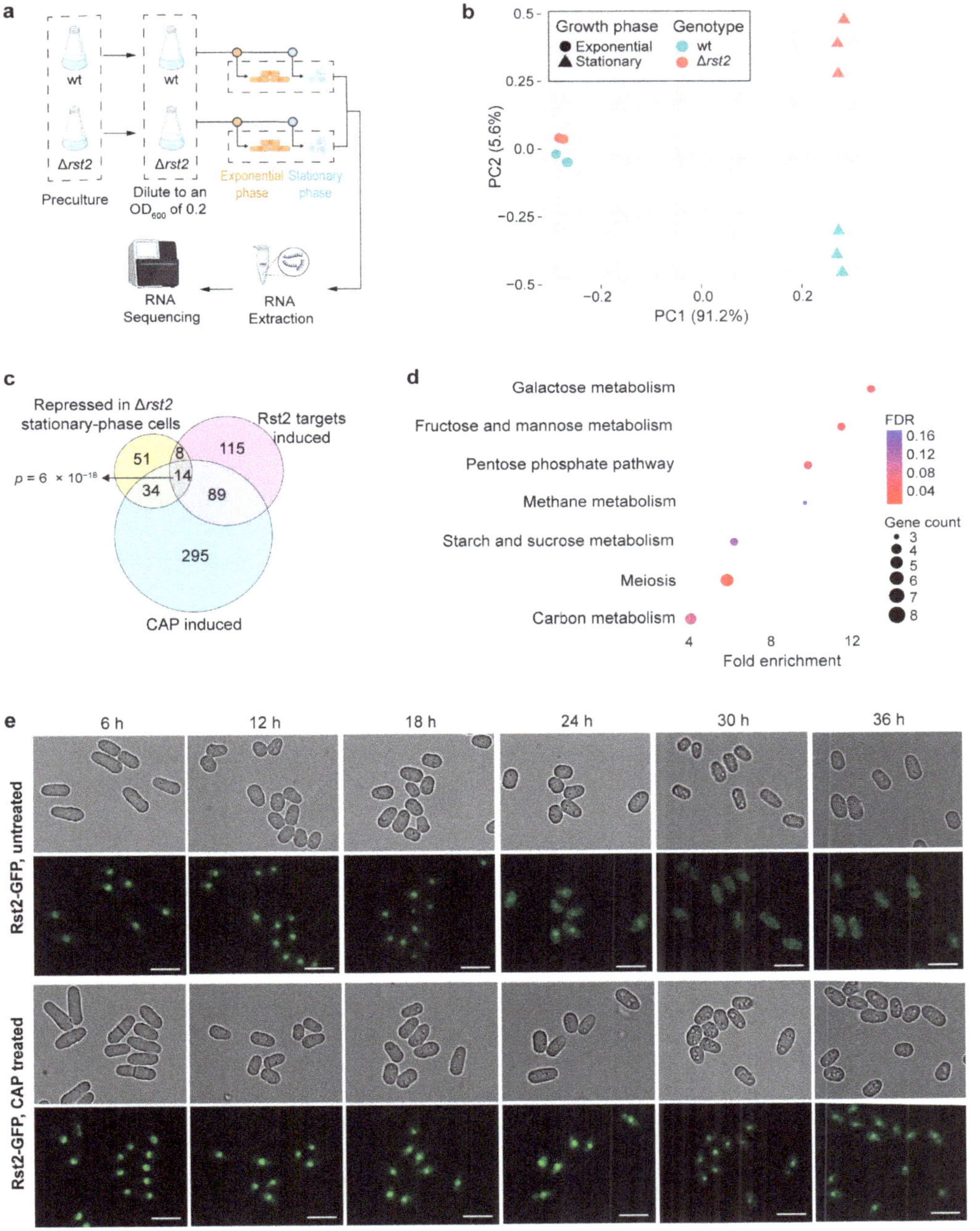


**Figure 5.** Rst2 contribution to gene regulation in response to mitochondrial translation inhibition during the stationary phase. (**a**) Experimental design of wild-type (wt) and Δ*rst2* strains used for RNA-

seq analysis. For experimental details, see the Materials and Methods Section. Three independent biological replicates were prepared for each condition. (**b**) PCA based on rlog-transformed normalized expression values of transcriptome responses from Δ*rst2* and wild-type (wt) cells collected during exponential and stationary phases. Each dot represents one biological replicate. Dot shapes indicate the growth phases (circle: exponential; triangle: stationary), and dot colors indicate the genotype (turquoise: wt; pink: Δ*rst2*). PC1 and PC2 account for 91.2% and 5.6% of the total variance, respectively. (**c**) Overlap analysis of genes repressed in Δ*rst2* cells during the stationary phase compared to CAP-induced genes during the stationary phase and to Rst2 target genes [25]. The *p*-value represents the significance of the overlap among the three gene sets. (**d**) KEGG pathway enrichment analysis of the genes repressed in Δ*rst2* cells during the stationary phase. The most enriched pathways are related to carbon metabolism and meiosis. Dot size represents gene count per pathway, and color indicates FDR values, as shown on the right. The analysis was performed using DAVID [36,37] with default parameters and visualized by R (version 4.4.1, R Foundation for Statistical Computing, Vienna) [38]. (**e**) Subcellular localization of Rst2 in untreated and CAP-treated cells. Wild-type cells expressing Rst2-GFP under its endogenous promoter were grown in YES medium to an initial $OD_{600}$ of 0.2 (0 h). Cells were then treated with CAP and collected at the indicated time points for fluorescence microscopy. Scale bar, 10 μm.

### *3.7. Rst2 Genetically Interacts with Genes Involved in Stress Protection and Nutrient Response*

To systematically uncover functional relationships of Rst2 with diverse biological processes, we generated a new Δ*rst2* deletion strain (*rst2::natMX6*), containing a marker suitable for a query mutant to assay genetic interactions with all 3420 non-essential prototrophic deletion mutants using the SGA method [57,58]. Genetic interactions were measured in glucose medium, both with and without CAP, based on colony size as a proxy for double-mutant fitness compared to the control mutant (see the Methods Section 2.6). We observed 246 negative and 145 positive genetic interactions for untreated double mutants, and 247 negative and 156 positive genetic interactions for CAP-treated double mutants (Supplementary Dataset S3). Significant overlaps of 74 positive ($p = 6.8 \times 10^{-59}$) and 171 negative ($p = 4.3 \times 10^{-146}$) interactions were identified between the untreated and CAP-treated conditions. Functional enrichment analysis using AnGeLi [59] revealed that these interacting genes in both untreated and CAP-treated conditions were significantly enriched for stress-related phenotypes, including the fission yeast phenotype ontology (FYPO) terms [42,60,61] 'increased sensitivity to chemical' ($p = 5.5 \times 10^{-15}$ and $p = 1.0 \times 10^{-13}$ for untreated and CAP-treated, respectively), 'sensitive to DNA-damaging agents' ($p = 4.7 \times 10^{-7}$ and $p = 9.5 \times 10^{-9}$), and 'sensitive to hydrogen peroxide' ($p = 7.9 \times 10^{-4}$ and $p = 4.6 \times 10^{-5}$). The term 'decreased mating efficiency' ($p = 2.7 \times 10^{-4}$ and $p = 9.5 \times 10^{-5}$) was also enriched, consistent with the enriched KEGG pathway 'meiosis' observed in Figures 3g and 5d. Moreover, in untreated cells, the interacting genes additionally showed enrichment for several GO terms [62] related to 'response to nutrient levels' ($p = 2.1 \times 10^{-3}$) and 'response to starvation' ($p = 3.7 \times 10^{-3}$).

Among the 171 negative genetic interactions shared between untreated and CAP-treated conditions, four genes (*gal1*, *ght8*, *SPAC212.04c*, and *SPBC1861.05*) were also repressed genes in Δ*rst2* stationary-phase cells—*gal1* and *ght8* are involved in sugar metabolism and transport, supporting a role of Rst2 in maintaining glucose-responsive gene expression during mitochondrial stress. In addition, 14 genes overlapped with genes that are induced in CAP-treated stationary-phase cells, including *ppr5*, encoding a mitochondrial pentatricopeptide repeat protein that negatively regulates mitochondrial translation [63], and *rec24*, encoding a meiotic recombination protein [64,65], consistent with the enriched KEGG pathways related to meiosis (Figures 3g and 5d). Collectively, our genetic interaction screen complements the transcriptomic data and highlights candidate genes that may work in parallel or together with Rst2 to promote the adaptation of stationary-phase cells to mitochondrial dysfunction.

## 4. Discussion

Mitochondrial translation is essential for cellular energy production and other mitochondrial functions [4]. Dysfunction of mitochondria is communicated to the nucleus through retrograde signaling, triggering adaptive gene expression programs [15,16]. Here, we investigated the transcriptome changes in response to impaired mitochondrial translation, induced by CAP treatment, in different metabolic and growth conditions. Our RNA-seq results indicated that CAP treatment primarily induces transcriptome changes in stationary-phase cells, whereas exponentially growing cells exhibit only a minor response (Figure 1b).

Surprisingly, in cells growing under a respiratory condition that requires high mitochondrial activity (glycerol medium), only 23 genes were differentially expressed upon inhibition of mitochondrial translation (Table 1). This limited response could indicate that the six-hour CAP treatment during exponential growth was insufficient to elicit a stronger transcriptional response. Such short-term CAP treatment is known to trigger only a modest reduction in the protein levels of mtDNA-encoded ETC subunits in the exponential phase, whereas in the stationary phase, after longer CAP treatment, the levels of these proteins are strongly repressed [47]. Nevertheless, CAP treatment in glycerol almost completely abolished cell growth (Figure S1). Thus, it is also possible that these cells are too impaired to launch a robust transcriptional response.

GO enrichment analysis of the differentially regulated genes in CAP-treated cells in the different conditions revealed that only the 432 genes induced in stationary-phase cells grown in glucose medium significantly overlapped with genes that were differentially expressed in response to various other stress conditions (Figure 1c). These findings suggest that mitochondrial translation inhibition in these stationary-phase cells triggers a response similar to other stresses, including the retrograde response. However, we note that the published stress-related gene sets used for comparison are from proliferating cells rather than stationary-phase cells. We included these datasets primarily as reference signatures to evaluate similarities between CAP-induced changes and previously characterized transcriptional programs. Furthermore, only these 432 genes were significantly enriched in processes and pathways related to carbon metabolism, particularly cytoplasmic pathways that provide alternatives to mitochondrial respiration (Figures 1d and 2). These transcriptome changes are consistent with a reprogramming of energy metabolism from respiration to fermentation to cope with the mitochondrial dysfunction.

Our results indicate that the CAP-induced transcriptional reprogramming of carbon metabolism involves Scr1- and Rst2-mediated regulation (Figure 3). Scr1 primarily acts as a transcriptional repressor under glucose-sufficient conditions, while Rst2 functions as a transcriptional activator when glucose is limited [25]. Both Δ*scr1* and Δ*rst2* mutants displayed normal cell viability and lifespans under control conditions (Figure 4), consistent with published results [25,56]. However, CAP-treated Δ*scr1* cells showed a dramatically shortened CLS, while Δ*rst2* cells featured an extended CLS compared to wild-type cells (Figure 4c). These results suggest that Scr1 and Rst2 exert opposing regulatory roles in response to inhibition of mitochondrial translation. Scr1 and Rst2 regulate a common set of target genes, while also each controlling distinct sets of genes (Figure 3a). The Scr1 and Rst2 target genes were, on average, significantly induced during the stationary phase in CAP-treated cells (Figure 3e), and they were enriched in carbon metabolism processes (Figure 3g). Together with published results, these findings support a model of functional convergence, where Scr1 and Rst2 may act as a transcriptional 'brake' and 'accelerator', respectively, to tune the expression of carbon metabolism genes in response to mitochondrial defects. Similar regulators exist in *S. cerevisiae*, where Mig1 is the ortholog of Scr1 that controls carbon source utilization [66,67], while Cat8 is the ortholog of Rst2 that activates metabolic genes under glucose-limited conditions [68]. These parallels suggest that coordination between glucose repression and metabolic adaptation via CCR regulators represents a conserved strategy across divergent yeast species.

Our findings further suggest that Rst2 primarily functions as a transcriptional activator for the induction of CAP-responsive metabolic genes in stationary-phase cells, based on RNA-seq (Figure 5a–d) and genetic interaction data. Rst2 was predominantly localized to the nucleus during the exponential phase and gradually moved to the cytoplasm during the late stationary phase (Figure 5e). In CAP-treated cells, however, Rst2 was retained in the nucleus during the stationary phase. This result supports the conclusion that Rst2 plays an important role in the transcriptional response of stationary-phase cells to mitochondrial dysfunction. The Rst2 localization pattern we observed in control cells seems to contradict published findings, where Rst2 is typically cytoplasmic under glucose-rich conditions and translocates to the nucleus only upon glucose depletion, regulated via AMPK-mediated phosphorylation [69]. However, a cytoplasmic localization of Rst2 has been observed in overnight cultures [70], which corresponds to our stationary-phase condition. Moreover, Rst2 localization may not be exclusively regulated by glucose levels, as illustrated by the following examples. Mutations in the Pka1 kinase that phosphorylates Rst2 in glucose-sufficient conditions lead to nuclear accumulation of Rst2 [71], and Rst2 nucleocytoplasmic shuttling is impaired by deletion of ETC complex subunits [70]. Furthermore, these studies typically examined acute changes (minutes to hours) of Rst2 dynamics in response to glucose levels, whereas we captured Rst2 localization over longer times in steady-state conditions. Another difference is that we used rich medium for consistency with the RNA-seq conditions, while the other studies used minimal medium. Together, these results highlight the influence of broader metabolic and signaling contexts, besides glucose levels, that affect Rst2 function and localization. A recent study identified high-occupancy target (HOT) regions in the *S. pombe* genome that are bound by multiple transcription factors, and Rst2 was enriched at these HOT regions [72]. This observation might reflect a broader involvement of Rst2 in global gene regulation.

## 5. Conclusions

We investigated whether and how the inhibition of mitochondrial translation affects genome regulation in different metabolic conditions and growth phases. During the stationary phase in glucose medium, CAP treatment triggered a transcriptional reprogramming of carbon metabolism from respiration to fermentation, along with changes resembling general stress and retrograde responses. The CCR regulators Scr1 and Rst2 controlled a shared set of metabolic genes in response to CAP treatment. Scr1 and Rst2 had opposing effects on the viability of CAP-treated stationary-phase cells. In these cells, which were in a glucose-limited state, Rst2 remained localized in the nucleus and functioned as a transcriptional activator of carbon metabolism genes. Genetic interactions pointed to functional relationships of Rst2 with stress and starvation responses. Together, our findings revealed a crosstalk between stress/retrograde response-like signatures and the CCR regulators Scr1 and Rst2 and highlighted a coordinated regulatory strategy that reprograms carbon metabolism in response to mitochondrial dysfunction. These findings deepen our understanding of mitochondrial–nuclear communication and the resulting transcriptome regulation in fission yeast and may also inform how other eukaryotic cells adapt their metabolism under mitochondrial stress.

**Supplementary Materials:** The following supporting information can be downloaded at: https://www.mdpi.com/article/10.3390/biom15101354/s1, Table S1: List of *S. pombe* strains used in this study. Table S2: Primers used in this study. Figure S1: Growth measurement with or without CAP treatment in glucose or glycerol media. Figure S2: GO terms and KEGG pathway enrichment analysis of 141 induced (a) and 250 repressed (b) genes following CAP treatment during the stationary phase in glycerol medium. Figure S3: Overlaps of CAP-induced genes with multiple stress-responsive gene sets. Figure S4: Chronological lifespan assays. Figure S5: The Nat$^R$ wild-type strain, unlike the Δ*ade6* strain, shows no phenotypic differences compared to the original wild-type strain. Figure S6: Growth assessment of the Rst2-GFP strain. Dataset S1: Transcriptomic changes after the inhibition of mitochondrial translation under different carbon sources and growth phases. Dataset S2: Transcriptomic changes in Δ*rst2* compared to wild-type cells. Dataset S3: SGA assay data. File S1: Original microscopy images.

**Author Contributions:** Conceptualization, Y.L. and J.B.; methodology, Y.L., S.H. and S.R.; formal analysis, Y.L.; investigation, Y.L.; writing—original draft preparation, Y.L.; writing—review and editing, Y.L., S.H. and J.B.; visualization, Y.L., S.H. and J.B.; supervision, J.B.; funding acquisition, Y.L. and J.B. All authors have read and agreed to the published version of the manuscript.

**Funding:** This work was funded by the Wellcome Discovery Award (No. 302608/Z/23/Z to J.B.) and the China Scholarship Council (No. 202306860040 to Y.L.).

**Institutional Review Board Statement:** Not applicable.

**Informed Consent Statement:** Not applicable.

**Data Availability Statement:** The RNA-seq data have been submitted to GEO under the accession number PRJNA1262735: https://www.ncbi.nlm.nih.gov/sra/PRJNA1262735 (accessed on 12 September 2025).

**Acknowledgments:** We thank Benjamin Murray Heineike and Manuel Lera-Ramírez for the helpful discussions, and Sam Yee Low, Qinglong Yang, and Xiao Yuan for technical assistance.

**Conflicts of Interest:** The authors declare no conflicts of interest. The funders had no role in the design of the study; in the collection, analyses or interpretation of the data; in the writing of the manuscript, or in the decision to publish the results.

## Abbreviations

The following abbreviations are used in this manuscript:

| | |
|---|---|
| mtDNA | Mitochondrial DNA |
| ETC | Electron transport chain |
| CCR | Carbon catabolite repression |
| AMPK | AMP-activated protein kinase |
| GO | Gene ontology |
| ANT | Antimycin A treatment |
| CESR | Core environmental stress response |
| COSG | Core oxidative stress gene |
| MNR | Mitonuclear retrograde |
| TCA | Tricarboxylic acid |
| CLS | Chronological lifespan |
| FYPO | Fission yeast phenotype ontology |

## References

1. Yan, C.; Duanmu, X.; Zeng, L.; Liu, B.; Song, Z. Mitochondrial DNA: Distribution, Mutations, and Elimination. *Cells* **2019**, *8*, 379. [CrossRef]
2. Gao, M.; Yi, J.; Zhu, J.; Minikes, A.M.; Monian, P.; Thompson, C.B.; Jiang, X. Role of Mitochondria in Ferroptosis. *Mol. Cell* **2019**, *73*, 354–363.e3. [CrossRef]

3. Mishra, P.; Chan, D.C. Metabolic regulation of mitochondrial dynamics. *J. Cell Biol.* **2016**, *212*, 379–387. [CrossRef]
4. Herbert, C.J.; Labarre-Mariotte, S.; Cornu, D.; Sophie, C.; Panozzo, C.; Michel, T.; Dujardin, G.; Bonnefoy, N. Translational activators and mitoribosomal isoforms cooperate to mediate mRNA-specific translation in *Schizosaccharomyces pombe* mitochondria. *Nucleic Acids Res.* **2021**, *49*, 11145–11166. [CrossRef] [PubMed]
5. Lightowlers, R.N.; Rozanska, A.; Chrzanowska-Lightowlers, Z.M. Mitochondrial protein synthesis: Figuring the fundamentals, complexities and complications, of mammalian mitochondrial translation. *FEBS Lett.* **2014**, *588*, 2496–2503. [CrossRef] [PubMed]
6. Kuzmenko, A.; Atkinson, G.C.; Levitskii, S.; Zenkin, N.; Tenson, T.; Hauryliuk, V.; Kamenski, P. Mitochondrial translation initiation machinery: Conservation and diversification. *Biochimie* **2014**, *100*, 132–140. [CrossRef]
7. Agirrezabala, X.; Frank, J. Elongation in translation as a dynamic interaction among the ribosome, tRNA, and elongation factors EF-G and EF-Tu. *Q. Rev. Biophys.* **2009**, *42*, 159–200. [CrossRef] [PubMed]
8. Kummer, E.; Ban, N. Mechanisms and regulation of protein synthesis in mitochondria. *Nat. Rev. Mol. Cell Biol.* **2021**, *22*, 307–325. [CrossRef] [PubMed]
9. Kummer, E.; Leibundgut, M.; Rackham, O.; Lee, R.G.; Boehringer, D.; Filipovska, A.; Ban, N. Unique features of mammalian mitochondrial translation initiation revealed by cryo-EM. *Nature* **2018**, *560*, 263–267. [CrossRef]
10. Luo, Y.; Su, R.; Wang, Y.; Xie, W.; Liu, Z.; Huang, Y. *Schizosaccharomyces pombe* Mti2 and Mti3 act in conjunction during mitochondrial translation initiation. *FEBS J.* **2019**, *286*, 4542–4553. [CrossRef]
11. Luo, Y.; Bähler, J.; Huang, Y. The Insertion Domain of Mti2 Facilitates the Association of Mitochondrial Initiation Factors with Mitoribosomes in *Schizosaccharomyces pombe*. *Biomolecules* **2025**, *15*, 695. [CrossRef] [PubMed]
12. Wang, Y.; Yan, J.; Zhang, Q.; Ma, X.; Zhang, J.; Su, M.; Wang, X.; Huang, Y. The *Schizosaccharomyces pombe* PPR protein Ppr10 associates with a novel protein Mpa1 and acts as a mitochondrial translational activator. *Nucleic Acids Res.* **2017**, *45*, 3323–3340. [CrossRef] [PubMed]
13. Luo, Y.; Wang, Y.; Huang, Y. *Schizosaccharomyces pombe* Ppr10 and Mpa1 together mediate mitochondrial translational initiation. *J. Biol. Chem.* **2021**, *297*, 100869. [CrossRef]
14. Zuo, L.; Zhou, T.; Chuang, C.-C. The Consequences of Damaged Mitochondrial DNA. In *Mitochondrial Mechanisms of Degeneration and Repair in Parkinson's Disease*; Buhlman, L.M., Ed.; Springer International Publishing: Cham, Switzerland, 2016; pp. 49–61.
15. Kotiadis, V.N.; Duchen, M.R.; Osellame, L.D. Mitochondrial quality control and communications with the nucleus are important in maintaining mitochondrial function and cell health. *Biochim. Biophys. Acta* **2014**, *1840*, 1254–1265. [CrossRef] [PubMed]
16. Malecki, M.; Bitton, D.A.; Rodríguez-López, M.; Rallis, C.; Calavia, N.G.; Smith, G.C.; Bähler, J. Functional and regulatory profiling of energy metabolism in fission yeast. *Genome Biol.* **2016**, *17*, 240. [CrossRef]
17. Desai, R.; East, D.A.; Hardy, L.; Faccenda, D.; Rigon, M.; Crosby, J.; Alvarez, M.S.; Singh, A.; Mainenti, M.; Hussey, L.K.; et al. Mitochondria form contact sites with the nucleus to couple prosurvival retrograde response. *Sci. Adv.* **2020**, *6*, eabc9955. [CrossRef] [PubMed]
18. Trendeleva, T.A.; Zvyagilskaya, R.A. Retrograde Signaling as a Mechanism of Yeast Adaptation to Unfavorable Factors. *Biochemistry* **2018**, *83*, 98–106. [CrossRef] [PubMed]
19. Hunt, R.J.; Bateman, J.M. Mitochondrial retrograde signaling in the nervous system. *FEBS Lett.* **2018**, *592*, 663–678. [CrossRef]
20. Richter, A.S.; Nägele, T.; Grimm, B.; Kaufmann, K.; Schroda, M.; Leister, D.; Kleine, T. Retrograde signaling in plants: A critical review focusing on the GUN pathway and beyond. *Plant Commun.* **2023**, *4*, 100511. [CrossRef]
21. Epstein, C.B.; Waddle, J.A.; Hale, W.T.; Davé, V.; Thornton, J.; Macatee, T.L.; Garner, H.R.; Butow, R.A. Genome-wide responses to mitochondrial dysfunction. *Mol. Biol. Cell* **2001**, *12*, 297–308. [CrossRef]
22. Guha, S.; López-Maury, L.; Shaw, M.; Bähler, J.; Norbury, C.J.; Agashe, V.R. Transcriptional and cellular responses to defective mitochondrial proteolysis in fission yeast. *J. Mol. Biol.* **2011**, *408*, 222–237. [CrossRef]
23. Broach, J.R. Nutritional control of growth and development in yeast. *Genetics* **2012**, *192*, 73–105. [CrossRef] [PubMed]
24. Bahn, Y.S.; Xue, C.; Idnurm, A.; Rutherford, J.C.; Heitman, J.; Cardenas, M.E. Sensing the environment: Lessons from fungi. *Nat. Rev. Microbiol.* **2007**, *5*, 57–69. [CrossRef]
25. Vassiliadis, D.; Wong, K.H.; Andrianopoulos, A.; Monahan, B.J. A genome-wide analysis of carbon catabolite repression in *Schizosaccharomyces pombe*. *BMC Genom.* **2019**, *20*, 251. [CrossRef] [PubMed]
26. Matsuzawa, T.; Fujita, Y.; Tohda, H.; Takegawa, K. Snf1-like protein kinase Ssp2 regulates glucose derepression in *Schizosaccharomyces pombe*. *Eukaryot. Cell* **2012**, *11*, 159–167. [CrossRef] [PubMed]
27. Hirota, K.; Hoffman, C.S.; Ohta, K. Reciprocal nuclear shuttling of two antagonizing Zn finger proteins modulates Tup family corepressor function to repress chromatin remodeling. *Eukaryot. Cell* **2006**, *5*, 1980–1989. [CrossRef]
28. Mukai, Y.; Matsuo, E.; Roth, S.Y.; Harashima, S. Conservation of histone binding and transcriptional repressor functions in a *Schizosaccharomyces pombe* Tup1p homolog. *Mol. Cell Biol.* **1999**, *19*, 8461–8468. [CrossRef]
29. Bähler, J.; Wu, J.Q.; Longtine, M.S.; Shah, N.G.; McKenzie, A., 3rd; Steever, A.B.; Wach, A.; Philippsen, P.; Pringle, J.R. Heterologous modules for efficient and versatile PCR-based gene targeting in *Schizosaccharomyces pombe*. *Yeast* **1998**, *14*, 943–951. [CrossRef]

30. Kim, D.U.; Hayles, J.; Kim, D.; Wood, V.; Park, H.O.; Won, M.; Yoo, H.S.; Duhig, T.; Nam, M.; Palmer, G.; et al. Analysis of a genome-wide set of gene deletions in the fission yeast *Schizosaccharomyces pombe*. *Nat. Biotechnol.* **2010**, *28*, 617–623. [CrossRef]
31. Anver, S.; Sumit, A.F.; Sun, X.M.; Hatimy, A.; Thalassinos, K.; Marguerat, S.; Alic, N.; Bähler, J. Ageing-associated long non-coding RNA extends lifespan and reduces translation in non-dividing cells. *EMBO Rep.* **2024**, *25*, 4921–4949. [CrossRef]
32. Luo, Y.; Xu, Y.; Ahmad, F.; Feng, G.; Huang, Y. Characterization of Shy1, the *Schizosaccharomyces pombe* homolog of human SURF1. *Sci. Rep.* **2024**, *14*, 21678. [CrossRef] [PubMed]
33. Lera-Ramírez, M.; Bähler, J.; Mata, J.; Rutherford, K.; Hoffman, C.S.; Lambert, S.; Oliferenko, S.; Martin, S.G.; Gould, K.L.; Du, L.L.; et al. Revised fission yeast gene and allele nomenclature guidelines for machine readability. *Genetics* **2023**, *225*, iyad143. [CrossRef] [PubMed]
34. Patro, R.; Duggal, G.; Love, M.I.; Irizarry, R.A.; Kingsford, C. Salmon provides fast and bias-aware quantification of transcript expression. *Nat. Methods* **2017**, *14*, 417–419. [CrossRef]
35. Lock, A.; Rutherford, K.; Harris, M.A.; Hayles, J.; Oliver, S.G.; Bähler, J.; Wood, V. PomBase 2018: User-driven reimplementation of the fission yeast database provides rapid and intuitive access to diverse, interconnected information. *Nucleic Acids Res.* **2019**, *47*, D821–D827. [CrossRef]
36. Sherman, B.T.; Hao, M.; Qiu, J.; Jiao, X.; Baseler, M.W.; Lane, H.C.; Imamichi, T.; Chang, W. DAVID: A web server for functional enrichment analysis and functional annotation of gene lists (2021 update). *Nucleic Acids Res.* **2022**, *50*, W216–W221. [CrossRef]
37. Huang, D.W.; Sherman, B.T.; Lempicki, R.A. Systematic and integrative analysis of large gene lists using DAVID bioinformatics resources. *Nat. Protoc.* **2009**, *4*, 44–57. [CrossRef]
38. R Development Core Team. R: A Language and Environment for Statistical Computing. 2024. Available online: https://www.r-project.org/ (accessed on 11 July 2024).
39. Shen, L. GeneOverlap: Test and visualize gene overlaps. In *Bioconductor*; Roswell Park Comprehensive Cancer Center: Buffalo, NY, USA, 2022. [CrossRef]
40. Wang, M.; Zhao, Y.; Zhang, B. Efficient Test and Visualization of Multi-Set Intersections. *Sci. Rep.* **2015**, *5*, 16923. [CrossRef]
41. Kahm, M.; Hasenbrink, G.; Lichtenberg-Fraté, H.; Ludwig, J.; Kschischo, M. grofit: Fitting Biological Growth Curves with R. *J. Stat. Softw.* **2010**, *33*, 1–21. [CrossRef]
42. Romila, C.A.; Townsend, S.; Malecki, M.; Kamrad, S.; Rodríguez-López, M.; Hillson, O.; Cotobal, C.; Ralser, M.; Bähler, J. Barcode sequencing and a high-throughput assay for chronological lifespan uncover ageing-associated genes in fission yeast. *Microb. Cell* **2021**, *8*, 146–160. [CrossRef]
43. Rallis, C.; Townsend, S.; Bähler, J. Genetic interactions and functional analyses of the fission yeast *gsk3* and *amk2* single and double mutants defective in TORC1-dependent processes. *Sci. Rep.* **2017**, *7*, 44257. [CrossRef] [PubMed]
44. Wagih, O.; Parts, L. gitter: A robust and accurate method for quantification of colony sizes from plate images. *G3 Genes Genomes Genet.* **2014**, *4*, 547–552. [CrossRef] [PubMed]
45. Wagih, O.; Usaj, M.; Baryshnikova, A.; VanderSluis, B.; Kuzmin, E.; Costanzo, M.; Myers, C.L.; Andrews, B.J.; Boone, C.M.; Parts, L. SGAtools: One-stop analysis and visualization of array-based genetic interaction screens. *Nucleic Acids Res.* **2013**, *41*, W591–W596. [CrossRef] [PubMed]
46. Hamburger, K.; Kramhft, B. The Effect of Chloramphenicol on Respiration, Fermentation and Growth in *Schizosaccharomyces pombe*. *Microbiology* **1980**, *120*, 279–282. [CrossRef]
47. Hu, Y.; Luo, Y.; Yin, D.; Zhao, L.; Wang, Y.; Rui, Y.; Zhang, P.; Wu, X.; Li, M.; Hidalgo, E.; et al. *Schizosaccharomyces pombe* MAP kinase Sty1 promotes survival of Δppr10 cells with defective mitochondrial protein synthesis. *Int. J. Biochem. Cell Biol.* **2022**, *152*, 106308. [CrossRef] [PubMed]
48. Vustin, M.M. The Biological Role of Glycerol in Yeast Cells. Yeast as Glycerol Producers. *Appl. Biochem. Microbiol.* **2021**, *57*, 907–916. [CrossRef]
49. Chen, D.; Toone, W.M.; Mata, J.; Lyne, R.; Burns, G.; Kivinen, K.; Brazma, A.; Jones, N.; Bähler, J. Global transcriptional responses of fission yeast to environmental stress. *Mol. Biol. Cell* **2003**, *14*, 214–229. [CrossRef]
50. Chen, D.; Wilkinson, C.R.; Watt, S.; Penkett, C.J.; Toone, W.M.; Jones, N.; Bähler, J. Multiple pathways differentially regulate global oxidative stress responses in fission yeast. *Mol. Biol. Cell* **2008**, *19*, 308–317. [CrossRef]
51. Lunt, S.Y.; Vander Heiden, M.G. Aerobic Glycolysis: Meeting the Metabolic Requirements of Cell Proliferation. *Annu. Rev. Cell Dev. Biol.* **2011**, *27*, 441–464. [CrossRef]
52. Kamrad, S.; Grossbach, J.; Rodríguez-López, M.; Mülleder, M.; Townsend, S.; Cappelletti, V.; Stojanovski, G.; Correia-Melo, C.; Picotti, P.; Beyer, A.; et al. Pyruvate kinase variant of fission yeast tunes carbon metabolism, cell regulation, growth and stress resistance. *Mol. Syst. Biol.* **2020**, *16*, e9270. [CrossRef]
53. Rutherford, K.M.; Lera-Ramírez, M.; Wood, V. PomBase: A Global Core Biodata Resource-growth, collaboration, and sustainability. *Genetics* **2024**, *227*, iyae007. [CrossRef]
54. Woolson, R.F. Wilcoxon Signed-Rank Test. In *Wiley Encyclopedia of Clinical Trials*; Wiley: Hoboken, NJ, USA, 2007; pp. 1–3.

55. Takenaka, K.; Tanabe, T.; Kawamukai, M.; Matsuo, Y. Overexpression of the transcription factor Rst2 in *Schizosaccharomyces pombe* indicates growth defect, mitotic defects, and microtubule disorder. *Biosci. Biotechnol. Biochem.* **2018**, *82*, 247–257. [CrossRef]
56. Kunitomo, H.; Higuchi, T.; Iino, Y.; Yamamoto, M. A zinc-finger protein, Rst2p, regulates transcription of the fission yeast ste11(+) gene, which encodes a pivotal transcription factor for sexual development. *Mol. Biol. Cell* **2000**, *11*, 3205–3217. [CrossRef]
57. Baryshnikova, A.; Costanzo, M.; Dixon, S.; Vizeacoumar, F.J.; Myers, C.L.; Andrews, B.; Boone, C. Synthetic genetic array (SGA) analysis in *Saccharomyces cerevisiae* and *Schizosaccharomyces pombe*. *Methods Enzymol.* **2010**, *470*, 145–179. [CrossRef]
58. Costanzo, M.; Kuzmin, E.; van Leeuwen, J.; Mair, B.; Moffat, J.; Boone, C.; Andrews, B. Global Genetic Networks and the Genotype-to-Phenotype Relationship. *Cell* **2019**, *177*, 85–100. [CrossRef]
59. Bitton, D.A.; Schubert, F.; Dey, S.; Okoniewski, M.; Smith, G.C.; Khadayate, S.; Pancaldi, V.; Wood, V.; Bähler, J. AnGeLi: A Tool for the Analysis of Gene Lists from Fission Yeast. *Front. Genet.* **2015**, *6*, 330. [CrossRef]
60. Harris, M.A.; Lock, A.; Bähler, J.; Oliver, S.G.; Wood, V. FYPO: The fission yeast phenotype ontology. *Bioinformatics* **2013**, *29*, 1671–1678. [CrossRef]
61. Sideri, T.; Rallis, C.; Bitton, D.A.; Lages, B.M.; Suo, F.; Rodríguez-López, M.; Du, L.L.; Bähler, J. Parallel profiling of fission yeast deletion mutants for proliferation and for lifespan during long-term quiescence. *G3 Genes Genomes Genet.* **2014**, *5*, 145–155. [CrossRef] [PubMed]
62. The Gene Ontology resource: Enriching a GOld mine. *Nucleic Acids Res.* **2021**, *49*, D325–D334. [CrossRef] [PubMed]
63. Kühl, I.; Dujeancourt, L.; Gaisne, M.; Herbert, C.J.; Bonnefoy, N. A genome wide study in fission yeast reveals nine PPR proteins that regulate mitochondrial gene expression. *Nucleic Acids Res.* **2011**, *39*, 8029–8041. [CrossRef]
64. Bonfils, S.; Rozalén, A.E.; Smith, G.R.; Moreno, S.; Martín-Castellanos, C. Functional interactions of Rec24, the fission yeast ortholog of mouse Mei4, with the meiotic recombination-initiation complex. *J. Cell Sci.* **2011**, *124*, 1328–1338. [CrossRef] [PubMed]
65. Steiner, S.; Kohli, J.; Ludin, K. Functional interactions among members of the meiotic initiation complex in fission yeast. *Curr. Genet.* **2010**, *56*, 237–249. [CrossRef] [PubMed]
66. Lin, X.; Zhang, C.Y.; Bai, X.W.; Song, H.Y.; Xiao, D.G. Effects of MIG1, TUP1 and SSN6 deletion on maltose metabolism and leavening ability of baker's yeast in lean dough. *Microb. Cell Factories* **2014**, *13*, 93. [CrossRef] [PubMed]
67. Alipourfard, I.; Bakhtiyari, S.; Datukishvili, N.; Haghani, K.; Di Renzo, L.; De Miranda, R.C.; Cioccoloni, G.; Basiratyan Yazdi, P.; Mikeladze, D. *Saccharomyces cerevisiae*, key role of MIG1 gene in metabolic switching: Putative fermentation/oxidation. *J. Biol. Regul. Homeost. Agents.* **2018**, *32*, 649–654.
68. Laera, L.; Guaragnella, N.; Ždralević, M.; Marzulli, D.; Liu, Z.; Giannattasio, S. The transcription factors ADR1 or CAT8 are required for RTG pathway activation and evasion from yeast acetic acid-induced programmed cell death in raffinose. *Microb. Cell* **2016**, *3*, 621–631. [CrossRef]
69. Higuchi, T.; Watanabe, Y.; Yamamoto, M. Protein kinase A regulates sexual development and gluconeogenesis through phosphorylation of the Zn finger transcriptional activator Rst2p in fission yeast. *Mol. Cell Biol.* **2002**, *22*, 1–11. [CrossRef]
70. Jiang, G.; Liu, Q.; Kato, T.; Miao, H.; Gao, X.; Liu, K.; Chen, S.; Sakamoto, N.; Kuno, T.; Fang, Y. Role of mitochondrial complex III/IV in the activation of transcription factor Rst2 in *Schizosaccharomyces pombe*. *Mol. Microbiol.* **2021**, *115*, 1323–1338. [CrossRef]
71. Gupta, D.R.; Paul, S.K.; Oowatari, Y.; Matsuo, Y.; Kawamukai, M. Multistep regulation of protein kinase A in its localization, phosphorylation and binding with a regulatory subunit in fission yeast. *Curr. Genet.* **2011**, *57*, 353–365. [CrossRef]
72. Skribbe, M.; Soneson, C.; Stadler, M.B.; Schwaiger, M.; Suma Sreechakram, V.N.; Iesmantavicius, V.; Hess, D.; Moreno, E.P.F.; Braun, S.; Seebacher, J.; et al. A comprehensive *Schizosaccharomyces pombe* atlas of physical transcription factor interactions with proteins and chromatin. *Mol. Cell* **2025**, *85*, 1426–1444.e8. [CrossRef]

biomolecules

MDPI

*Article*

# Diapause and Anoxia-Induced Quiescence Are Unique States in Embryos of the Annual Killifish, *Austrofundulus limnaeus*

**Patrick R. Clouser [1,*], Claire L. Riggs [2,3], Amie L. T. Romney [1] and Jason E. Podrabsky [1,*]**

1 Department of Biology, Center for Life in Extreme Environments, Portland State University, P.O. Box 751, Portland, OR 97201, USA; arom2@pdx.edu

2 Division of Rheumatology, Inflammation, and Immunity, Brigham and Women's Hospital, Boston, MA 02115, USA; clriggs@bwh.harvard.edu

3 Department of Medicine, Harvard Medical School, Boston, MA 02115, USA

* Correspondence: pat6@pdx.edu (P.R.C.); jpod@pdx.edu (J.E.P.)

**Abstract:** Diapause is a state of developmental and metabolic dormancy that precedes exposure to environmental stresses. Yet, diapausing embryos are typically stress-tolerant. Evidence suggests that diapausing embryos "prepare" for stress as part of a gene expression program as they enter dormancy. Here, we investigate if diapause II embryos of the annual killifish *Austrofundulus limnaeus*, which can survive for hundreds of days of anoxia, can mount a transcriptomic response to anoxic insult. Bulk RNAseq was used to characterize the transcriptomes of diapause II embryos exposed to normoxia, 4 h and 24 h anoxia, and 2 h and 24 h normoxic recovery from anoxia. Differential expression and gene ontology analyses were used to probe for pathways that may mitigate survival. Transcriptional factor analysis was used to predict potential mediators of this response. Diapausing embryos exhibited a robust transcriptomic response to anoxia and recovery that returns to near baseline conditions after 24 h. Anoxia induced an upregulation of genes involved in the integrated stress response, lipid metabolism, p38mapk kinase signaling, and apoptosis. Developmental and mitochondrial genes decreased. We conclude that diapause II embryos mount a robust transcriptomic stress response when faced with anoxic insult. This response is consistent with mediating expected challenges to cellular homeostasis in anoxia.

**Keywords:** dormancy; stress; transcriptomics; development; killifish

## 1. Introduction

Metabolic dormancy is deployed across the tree of life to synchronize activity, reproduction, and development with favorable environmental conditions [1–3]. Diapause is a state of developmental and metabolic dormancy that occurs during embryonic development in a variety of animal species [3]. Diapause is considered an endogenously promoted dormancy, which means that internal signals—typically considered part of the developmental program—control entrance into dormancy. Thus, diapause dormancy occurs under conditions conducive to normal development, although environmental signals and factors are typically involved in the induction of diapause. Stress tolerance, sometimes even extreme stress tolerance, almost always co-occurs with diapause [4]. Due to the general physiological characteristics of most diapausing embryos, such as severely reduced rates of ATP turnover and protein synthesis, it is somewhat paradoxical that they exhibit increased tolerance to environmental stresses that require a robust and complex physiological and molecular response in active animals. Thus, it is assumed that diapausing embryos "prepare" for stress as part of a gene expression program as they enter dormancy.

 https://doi.org/10.3390/biom15040515

Indeed, there is a great deal of evidence supporting the accumulation of molecular and chemical chaperones and the alteration of metabolic pathways during the entry phase of diapause [5–10]. However, few studies have profiled global gene expression responses to stress in diapausing embryos. In this study, we challenge diapausing annual killifish embryos with anoxic stress to test the hypothesis that they are already poised to tolerate stress and will lack a typical transcriptional response.

While there are many species of annual killifishes that produce diapausing embryos, the most well-studied species in terms of the physiological, biochemical, and molecular aspects of diapause and stress tolerance is *Austrofundulus limnaeus* [11,12]. Populations of *A. limnaeus* survive in ephemeral ponds by producing embryos that can enter embryonic diapause at three distinct developmental stages, termed diapause I, II, and III [13,14]. Entry into diapause II, which occurs midway through development, is a unique developmental trajectory with patterns of gene expression that diverge from actively developing embryos many days prior to the cessation of development [15]. During this period, embryos experience an increase in tolerance to environmental stresses such as oxygen limitation and osmotic stress [11,16–19]. The gene expression program associated with entrance into diapause II is presumed to prepare the embryos for survival of the harsh conditions associated with the dry season in their native habitat, because once in diapause, embryos have a severely reduced metabolic rate and experience over a 90% reduction in the rate of protein synthesis [20,21].

Embryos of *A. limnaeus* are the most anoxia-tolerant vertebrates [11,19,22–24]. Diapause II embryos can survive for hundreds of days without oxygen and have an $LT_{50}$ of 62 days of anoxia at 25 °C [25]. Embryos of *A. limnaeus* respond quickly to removal of oxygen. When diapause II embryos are exposed to anoxia, heat dissipation decreases slightly but not significantly over the first 16 h, suggesting that metabolism remains relatively constant across the transition into anoxia [26]. However, ATP levels fall by 50%, and various indices of energetic status suggest a severely energy-limited system after only 14 h of anoxia [26]. Amazingly, embryos continue to survive for months with severely reduced ATP levels. Thus, we hypothesized that diapausing embryos of *A. limnaeus* will be prepared for stress and have a muted or non-existent stress response during initial exposure to anoxia. Alternatively, if they do respond to anoxic stress, it will most likely occur within the first 24 h of exposure and before the large-scale reductions in ATP are complete. Here we report that diapausing embryos respond with a strong transcriptomic response within 24 h of anoxia. This paper highlights the players that likely orchestrate the early responses to oxygen limitation in this species and may explain the extreme stress tolerance in these embryos.

## 2. Materials and Methods

### 2.1. Embryo Rearing and Staging

Embryos were incubated at 25 °C in the dark, and embryo medium was changed daily. At 4 days post-fertilization (dpf), the embryos were treated with two 5 min washes in 0.03% sodium hypochlorite (Clorox bleach, Clorox company, Oakland, CA, USA), as previously described [27]. Following bleaching, the embryos were incubated in embryo medium containing 10 µg/L gentamicin sulfate. Embryos were staged according to Wourms' staging regime; embryos typically enter diapause II at 24 days post-fertilization (dpf) when incubated at 25 °C [14,28].

### 2.2. Anoxic Exposure and Aerobic Recovery

Diapause II embryos at 32 dpf were exposed to 24 h of anoxia at 25 °C in the dark in a Bactron III anaerobic chamber (Sheldon Manufacturing, Cornelius, OR, USA), as previously described [25,29]. This methodology immediately exposes embryos to an anoxic environment. Aerobic recovery was initiated by removing the embryos from the chamber and rinsing them three times in normoxic embryo medium. Recovering embryos were incubated at 25 °C in the dark until sampling [25]. Embryos were sampled at the following time points: normoxia (t = 0), 4 h and 24 h anoxia, and after 2 h and 24 h of aerobic recovery (Figure 1). Experimental trials were repeated on 4 separate spawning events (n = 4). Each time point/sample consisted of groups of 20 pooled embryos. For qPCR validation of RNAseq results, only the normoxic and 24 h anoxia time points were replicated in diapause II embryos that were 55 dpf (n = 3–4).

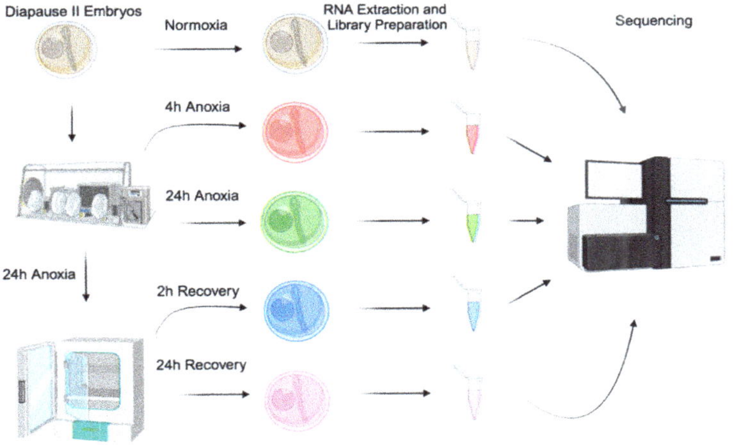


**Figure 1.** Experimental design and workflow. Experimental conditions include the following: normoxia (t = 0, tan), 4 h of anoxia (red), 24 h of anoxia (green), 2 h of normoxic recovery (green), and 24 h of normoxic recovery (pink). All samples are a pool of 20 embryos, and 4 different spawning events were sampled (n = 4).

## 2.3. Sampling Embryos for RNA Sequencing and qPCR

To sample, embryos were blotted dry on a 100 µm nylon mesh screen, and excess embryo medium was wicked away on a paper towel. Groups of 20 embryos were transferred to a pre-weighed microcentrifuge tube and flash-frozen in liquid nitrogen. This procedure was performed within the anaerobic chamber for anoxic samples that were immediately flash-frozen upon removal from the anaerobic environment [25]. Frozen samples were stored at −80 °C until RNA extraction.

## 2.4. RNA Extraction

Total RNA was extracted from each sample with TRIzol reagent (Invitrogen, Carlsbad, CA, USA), as previously described [25,30–33]. RNA was precipitated by overnight incubation in a high-salt solution at −20 °C according to the manufacturer's instructions [32], and samples were resuspended in 1 mM sodium citrate (pH = 6.5). Concentration and purity of RNA were determined by UV spectroscopy using an a NanoQuant plate and an Infinite Pro M200 plate reader (Tecan, San Jose, CA, USA; Table S1). RNA integrity was assessed using agarose gel electrophoresis. RNA samples were stored at −80 °C until library preparation. The RNA samples used in this study are the same as those used to previously profile small noncoding RNAs in response to anoxia [25].

### 2.5. Poly-A RNA Sequencing

cDNA libraries for sequencing were prepared using the Illumina TruSeq RNA Sample Preparation Kit (v2, Illumina, San Diego, CA, USA) according to the manufacturer's instructions with a starting input of 1 µg of total RNA. The quality of the purified cDNA libraries was assessed using an Agilent 2100 Bioanalyzer equipped with a DNA 1000 chip (Agilent Technologies, Santa Clara, CA, USA). Quantitative PCR was used to measure cDNA content. Libraries were sequenced (100 nt paired-end reads) on an Illumina HiSeq 2000 at the Oregon Health & Science University's Massively Parallel Sequencing Shared Resource Facility.

### 2.6. Analysis of Sequence Data

Sequence data were analyzed as previously reported for transcriptomic analysis in *A. limnaeus* [34]. The bioinformatic pipeline was performed on a Linux high-performance computing resource at PSU. Read data quality was assessed using FastQC (version 0.10.1) to ensure high-quality sequencing results for each sample [35]. Poor-quality sequence reads and adaptors were removed with Trimmomatic (v 0.33) using the following settings: "ILLUMINACLIP:2:30:7:1:true", "SLIDINGWINDOW:5:15", "LEADING:20", "TRAILING:20", and "MINLEN: 25" [36]. Resultant reads were mapped to the *A. limnaeus* genome with Bowtie2 (v 2.5.1) using preset parameters for very fast local alignments [37]. Only reads that aligned with 0 mismatches were used. Gene counts were generated with the R/Bioconductor software package GenomicAlignments (ver 1.42.0) in union mode with the *summarizeOverlaps* function [38] in conjunction with the NCBI *A. limnaeus* genome annotation Release 100 [39].

### 2.7. Differential Gene Expression Analysis

Differential expression analysis was performed using the R package DEseq2 (v 1.46.0) [40]. Gene count data were initially prefiltered to remove genes with row sum totals across all conditions with less than 5 counts and used to create a DEseq data set object. Principal component analysis was also performed with DEseq2 to assess the variance between all experimental conditions. Differentially expression data were acquired at the gene level for all experimental conditions by pairwise comparison to normoxic (t = 0) samples. Transcripts with a *p*-value less than 0.05 (Benjamini–Hochberg corrected for multiple comparisons) and a $\log_2$ fold change $\pm$ 0.58 were considered differentially expressed [40].

### 2.8. Quantitative Real-Time Polymerase Chain Reaction (qPCR)

The original RNA samples used to generate the RNAseq data were lost due to a freezer failure, so new RNA samples were generated as described above using diapause II embryos exposed to normoxia and 24 h of anoxia. Total RNA in the samples was quantified using a NanoDrop One spectrophotometer (ThermoFisher, Scientific, Waltham, MA, USA). Single-stranded cDNA libraries were generated using a Qiagen reverse transcriptase kit (RT-020, Qiagen, Hilden, Germany) using 85 ng of total RNA as the starting material in a 20 µL reaction. Samples were incubated for 10 min at 25 °C, 30 min at 50 °C, and 5 min at 85 °C in a thermocycler (Geneamp 2700 thermocycler, Applied Biosystems, Waltham, MA, USA) and then immediately placed on ice. qPCR (Applied Biosystems Vii7 Real-Time PCR system, Applied Biosystems, Waltham, MA, USA) was performed on 10 µL reactions in a 386-well plate format. Each reaction consisted of 5 µL of Luna universal PCR master mix (New England biolabs, Ipswich, MA, USA), 0.25 µL each of forward and reverse primers (Integrated DNA Technologies, Coralville, IA, USA), 3.5 µL nuclease-free water, and 1 µL of ss cDNA template. Reactions were subjected to an initial denaturation step at 95 °C for 60 s, followed by 40 cycles of denaturation for 15 s at 95 °C and extension for 30 s at 60 °C.

Melting curves were performed after each reaction to confirm a single PCR product. At least 2 technical replicates were performed for each sample, on 3–4 biological replicates. Relative fold-change values between normoxia and 24 h anoxia were calculated using the ΔCt method with the mean Ct values for normoxic samples serving as the reference for each anoxic sample.

*2.9. Ortholog Mapping and Gene Ontology Analysis*

Orthologous proteins were identified between *A. limnaeus* and *Homo sapiens* using NCBI BLASTp and the R package OrthoFinder (v 2.5.5) [41]. Sequences for all *A. limnaeus* proteins were aligned to the GHR38 human protein database retrieved from ENSMBL, keeping the top returned result with an e-value less than 0.0001. The same *A. limnaeus* protein sequences were also analyzed against the human genome with OrthoFinder [41]. The two returned lists were then merged, and any duplicate *A. limnaeus* proteins found associated to the same transcript were removed.

Gene ontologies analyses were generated using the R library clusterProfiler (v 4.14.4) [42], with the *H. sapiens* database from annotationDBi (v 1.68.0). Differential expression data from DEseq2 for each of the experimental conditions were matched to the ENSMBL ID of the *H. sapiens* ortholog list described above. Overrepresentation analysis (ORA) was used to return the associated statistically significant gene ontology terms for biological process, molecular function, and cellular component ($p$-values < 0.05). The simplify function in clusterProfiler was used to remove redundant terms (cutoff = 0.5, measure = "Wang"). In addition, KEGG pathways that were enriched in the differentially expressed genes were also identified within clusterProfiler. Relationships between the sets of differentially expressed genes in each treatment (and their associated gene ontologies and KEGG pathways) were compared using clusterProfiler's compareCluster function.

*2.10. Transcription Factor Analysis Using TFLink*

Differentially expressed genes were interrogated for patterns that may implicate the action of known transcription factors (TFs) using the TFLink website [43]. Transcription factors that had significant associations with any of the up- or downregulated genes (with human orthologs) were identified for each experimental time point ($p$ adjust value < 0.05).

*2.11. Data Visualization*

All graphs and tables were created using R packages ggplot2 (v 3.5.1) and clusterProfiler (v 4.14.4), GraphPad Prism (v 10.4.1), and Biorender (www.biorender.com).

## 3. Results

*3.1. Sequencing Library Quality and Coverage*

Illumina sequencing yielded an average of 49.1 million paired reads per cDNA library. Of these reads, an average of 93% were mapped to the *A. limnaeus* genome with zero mismatches after trimming and filtering. This depth of sequencing represents an average of 35X coverage of the genome in the expressed transcriptome. A total of 26,401 unique gene sequences were identified in the data set. A full table of sequencing results for each cDNA library can be found in Table S1.

*3.2. Characterization of the Normoxic Diapause II Transcriptome Profile*

Normoxic diapause II embryos (t = 0) had an average FPKM of 49 ± 394 (mean ± S.D.) reads per transcript. Transcripts equal to and above one-half standard deviation of the mean (245 reads) were denoted as "highly expressed" in normoxic diapause II embryos. This gene list consists of 271 transcripts, 245 of which map to *Homo sapiens* orthologs (Table S2). An enrichment analysis of all gene ontology terms (biological processes, molecular functions, and cellular component) based on human annotations that were enriched in this gene set are listed in Figure 2. Enriched cellular components of note include cytoplasmic ribosomes, polysomes, focal adhesions, several mitochondrial components, the spliceosome complex, and the GAIT complex (Figure 2). Of note, the third most abundant transcript expressed in diapause II embryos is a long non-coding RNA (LOC106513764) with unknown function.

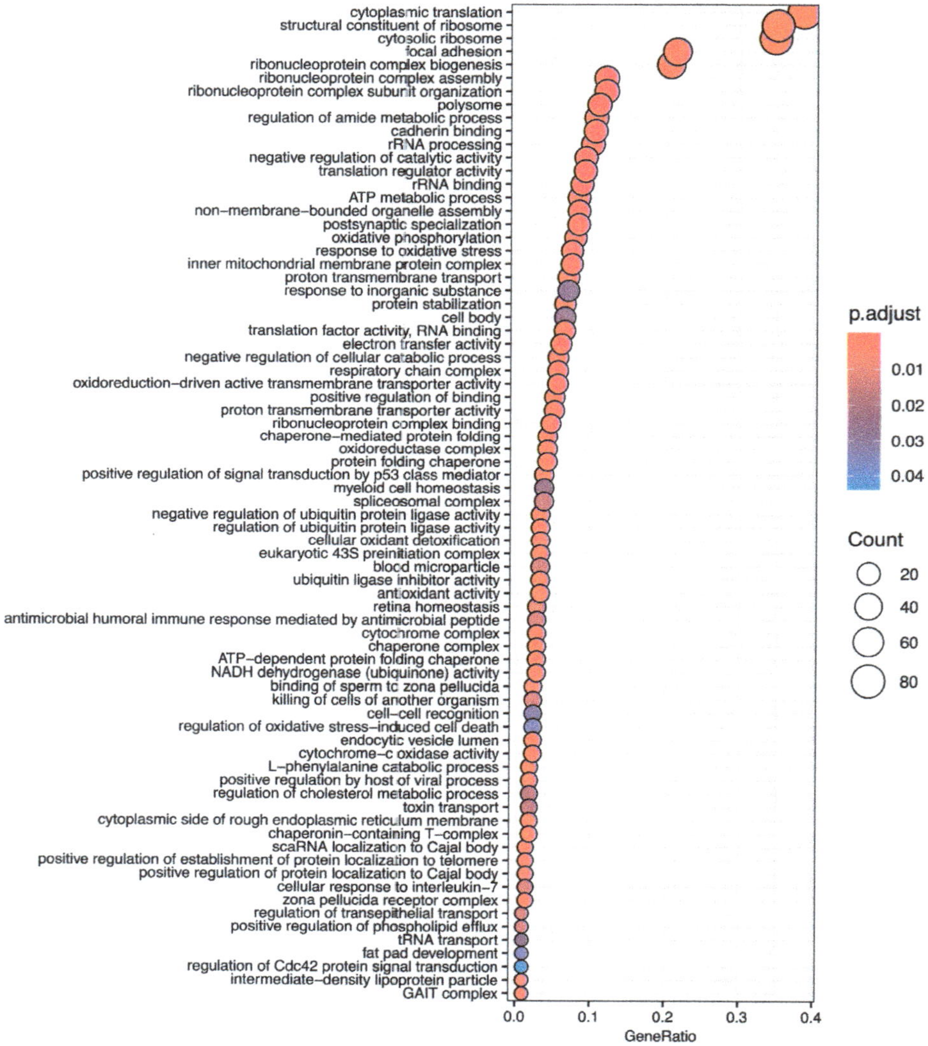


**Figure 2.** Gene ontology terms enriched in highly expressed genes in diapause II embryos of *Austrofundulus limnaeus*. Notable biological process terms are associated with protein expression, mitochondrial metabolic functions, and negative regulation of catabolism. Top molecular function terms of note include ribosomal and translational regulation, electron transport, ubiquitin ligase inhibitor activity, and protein folding. Terms listed here are limited to the highest possible term in the hierarchy.

### 3.3. Overall Characterization of Experimental Treatments

Principal component analysis reveals a separation of gene expression patterns as embryos respond to and recover from anoxia (Figure 3). Normoxic (t = 0) embryos and embryos after 24 h of aerobic recovery group together, suggesting a return to "control" gene expression patterns within 24 h of recovery. Gene expression patterns after 24 h of anoxia and at 2 h of aerobic recovery group also group together. The most distinct group of samples are those at 4 h of anoxia.

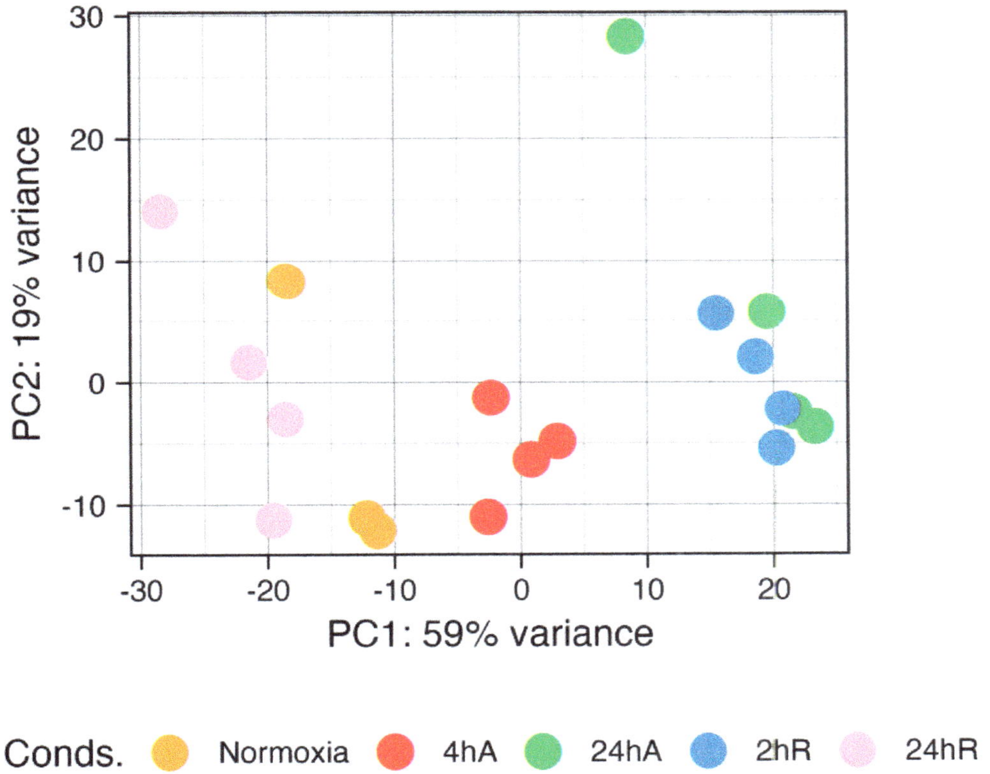


**Figure 3.** Principal component analysis of gene expression data for diapause II embryos of *Austrofundulus limnaeus* exposed to anoxia and recovery from anoxia. Gene expression patterns are similar in the 24 h of anoxia (24 hA) and 2 h (2 hR) embryos. Normoxic (t = 0) and 24 h of aerobic recovery (24 hR) embryos also share similar gene expression patterns. The 4 h of anoxia samples (4 hA) are the most unique. The proportion of the variance captured by each principal component suggests that the first two principal components explain around 80% of the variation in the data.

### 3.4. Differential Gene Expression

Differentially expressed transcripts were split into five categories: those with identified human orthologs, uncharacterized proteins in NCBI, NCBI predicted proteins, tRNAs, and noncoding RNAs as described by the NCBI's annotation of the *A. limnaeus genome* (Figure 4). A full list of differentially expressed transcripts is provided in Table S3 (up-regulated) and Table S4 (down-regulated). Differential gene expression patterns determined by DEseq2 and qPCR were consistent across platforms for a handful of genes validated by qPCR (Table S5). The greatest number of differentially expressed transcripts, relative to 0 h anoxia, were observed after 24 h of anoxia, with a total of 544 genes differentially expressed (up and down) (Figure 4A,B). Interestingly, only 37 transcripts were in higher abundance after 4 h of anoxia, and the majority of those were shared with upregulated transcripts after 24 h of anoxia and 2 h of recovery (Figures 4 and 5). This relatively small number of immediately up-regulated transcripts contains genes with some interesting functions, such as phosphoenolpyruvate carboxykinase, lipase E, iodothyronine deiodinase 2, molecular chaperones and heat shock proteins, and several genes of unknown function (Figure 5A). Only one transcript from an uncharacterized locus in *A. limnaeus* was uniquely upregulated after 4 and 24 h of anoxia, LOC106533231. This locus codes for a gene with significant similarity to human *kmt5c*, a histone methyltransferase. Over 75% of the transcripts that increased in abundance are shared between 24 h of anoxia and 2 h of aerobic recovery, and about 47% of transcripts that decreased in abundance are shared across these two time points. Interestingly, only one transcript was still upregulated after 24 h of recovery from anoxia, a locus with unknown function named protein TsetseEP-like (Figure 5D).

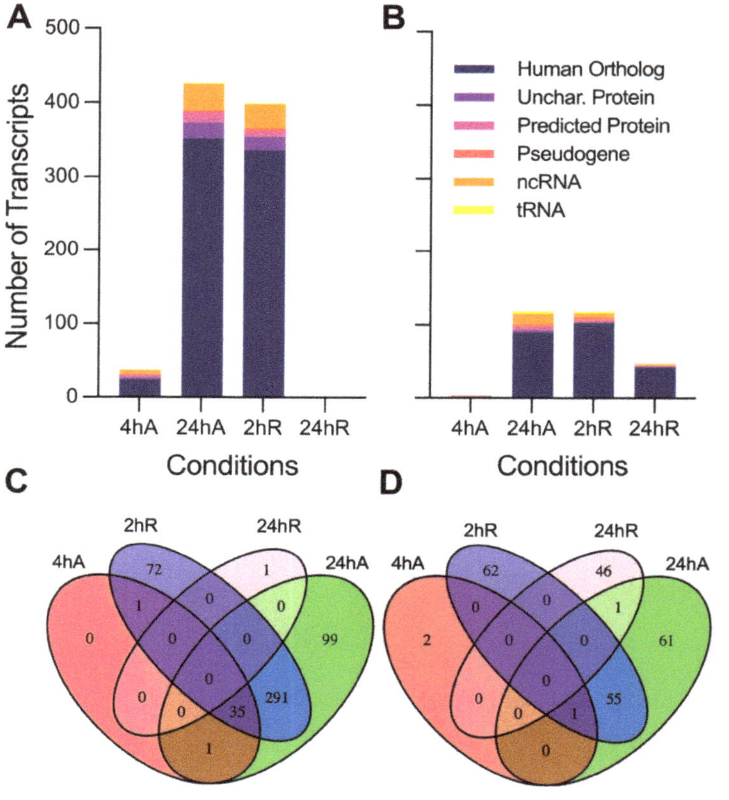


**Figure 4.** Differential gene expression patterns for diapause II embryos of *Austrofundulus limnaeus* exposed to anoxia and recovery from anoxia. Transcripts that increase in abundance (**A**) during anoxia and recovery from anoxia outnumber transcripts that decrease in abundance (**B**). Venn diagrams indicate the number of transcripts shared across experimental treatment groups that are (**C**) upregulated and (**D**) downregulated across the 24 h or exposure to and recovery from anoxia. Differential expression was determined by an adjusted $p$-value < 0.05 and a $\log_2$ fold change of $\pm 0.58$ (1.5-fold change).

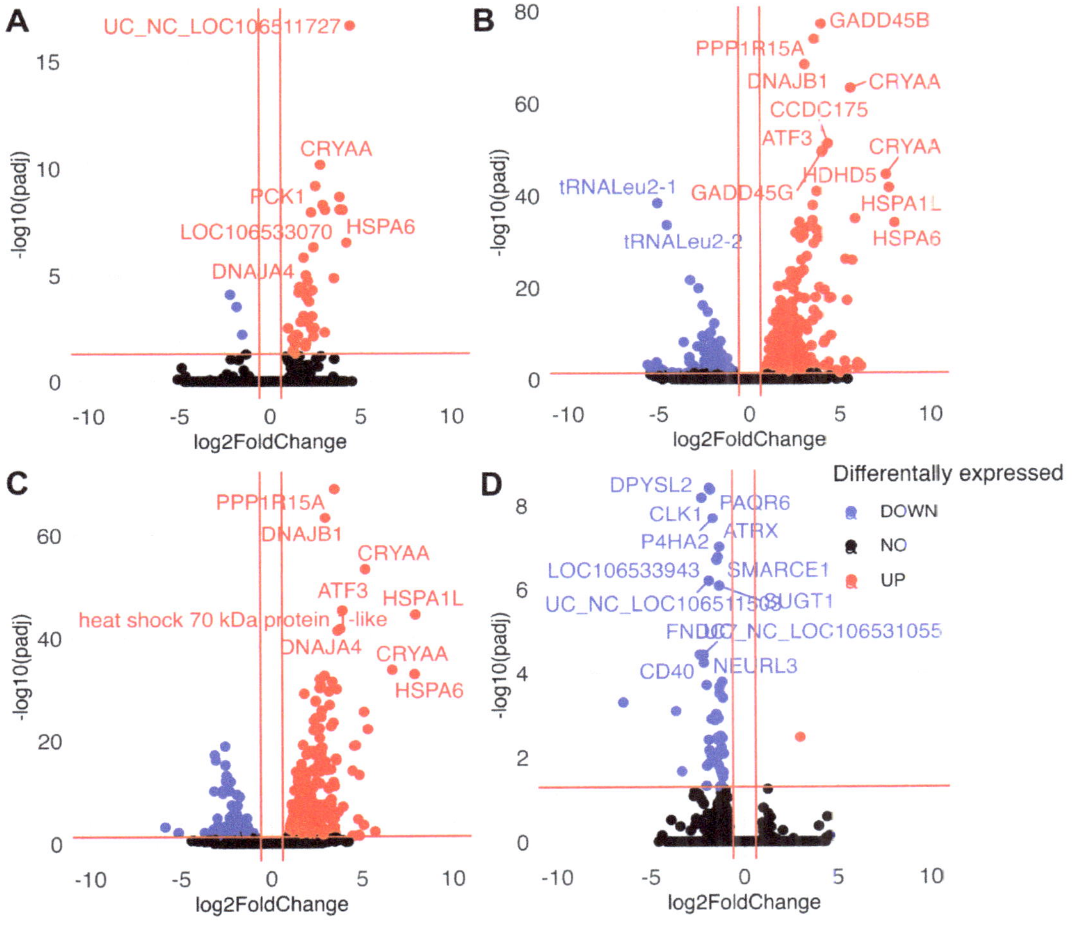


**Figure 5.** Volcano plots indicating differentially abundant transcripts in diapause II embryos of *Austrofundulus limnaeus* exposed to anoxia and recovery from anoxia. Transcripts differentially expressed after (**A**) 4 h of anoxia, (**B**) 24 h of anoxia, (**C**) 2 h of aerobic recovery from 24 h of anoxia, and (**D**) 24 h of aerobic recovery from 24 h of anoxia. Data points in blue significantly decrease in abundance, while those in red increase in abundance. Black data points are not statistically significant. Differential expression was determined by an adjusted $p$-value < 0.05 and a $\log_2$ fold change of $\pm 0.58$ (1.5-fold change).

Somewhat surprisingly, far fewer transcripts decreased in abundance in response to anoxia compared to those that increased (Figures 4B and 5A). Notably, only three transcripts immediately decreased in abundance in response to anoxia, transcripts for regulator of G-protein signaling-2 (*rgs-2*), taxilin γ (*txlng*), and a pseudo 16S transcript from the mitochondrial genome (Figure 5A). Downregulated RNAs also contained a higher proportion of noncoding RNAs and tRNAs compared to upregulated RNA species, with three mitochondrially encoded transcripts, *tRNALeu2-1*, *tRNALeu2-2*, and a segment of the pseudo16S RNA, being strongly decreased (Figure 5A–C). Interestingly, transcripts that are suppressed by anoxia do not appear to return to normoxic values, with almost 40% of the genes still in lower abundance compared to controls after 24 h of recovery (Figures 4B and 5D). Further, downregulated transcripts appear to have a more time-point specific expression pattern compared to those that are upregulated, as evidenced by the 46 transcripts uniquely downregulated only at 24 h of recovery (Figures 4D and 5D).

### 3.5. Gene Ontology Overrepresentation Analysis of Differentially Abundant Transcripts

Ortholog mapping resulted in 85% of the *A. limnaeus* proteins being mapped to human orthologs. For the differentially expressed transcripts, over 83% of transcripts were mapped to human orthologs, with an additional 9% mapping to non-coding RNAs described in the *A. limnaeus* genome. Figure 6 presents dot plots of KEGG pathways, and Figure 7 shows gene ontology terms overrepresented in lists of transcripts that increased in abundance in experimental compared to control conditions (t = 0). These pathways are highly interrelated and appear to build over time, as illustrated in Figure 8. A full list of upregulated transcripts that contribute to the KEGG and gene ontology terms are provided in Tables S6 and S7, respectively. KEGG pathway overrepresentation appears to increase over the time course of exposure to anoxia and during early recovery, suggesting a sequence of immediately acting and later-acting pathways that build on each other. The most significantly overrepresented KEGG pathways that initially respond to anoxia (4 hA) are associated with protein processing in the ER, estrogen signaling, the FoxO pathway, and a variety of cancers (Figure 6). By 24 h of anoxia, additional pathways associated with AMPK signaling, TNF-signaling, longevity, MAPK signaling, insulin signaling, and thyroid hormone signaling are overrepresented as well. Finally, during the initial recovery from anoxia, pathways associated with interleukin-17 signaling, necroptosis, NF-kappaB signaling, and HIF-1 signaling area additionally overrepresented (Figure 6). Gene ontology overrepresentation provides a slightly different pattern of functional analysis, with biological process terms being more uniquely overrepresented in different treatment groups (Figure 7). Not surprisingly, protein folding is strongly overrepresented in all groups, as is response to steroid hormones. Cell signaling pathways that are overrepresented during initial exposure to anoxia are p38MAPK cascade and cellular response to cAMP. Only during early recovery does cellular response to hypoxia become overrepresented.

Far fewer transcripts decreased in abundance in response to anoxia and recovery from anoxia, resulting in a less diverse set of overrepresented pathways and gene ontology terms (Figure 9, Table S8). Interestingly, genes that decreased in abundance do not appear to be interrelated but instead are specific to a single time point (Figure 10). After 24 h of anoxia, KEGG pathways associated with mitochondrial electron transport and ATP synthesis are overrepresented (Table S9). During early recovery from anoxia, KEGG pathways associated with development, stem cell proliferation, body morphogenesis, vasculogenesis, and vitamin D signaling are all overrepresented in the downregulated gene set. Gene ontology terms associated with biological process suggest a downregulation of G-protein coupled receptor activity and reduced electron transport processes in the mitochondrial membrane, as well as retinoic acid signaling and metabolism.

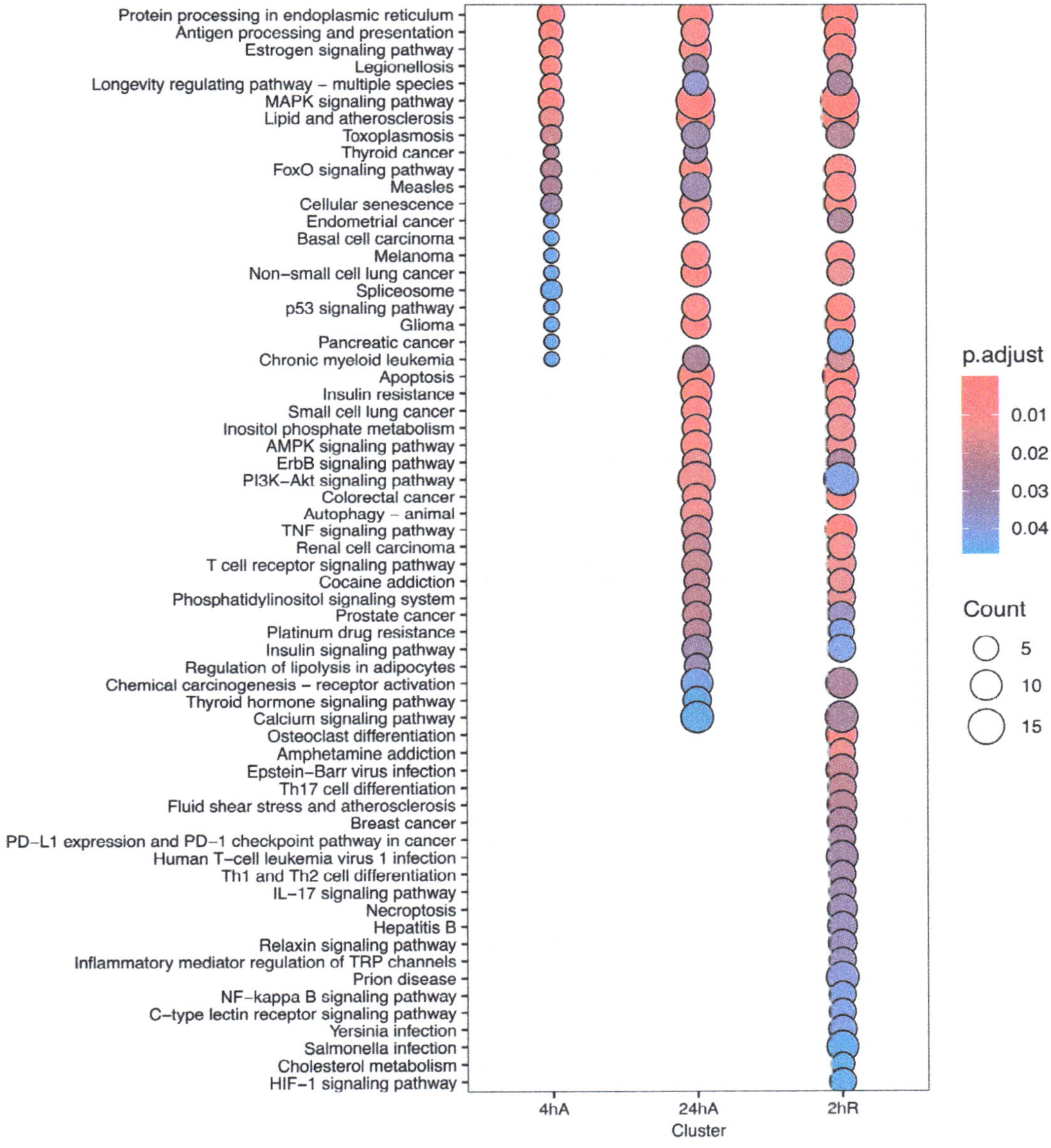


**Figure 6.** KEGG pathway analysis for genes that increase in abundance in response to anoxia and recovery from anoxia in diapause II embryos of *Austrofundulus limnaeus*.

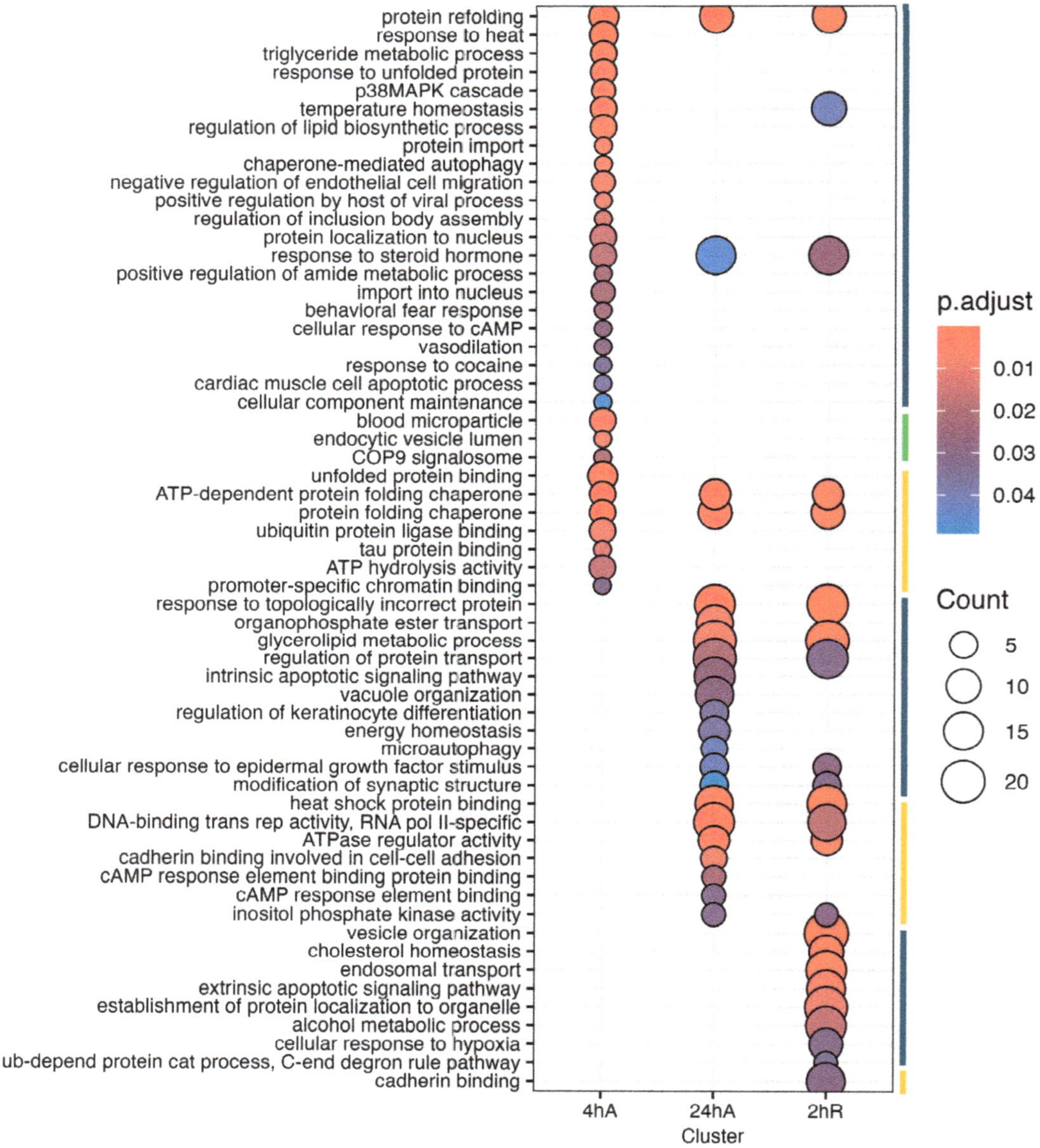


**Figure 7.** Gene ontology enrichment analysis for genes that increase in abundance in response to anoxia and recovery from anoxia in diapause II embryos in *Austrofundulus limnaeus*. The colored bar on the right side of the graph represents ontology family: blue = biological process, yellow = molecular function, and green = cellular component.

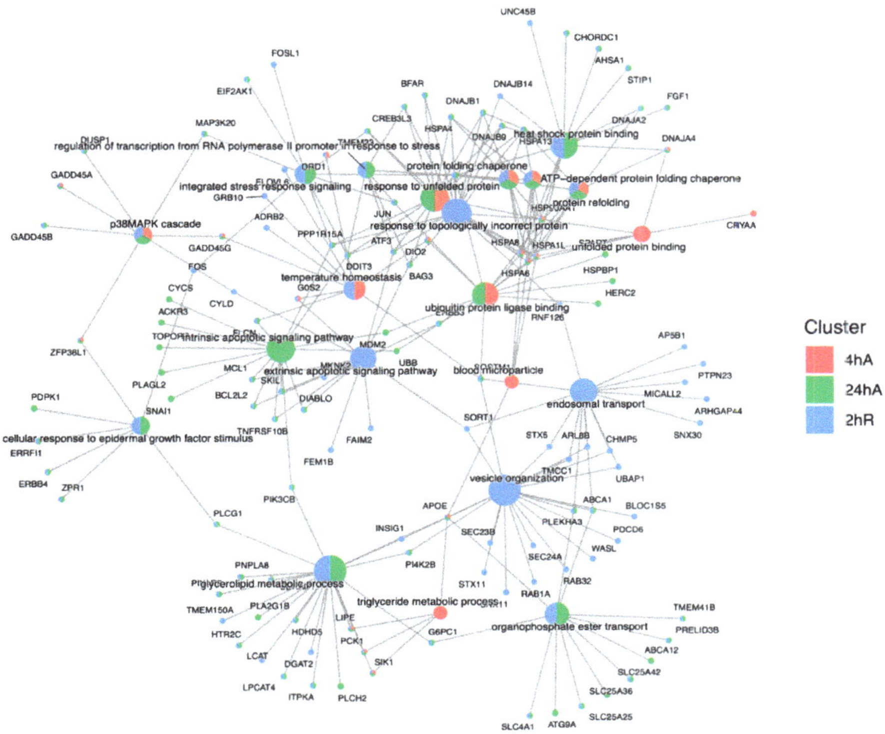


**Figure 8.** Cnet plot of biological process gene ontology enrichment analysis for genes that increase in abundance in response to anoxia and recovery from anoxia in diapause II embryos of *Austrofundulus limnaeus*. Central nodes represent ontology terms, while peripheral nodes are genes associated with the ontology term.

### 3.6. Transcription Factor Prediction

The set of differentially expressed genes initially responding to anoxia (4 h anoxia time point) are enriched for potential targets of 13 transcription factors (Table 1, adjusted $p$ values < 0.05). The TF with the highest significance is peroxisome proliferator-activated receptor α (PPARA) which is associated with the regulation of seven of the upregulated transcripts at 4 h of anoxia (*lipE, gadd45g, gadd45a, g0s2, pck1, G0S2, CBFA2T3*). CBFA2T3, a target of PPARA, is a transcriptional coregulator associated with the modulation of RGS2, TRIM29, TXLNG, PCK1, DNAJA4, ZFP36L1, PAQR5, GADD45G, and HSPA8. Thus, PPARA and CBFA2T3 can account for a large portion of the differentially expressed genes in the initial response to anoxia. The TFlink database predicts 705 significant target interactions based on the gene sets associated with 24 h of anoxia; among the top returns are *TFAP4, TBP, KLF5, SMAD1,* and *E2F1* (Table 1). Gene expression patterns associated with the initial recovery from anoxia are associated with 752 significant transcription factors, with the most statistically significant being *SIN3A*. A full list of predicted transcription factors can be found in Table S10.

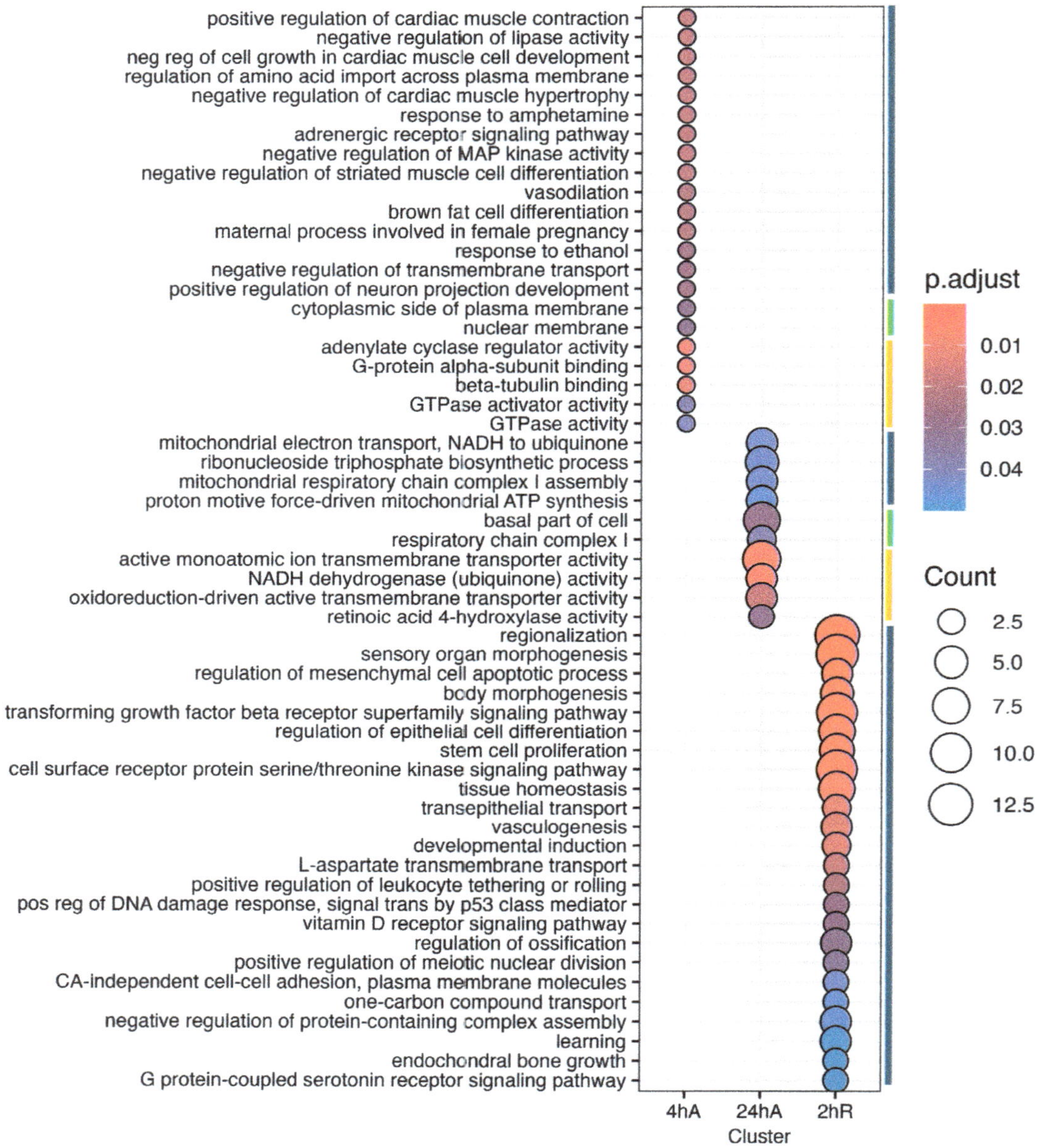


**Figure 9.** Gene ontology enrichment analysis for genes that decrease in abundance in response to anoxia and recovery from anoxia in diapause II embryos of *Austrofundulus limnaeus*. The colored bar on the right side of the graph represents ontology family: blue = biological process, yellow = molecular function, and green = cellular component.

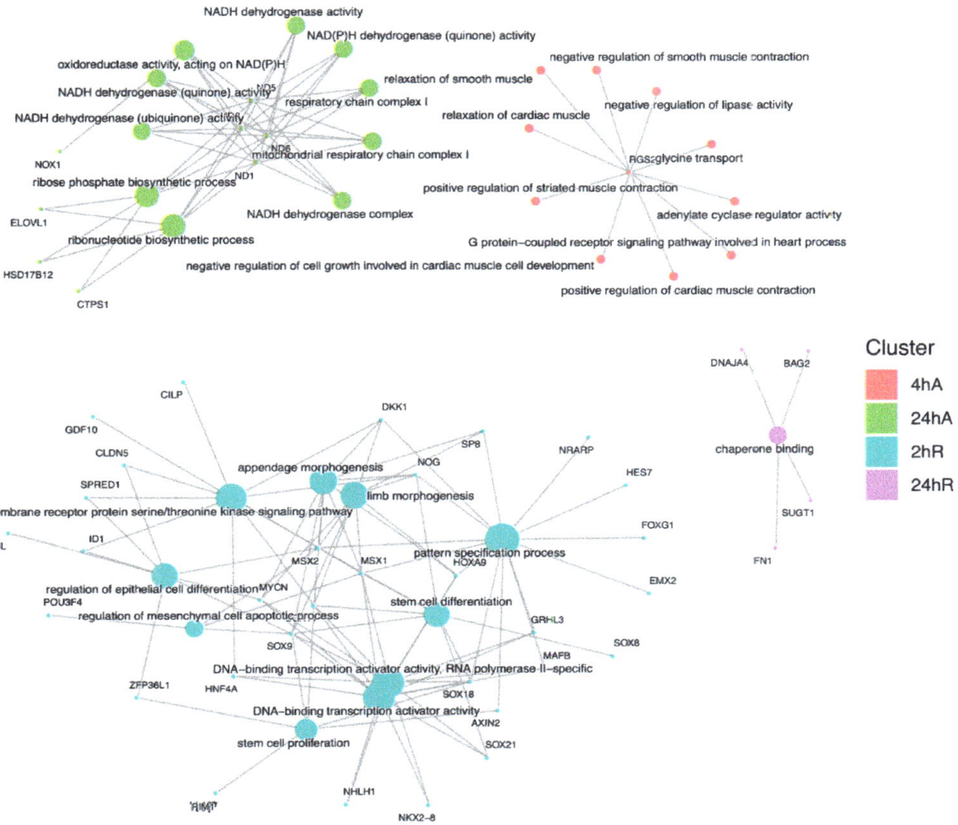


**Figure 10.** Cnet plot of biological process gene ontology enrichment analysis for genes that decrease in abundance in response to anoxia and recovery from anoxia in diapause II embryos of *Austrofundulus limnaeus*. Central nodes represent ontology terms, while peripheral nodes are genes associated with the ontology term.

**Table 1.** Transcription factors identified by TFLink that are associated with patterns of differential gene expression in response to anoxia and recovery from anoxia in diapause II embryos of *Austrofundulus limnaeus*. Number of associated genes is the number of genes from the gene list that were associated with the transcription factor.

| Transcription Factor | Number of Associated Genes | Gene List | *p* Value | *p* Adjust |
|---|---|---|---|---|
| | | 4 h anoxia | | |
| PPARA | 5 | 25 | $2.13 \times 10^{-8}$ | $2.00 \times 10^{-5}$ |
| POU4F1 | 1 | 25 | $1.52 \times 10^{-6}$ | $4.74 \times 10^{-4}$ |
| ZNF746 | 1 | 25 | $1.52 \times 10^{-6}$ | $4.74 \times 10^{-4}$ |
| HSF2 | 4 | 25 | $7.71 \times 10^{-6}$ | $1.83 \times 10^{-3}$ |
| POU3F1 | 1 | 25 | $1.51 \times 10^{-5}$ | $2.84 \times 10^{-3}$ |
| TFAP2B | 1 | 25 | $2.67 \times 10^{-5}$ | $3.54 \times 10^{-3}$ |
| DICER1 | 2 | 25 | $4.43 \times 10^{-5}$ | $7.28 \times 10^{-3}$ |

**Table 1.** *Cont.*

| Transcription Factor | Number of Associated Genes | Gene List | *p* Value | *p* Adjust |
|---|---|---|---|---|
| **24 h anoxia** | | | | |
| TFAP4 | 366 | 409 | $4.80 \times 10^{-14}$ | $1.17 \times 10^{-11}$ |
| TBP | 376 | 409 | $1.23x\ 10^{-13}$ | $2.50 \times 10^{-11}$ |
| KLF5 | 347 | 409 | $2.18 \times 10^{-12}$ | $3.82 \times 10^{-10}$ |
| SMAD1 | 323 | 409 | $5.72 \times 10^{-12}$ | $8.74 \times 10^{-10}$ |
| E2F1 | 383 | 409 | $8.09 \times 10^{-12}$ | $1.11 \times 10^{-9}$ |
| SNAI2 | 302 | 409 | $1.07 \times 10^{-11}$ | $1.30 \times 10^{-9}$ |
| SP5 | 351 | 409 | $1.75 \times 10^{-11}$ | $1.94 \times 10^{-9}$ |
| SMARCA4 | 401 | 409 | $1.98 \times 10^{-11}$ | $2.01 \times 10^{-9}$ |
| EED | 348 | 409 | $2.30 \times 10^{-11}$ | $2.16 \times 10^{-9}$ |
| CEBPA | 381 | 409 | $3.19 \times 10^{-11}$ | $2.79 \times 10^{-9}$ |
| **2 h recovery** | | | | |
| SIN3A | 375 | 405 | $1.06 \times 10^{-14}$ | $2.58 \times 10^{-12}$ |
| TAF1 | 369 | 405 | $1.92 \times 10^{-14}$ | $3.72 \times 10^{-12}$ |
| TBP | 374 | 405 | $2.15 \times 10^{-14}$ | $3.72 \times 10^{-12}$ |
| CHD2 | 339 | 405 | $3.54 \times 10^{-14}$ | $5.35 \times 10^{-12}$ |
| SP4 | 322 | 405 | $3.45 \times 10^{-13}$ | $4.63 \times 10^{-11}$ |
| KLF9 | 339 | 405 | $4.06 \times 10^{-13}$ | $4.91 \times 10^{-11}$ |
| SP1 | 390 | 405 | $4.87 \times 10^{-13}$ | $5.35 \times 10^{-11}$ |
| TFAP4 | 360 | 405 | $7.99 \times 10^{-13}$ | $8.05 \times 10^{-11}$ |
| MYCN | 381 | 405 | $9.70 \times 10^{-13}$ | $9.02 \times 10^{-11}$ |
| E2F1 | 381 | 405 | $1.19 \times 10^{-12}$ | $1.03 \times 10^{-10}$ |
| **24 h recovery** | | | | |
| XRCC6 | 1 | 42 | $4.35 \times 10^{-6}$ | $4.24 \times 10^{-3}$ |
| TRAF6 | 1 | 42 | $1.30 \times 10^{-5}$ | $6.36 \times 10^{-3}$ |

## 4. Discussion

The transcriptome of normoxic diapause II embryos is enriched for multiple gene ontologies associated with stress tolerance and protein synthesis, suggesting that preparation for stress may be part of the developmental program associated with entrance into diapause. While additional studies, including comparisons to actively developing embryos, are needed to fully evaluate the unique aspects of the diapause II transcriptome, there are some general trends that are worth highlighting. First, the transcript with the highest abundance in normoxic diapause II embryos is *HSPA8*. This finding agrees with previous research showing elevated levels of HSP70 proteins in diapause II embryos of *A. limnaeus* [44]. *HSPA8* is a constitutively expressed heat shock 70 transcript, which has a variety of functions in addition to its ability to act as a molecular chaperone [45], including a role in the normal development of anterior structures in zebrafish [46]. The expression of *HSPA8* prior to insult may be an indication of diapausing embryos' preparedness to encounter and deal with environmental stress while in diapause, or it may indicate stabilization of a transcript that is needed to support immediate post-diapause II development. The latter conclusion is also supported by the domination of transcripts that encode for major components and regulators of cytoplasmic protein synthesis, RNA splicing, and oxidative phosphorylation. The high abundance of transcripts encoding ribosomal proteins and various translation initiation and elongation factors indicate that diapause II embryos may be prepared to synthesize and assemble the translation machinery in response to termination of diapause or to external stresses. Three mitochondrial transcripts, *co1*, *co3*, and *atp6*, are among the top 12 expressed genes in the diapause II transcriptome, which suggests active or at least stabilized mitochondrial physiology in dormant embryos. Thus, some aspects of the highly expressed transcriptome in diapause II embryos suggest preparation for stress, but perhaps the majority of transcripts appear to poise embryos for responses to internal or external signals with a robust gene expression response.

It is important to note that these experiments were conducted in a controlled laboratory environment that cannot mimic the natural conditions experienced by annual killifish embryos. In the wild, embryos would likely be exposed to multiple simultaneous stressors, and what appears as an inducible response in the lab may be a constitutive response in the field. However, this study was focused on creating controlled conditions that allow for the genetic programs that regulate diapause dormancy to be viewed in relative isolation from those that respond to stressors such as anoxia that induce quiescence. Thus, from a mechanistic point of view, these experiments are necessary to clearly outline the discrete responses associated with each state of dormancy and will be highly useful for future studies of embryos responding to natural conditions.

Despite being in a profound state of metabolic depression, with severely reduced rates of protein synthesis [21,47,48], diapause II embryos respond to anoxia with over 500 differentially expressed transcripts, 75% of which increase in abundance. This is a surprising result given the metabolic and energetic status of diapausing embryos and the fact that they do not experience an increase in heat production during exposure to anoxia [26,48]. RNA synthesis can account for 1–15% of the basal metabolic rate of a cell/organism [49–51], and thus unnecessary transcription would presumably come at a significant cost to embryos already in an energy limited state. It is possible that this transcriptomic response is at least partially responsible for the 50% drop in ATP levels during the first 14 h of anoxic exposure [26]. Given that RNAseq measures standing stocks of transcripts, these data could be due to de novo synthesis or differential stabilization of transcripts. Stabilization of transcripts has been observed in other anoxia-tolerant systems [52] and thus likely explains at least some of the patterns presented here. However, a robust transcriptional response is a conserved aspect of the cellular stress response [53], and increases in transcripts have been reported in response to anoxia in anoxia-tolerant turtles [54]. The fact that many of the genes upregulated in this study are known to be important in the cellular stress response adds further support for a role of de novo synthesis. Thus, evidence suggests that de novo synthesis of genes is likely; however, an assay specifically measuring new transcripts is needed to be conclusive. How these embryos can mount such a robust response in the face of severe limitations on ATP production is a topic worthy of future exploration.

Recovery from exposure to 24 h of anoxia appears to be an active process with many shared characteristics of anoxic embryos. In early recovery, there are 399 upregulated transcripts and 188 downregulated transcripts; 73% of the upregulated transcripts and 47% of the downregulated transcripts are shared with embryos after 24 h of anoxia. The shared transcripts between early recovery and 24 h of anoxia may indicate the importance of these genes in supporting an adaptive response to reoxygenation, a time when oxidative damage is known to have serious deleterious effects in most organisms [55]. After 24 h of recovery from anoxia, there is only one upregulated transcript, while 47 are downregulated. The upregulated transcript is predicted to be *tsetseEP-like*, a protein with unknown function in vertebrates. Only a single gene ontology term is still overrepresented after 24 h of recovery from anoxia, the downregulation of chaperone binding—perhaps indicating a return to non-stressful conditions, as is also supported by the PCA analysis.

The initial response to acute anoxia (4 h time point) includes a predictable upregulation of transcripts associated with supporting proper protein folding and preventing protein denaturation. The most upregulated transcripts encode for a number of heat shock proteins of the hsp30, 40, 70, and 90 kDa classes and account for five of the top 10 gene ontology terms associated with upregulated genes at 4 h of anoxia. These proteins mediate the folding and compartmentalization of nascent proteins, maintain proteins in functional conformations, protect proteins from irreversible denaturation during stress, aid in their refolding, and assist in the destruction of non-salvageable denatured proteins [4]. After 24 h of anoxia, the initial heat shock response observed at 4 h of anoxia expands to include the endoplasmic reticulum (ER) unfolded protein response (UPR) and the integrated stress response (ISR). The genes associated with the integrated stress response include *MAP3K20*, *DDIT3*, *HSPA5*, *JUN*, *PPP1R15*, *TMEM33*, and *EIF2AK1*. The ISR involves the phosphorylation of *eIF2α*, which causes an immediate decrease in protein synthesis, but interestingly also leads to the increase in translation of the transcription factor ATF4. ATF4 expression is involved with balancing two contradictory outcomes of the ISR, pro-survival and cellular apoptosis [56]. The expression of these upregulated genes is consistent with an activation of the ISR to support survival and recovery [57]. Interestingly, the transcript for *PPP1R15A* increases in abundance after 24 h of anoxia. This protein dephosphorylates eIF2α to regulate translation and modulate the ISR [58], suggesting an initial upregulation of the ISR after 4 h of anoxia and an attenuation of this early response by 24 h of anoxia to perhaps allow for protein synthesis needed to respond to anoxic insult. Finally, the ISR is modified during recovery from anoxia by the additional up-regulation of two key genes, *FOS* and *JUN*. Both *FOS* and *JUN* are members of the AP-1 complex that is activated by p38MAPK in response to oxidative stress [59]. AP-1 proteins may also interact with CHOP (encoded by the *DDIT3* gene) to regulate AP-1 targets in response to stress [60]. When the AP-1 complex interacts with CHOP, cell proliferation and survival are supported through increased expression of BCL-2 and cyclins D and B in human lung tissue exposed to hypoxia [61]. Thus, it appears that AP-1 is likely interacting with the activated ISR to promote return to cell proliferation during recovery from anoxia.

Interestingly, only two protein-coding transcripts, regulator of G-protein signaling 2 (*RGS2*) and taxilin γ (*TXLNG*), were significantly downregulated within the first 4 h of anoxia. RGS2 is a protein that inhibits MAPK3 and AKT signaling in zebrafish [62] and contributes to translational arrest in dormant cancer cells [63]. RGS2 has been shown to inhibit translation through direct interactions with eIF2B as well by interfering with the eIF2–eIF2B GTPase cycle required for initiation of translation [64].Thus, elevated RGS2 in normoxic diapause II embryos may be acting as a brake on protein synthesis to aid in the maintenance of diapause, and this brake may need to be removed for the embryos to respond to anoxic insult. Taxilin γ (TXLNG) can inhibit ATF4 activity and the inflammatory effects of TNFα [65]. *TXLNG* decreased in mouse brain slices in response to hypoxia, which led to activation of ATF4 and the ER unfolded protein response [66]. Thus, we hypothesize that early downregulation of *RGS2* and *TXLNG* in response to anoxia may contribute to the activation of ATF4 and the induction of the ISR as discussed above, making ATF4 a key mediator of the response to anoxia in this species.

p38MAPK appears to be activated within the first few hours of anoxia based on the increase in the relative abundance of several of its key target genes, including *ZFP36L1*, *GADD45A*, and *GADD45G*. By 24 h of anoxia, a total of six genes associated with p38MAPK signaling are upregulated, with *GADD45B*, *DUSP1*, and *MAP3K20* being added to the list. The later upregulation of *DUSP1*, a phosphatase and inhibitor of MAPK [67] that is induced through activation of *P53* in colon cancer cells, may suggest attenuation of the MAPK response after 24 h of anoxia. Conversely, upregulation of *MAP3K20* (*MLK7*), a protein kinase that activates p38MAPK and JNK pathways, would support continued p38MAPK activity if increased transcripts indeed lead to protein expression [68]. The complex nature of MAPK signaling and regulation will require additional experiments to confirm a true role for p38MAPK in supporting anoxia tolerance. It is worth noting that p38MAPK is downregulated in anoxic turtle brains [69], and thus increased activity of p38MAPK in *A. limnaeus* could be a unique adaptation in this species. p38MAPK has been associated with cell cycle regulation and apoptosis pathways through interactions with the transcription factor E2F and anti-apoptotic genes *MCL1* and *BCL2* in other systems [70], and we predict that future studies will elucidate a major role for this protein in the response to anoxia in killifish embryos.

In many tissue types, activation of p38MAPKs during stress pushes cells towards apoptosis, but in some instances, activation promotes cell survival [71,72]. Many of the genes that increased in abundance in response to anoxia are involved with regulation of apoptosis, and many of them are anti-apoptotic (for example MCL-1 [67]). There also appear to be connections between p38MAPK signaling, the ISR, and regulation of apoptosis [73,74]. Embryos of *A. limnaeus* survive short-term exposures to anoxia with little or no apoptosis, suggesting mechanisms that can block apoptosis in response to oxygen stress in comparison to mammals and other oxygen-sensitive species [75,76]. Further, apoptosis is often activated when cells or organisms experience reoxygenation stress [77–80], but in diapause II embryos, the expression of apoptotic genes remains stable during the initial hours of recovery form anoxia. This suggests that the complement of apoptosis-regulating transcripts is likely appropriate for avoiding apoptosis during recovery. Thus, the pathways involved in regulating apoptosis in these embryos likely reflect a complex balance between pro- and anti-apoptotic signals. The obvious outcome of these interactions is a lack of apoptosis, making diapause II embryos of *A. limnaeus* a potentially powerful model for probing crosstalk and interactions between pro- and anti-apoptotic signaling in response to oxygen limitation.

Major alterations in the transcriptome for lipid homeostasis occur during exposure to anoxia and during initial recovery in diapause II embryos of *A. limnaeus*. Several lipid metabolism ontologies, including glycerolipid metabolic process, cholesterol homeostasis, vesicle organization, and endosomal transport, are overrepresented in upregulated transcripts of embryos during exposure to anoxia and during the initial hours of recovery. The key genes associated with these ontologies are *APOE, LIPE, SIK1,* and *PCK1*. Salt-inducible kinase 1 (*SIK1*) can inhibit liver lipogenesis in humans [81] and is highly expressed in the ventricle tissue of western painted turtles in anoxia [82]. Lipase e (hormone sensitive) and apolipoprotein E are both involved with the catabolism of lipids, including triacylglycerols, and cholesteryl and retinyl ester bonds [83–85]. Furthermore, the cytosolic form of phosphoenolpyruvate carboxykinase 1 (encoded by the gene *PCK1*), is a key enzyme involved in mediating transfer of metabolites and reducing equivalents between the cytosol and mitochondrial compartments to support processes such as gluconeogenesis and glyceroneogenesis [86]. There are also a number of genes involved in vesicular transport and endocytosis [87,88] that are upregulated after 2 h of aerobic recovery. Proper regulation of endosomal membrane transport is known to be regulated during stress by the p38MAPK pathway [89]. Genes that regulate autophagosome dynamics, mitophagy, and the proper handling of ubiquitylated proteins are also upregulated during recovery from anoxia [90–92].

There is a notable absence of a transcriptomic signature associated with hypoxia inducible factor (HIF) signaling in diapause II embryos of *A. limnaeus* exposed to anoxia. In fact, HIF signaling pathways are not observed until aerobic recovery from anoxia. We hypothesize that the upregulation of *ZFP36L1* may play a role in blocking HIF signaling in this system. ZFP36L1 is known to target mRNA 3′UTRs and lead to their degradation; *HIF1A* is a known target of *ZFP36L1* [93].

Gene ontology terms associated with transcripts that decrease in abundance during exposure to anoxia and the initial recovery from anoxia are associated with embryonic development, mitochondrial metabolism, and negative regulation of lipase activity. For example, both *SOX8* and *SOX9* are of key importance to embryonic development and are downregulated during anoxia. Many more transcripts that are involved in embryonic development are downregulated during recovery. This list of transcripts includes multiple transcription factors related to development, including *POU3F4*, *SOX7*, *SOX8*, *SOX9*, *SOX18*, *SP8*, and *HOXA9* [94–97]. The downregulation of these developmental transcription factors may be vital in ensuring normal development while recovering from anoxic stress and ensuring maintenance of diapause. VDR signaling plays a pivotal role in regulation of active development and entrance into diapause in embryos of *A. limnaeus* [15]. The expression of *PIM1* during anoxia with a significant decrease during recovery is an interesting finding given the known interaction between PIM1 and the VDR [98]. *PIM1* is a serine/threonine kinase that supports stem cell proliferation and is protective against ROS induced cell death [99], likely through phosphorylation/inhibition of BAD. The roles of VDR signaling during diapause and in response to anoxia are currently unexplored, but it is intriguing to explore a possible role for *PIM1* and the VDR in the response of diapause embryos to stress.

Transcription factor analysis predicts PPARA as a potential master regulator of the response to anoxia in diapause II embryos of *A. limnaeus*. PPARA is a nuclear hormone receptor known to mediate adaptive responses to fasting and to regulate a variety of metabolic pathways [100,101]. Interestingly, a transcription factor with nine targets in the gene list, CBFA2/RUNX1 partner transcriptional co-repressor 3 (CBFA2T3), is a PPARA-regulated gene that modulates metabolic stress in mouse liver [102]. Thus, between PPARA and CBFA2T3, many of the genes that change significantly during the early response to anoxia (4 h anoxia) can be accounted for, suggesting a large role for PPARA in controlling the immediate response to anoxia. To our knowledge, this is the first implication of PPARA in regulating a response to anoxia.

## 5. Conclusions

This report represents the first transcriptomic analysis of diapausing vertebrate embryos exposed to stress. Despite being in a state of developmental dormancy, diapausing embryos respond to anoxic stress with a robust transcriptomic response that begins shortly after exposure and develops over 24 h. Within hours of anoxic exposure, HSPs and other molecular chaperones are transcriptionally activated, there is a shift expression of genes involved in lipid metabolism, and the p38MAPK signaling pathway is activated. This initial response may be at least partially coordinated by the downregulation of two transcripts, *RGS2* and *TXLNG*, which likely play a significant role in the regulation of translation. This initial response is the foundation of the continued response that develops over the next 20 h. Recovery from anoxia appears to require downregulation of transcripts involved in supporting active development and morphogenesis and may be a response to keep the embryo in diapause. The overarching response to anoxia is an impressive stabilization of 97% of the transcriptome in response to large-scale declines in ATP. In terms of the question of whether diapause II embryos are pre-adapted to survive anoxic stress, the answer appears to be yes, but only partially. While the normoxic transcriptome appears to contain an enrichment of expressed genes that are known to participate in the response to stressful insults, the 3% of transcripts that do change in response to anoxia highlight some interesting and likely essential pathways that must augment any preparations made prior to entrance into dormancy. Perhaps pre-diapause preparation supports stress tolerance in general, and the ability to mount a stress-specific response is retained to tune the response to the particular stress experienced. Future studies may be able to tease these two programs apart by exposing embryos to a variety of stresses and looking for common and unique responses.

**Supplementary Materials:** The following supporting information can be downloaded at https://www.mdpi.com/article/10.3390/biom15040515/s1: Table S1: cDNA library results; Table S2: Highly expressed T0 transcripts; Table S3: Gene ontology terms enriched in highly expressed genes in diapause II embryos of *Austrofundulus limnaeus*; Table S4: Differential gene expression patterns for diapause II embryos of *Austrofundulus limnaeus* exposed to anoxia and recovery from anoxia; Table S5: qPCR validation of RNAseq differential expression analysis; Table S6: KEGG pathway analysis for genes that increase in abundance in response to anoxia and recovery from anoxia in diapause II embryos of *Austrofundulus limnaeus*; Table S7: Gene ontology enrichment analysis for genes that increase in abundance in response to anoxia and recovery from anoxia in diapause II embryos in *Austrofundulus limnaeus*; Table S8: Gene ontology enrichment analysis for genes that decrease in abundance in response to anoxia and recovery from anoxia in diapause II embryos of *Austrofundulus limnaeus*; Table S9: KEGG pathway analysis for genes that decrease in abundance in response to anoxia and recovery from anoxia in diapause II embryos of *Austrofundulus limnaeus*; Table S10: Transcription factor analysis of differentially expressed genes.

**Author Contributions:** Conceptualization, J.E.P. and C.L.R.; methodology, C.L.R.; validation, P.R.C., C.L.R. and A.L.T.R.; formal analysis, P.R.C. and A.L.T.R.; investigation, P.R.C., J.E.P., C.L.R and A.L.T.R.; resources, J.E.P.; data curation, P.R.C. and A.L.T.R.; writing—original draft preparation, P.R.C.; writing—review and editing, P.R.C., J.E.P., C.L.R. and A.L.T.R.; visualization, P.R.C.; supervision, J.E.P.; project administration, J.E.P.; funding acquisition, J.E.P. All authors have read and agreed to the published version of the manuscript.

**Funding:** This research was funded by the US National Science Foundation grant numbers IOS 1354549 and 2025832.

**Institutional Review Board Statement:** Embryos of *Austrofundulus limnaeus* were collected and cared for according to established methods [25] from a stock of fish maintained in the lab since 1995. All procedures were conducted in accordance with rules and regulations outlined by the Office of Laboratory Animal Welfare at the US National Institutes of Health and consistent with guidance provided by the American Veterinary Association. Procedures were reviewed and approved prior to the start of work by the Portland State University (PSU) Institutional Animal Care and Use Committee (IACUC) protocols (PSU IACUC protocol #33 on 2 December 2015 and #92 on 3 November 2024).

**Informed Consent Statement:** Not applicable.

**Data Availability Statement:** The sequencing data files, in FASTQ format, are available on the NCBI's website. https://www.ncbi.nlm.nih.gov/Traces/study/?acc=PRJNA272154&o=acc_s:a (accessed on 31 March 2025), is the web address to directly access our lab's uploaded sequencing files. At the time of submission, metadata for the samples presented in this paper were not assigned on the NCBI website, but these metadata and accession numbers are contained in Table S1. Metadata for samples presented in the paper will be updated on the NCBI website in a timely manner.

**Acknowledgments:** The authors would like to acknowledge Gazal Kalyan (Portland State University, United States of America) for her support in computational analysis methods and strategies.

**Conflicts of Interest:** The authors declare no conflicts of interest. The funders had no role in the design of the study; in the collection, analyses, or interpretation of data; in the writing of the manuscript; or in the decision to publish the results.

## Abbreviations

The following abbreviations are used in this manuscript:

| | |
|---|---|
| dpf | days post fertilization |
| NCBI | The National Center for Biotechnology Information |
| ORA | overrepresentation analysis |
| TFs | transcription factors |
| ER | endoplasmic reticulum |
| UPR | unfolded protein response |
| ISR | integrated stress response |

## References

1. Storey, K.B.; Storey, J.M. Metabolic rate depression and biochemical adaptation in anaerobiosis, hibernation and estivation. *Q. Rev. Biol.* **1990**, *65*, 145–174. [CrossRef] [PubMed]
2. Storey, K.B. Suspended animation: The molecular basis of metabolic depression. *Can. J. Zool.* **1988**, *66*, 124–132. [CrossRef]
3. Hand, S.C. Metabolic dormancy in aquatic invertebrates. In *Advances in Comparative and Environmental Physiology*; Gilles, R., Ed.; Springer: Berlin/Heidelberg, Germany, 1991; pp. 1–50. [CrossRef]
4. MacRae, T.H. Gene expression, metabolic regulation and stress tolerance during diapause. *Cell. Mol. Life Sci.* **2010**, *67*, 2405–2424. [CrossRef]
5. Denlinger, D.L. Insect diapause: From a rich history to an exciting future. *J. Exp. Biol.* **2023**, *226*, jeb245329. [CrossRef]
6. Sim, C.; Kang, D.S.; Kim, S.; Bai, X.; Denlinger, D.L. Identification of FOXO targets that generate diverse features of the diapause phenotype in the mosquito *Culex pipiens*. *Proc. Natl. Acad. Sci. USA* **2015**, *112*, 3811–3816. [CrossRef]
7. Rinehart, J.P.; Yocum, G.D.; Denlinger, D.L. Developmental upregulation of inducible hsp70 transcripts, but not the cognate form, during pupal diapause in the flesh fly, *Sarcophaga crassipalpis*. *Insect Biochem. Mol. Biol.* **2000**, *30*, 515–521. [CrossRef]
8. Rinehart, J.P.; Denlinger, D.L. Heat-shock protein 90 is down-regulated during pupal diapause in the flesh fly, *Sarcophaga crassipalpis*, but remains responsive to thermal stress. *Insect Mol. Biol.* **2000**, *9*, 641–645. [CrossRef]
9. Tammariello, S.P.; Rinehart, J.P.; Denlinger, D.L. Desiccation elicits heat shock protein transcription in the flesh fly, *Sarcophaga crassipalpis*, but does not enhance tolerance to high or low temperatures. *J. Insect Physiol.* **1999**, *45*, 933–938. [CrossRef]
10. Yocum, G.; Joplin, K.; Denlinger, D. Upregulation of a 23 kDa small heat shock protein transcript during pupal diapause in the flesh fly, *Sarcophaga crassipalpis*. *Insect Biochem. Mol. Biol.* **1998**, *28*, 677–682. [CrossRef]
11. Podrabsky, J.; Riggs, C.; Wagner, J. Tolerance of Environmental Stress. In *Annual Fishes. Life History Strategy, Diversity, and Evolution*; Berois, N., García, G., De Sá, R., Eds.; CRC Press, Taylor & Francis: Boca Raton, FL USA, 2016; pp. 159–184. [CrossRef]
12. Podrabsky, J.; Romney, A.; Culpepper, K. Alternative Developmental Pathways. In *Annual Fishes. Life History Strategy, Diversity, and Evolution*; Berois, N., García, G., De Sá, R., Eds.; CRC Press, Taylor & Francis: Boca Raton, FL USA, 2016; pp. 63–73.
13. Wourms, J.P. The developmental biology of annual fishes III. Pre-embryonic and embryonic diapause of variable duration in the eggs of annual fishes. *J. Exp. Zool.* **1972**, *182*, 389–414. [CrossRef]
14. Podrabsky, J.; Riggs, C.; Romney, A.; Woll, S.; Wagner, J.; Culpepper, K.; Cleaver, T. Embryonic development of the annual killifish *Austrofundulus limnaeus*: An emerging model for ecological and evolutionary developmental biology research and instruction. *Dev. Dyn.* **2017**, *246*, 779–801. [CrossRef] [PubMed]
15. Romney, A.; Davis, E.; Corona, M.; Wagner, J.; Podrabsky, J. Temperature dependent vitamin D signaling regulates developmental trajectory associated with diapause in an annual killifish. *Proc. Natl. Acad. Sci. USA* **2018**, *115*, 12763–12768. [CrossRef]
16. Podrabsky, J.; Wilson, N. Hypoxia and anoxia tolerance in the annual killifish *Austrofundulus limnaeus*. *Integr. Comp. Biol.* **2016**, *56*, 500–509. [CrossRef] [PubMed]
17. Podrabsky, J.E.; Lopez, J.P.; Fan, T.W.M.; Higashi, R.; Somero, G.N. Extreme anoxia tolerance in embryos of the annual killifish *Austrofundulus limnaeus*: Insights from a metabolomics analysis. *J. Exp. Biol.* **2007**, *210*, 2253–2266. [CrossRef] [PubMed]
18. Machado, B.E.; Podrabsky, J.E. Salinity tolerance in diapausing embryos of the annual killifish *Austrofundulus limnaeus* is supported by exceptionally low water and ion permeability. *J. Comp. Physiol. B* **2007**, *177*, 809–820. [CrossRef]
19. Podrabsky, J.E.; Riggs, C.L.; Duerr, J.M. Anoxia Tolerance During Vertebrate Development—Insights from Studies on the Annual Killifish *Austrofundulus limnaeus*. In *Anoxia*; Padilla, P., Ed.; InTech: Rijeka, Croatia, 2012; pp. 3–24. [CrossRef]
20. Hand, S.C.; Podrabsky, J.E. Bioenergetics of diapause and quiescence in aquatic animals. *Thermochim. Acta* **2000**, *349*, 31–42. [CrossRef]
21. Podrabsky, J.E.; Hand, S.C. Depression of protein synthesis during diapause in embryos of the annual killifish *Austrofundulus limnaeus*. *Physiol. Biochem. Zool.* **2000**, *73*, 799–808. [CrossRef]

22. Polačik, M.; García, D.; Arezo, M.J.; Papa, N.; Schlueb, H.; Blanco, D.; Podrabsky, J.E.; Vrtílek, M. Embryonic development of natural annual killifish populations of the genus *Austrolebias*: Evolutionary parallelism and the role of environment. *Freshw. Biol.* **2023**, *68*, 1726–1738. [CrossRef]
23. Polačik, M.; Vrtílek, M.; Reichard, M.; Žák, J.; Blažek, R.; Podrabsky, J. Embryo ecology: Developmental synchrony and asynchrony in the embryonic development of wild annual fish populations. *Ecol. Evol.* **2021**, *11*, 4945–4956. [CrossRef]
24. Zajic, D.; Nicholson, J.; Podrabsky, J. No water, no problem: Stage-specific metabolic responses to dehydration stress in annual killifish embryos. *J. Exp. Biol.* **2020**, *223*, jeb231985. [CrossRef]
25. Riggs, C.; Podrabsky, J. Small noncoding RNA expression during extreme anoxia tolerance of annual killifish (*Austrofundulus limnaeus*) embryos. *Physiol. Genom.* **2017**, *49*, 505–518. [CrossRef]
26. Podrabsky, J.E.; Menze, M.A.; Hand, S.C. Rapid Communication: Long-term survival of anoxia despite rapid ATP decline in embryos of the annual killifish *Austrofundulus limnaeus*. *J. Exp. Zool. A Ecol. Genet. Physiol.* **2012**, *317*, 524–532. [CrossRef]
27. Podrabsky, J.E. Husbandry of the annual killifish *Austrofundulus limnaeus* with special emphasis on the collection and rearing of embryos. *Env. Biol. Fishes* **1999**, *54*, 421–431. [CrossRef]
28. Wourms, J.P. The developmental biology of annual fishes I. Stages in the normal development of *Austrofundulus myersi* Dahl. *J. Exp. Zool.* **1972**, *182*, 143–168. [CrossRef] [PubMed]
29. Meller, C.L.; Meller, R.; Simons, R.P.; Podrabsky, J.E. Patterns of ubiquitylation and SUMOylation associated with exposure to anoxia in embryos of the annual killifish *Austrofundulus limnaeus*. *J. Comp. Physiol. B* **2014**, *184*, 235–247. [CrossRef]
30. Chomczynski, P.; Sacchi, N. Single-step method of RNA isolation by acid guanidinium thiocyanate-phenol-chloroform extraction. *Anal. Biochem.* **1987**, *162*, 156–159. [CrossRef]
31. Chomczynski, P.; Sacchi, N. The single-step method of RNA isolation by acid guanidinium thiocyanate–phenol–chloroform extraction: Twenty-something years on. *Nat. Protoc.* **2006**, *1*, 581–585. [CrossRef]
32. Chomczynski, P.; Mackey, K. Short technical reports. Modification of the TRI reagent procedure for isolation of RNA from polysaccharide-and proteoglycan-rich sources. *Biotechniques* **1995**, *19*, 942–945. [PubMed]
33. Wagner, J.T.; Podrabsky, J.E. Gene expression patterns that support novel developmental stress buffering in embryos of the annual killifish *Austrofundulus limnaeus*. *EvoDevo* **2015**, *6*, 2. [CrossRef]
34. Romney, A.; Podrabsky, J. Transcriptomic analysis of maternally provisioned cues for phenotypic plasticity in the annual killifish, *Austrofundulus limnaeus*. *EvoDevo* **2017**, *8*, 6. [CrossRef]
35. Andrews, S. FastQC: A Quality Control Tool for High Throughput Sequence Data. Available online: https://www.bioinformatics.babraham.ac.uk/projects/fastqc/ (accessed on 18 June 2023).
36. Bolger, A.A.; Lohse, M.; Usadel, B. Trimmomatic: A flexible trimmer for Illumina Sequence Data. *Bioinformatics* **2014**, *30*, 2114–2120. [CrossRef]
37. Langmead, B.; Salzberg, S.L. Fast gapped-read alignment with Bowtie 2. *Nat. Methods* **2012**, *9*, 357–359. [PubMed]
38. Gentleman, R.C.; Carey, V.J.; Bates, D.M.; Bolstad, B.; Dettling, M.; Dudoit, S.; Ellis, B.; Gautier, L.; Ge, Y.; Gentry, J.; et al. Bioconductor: Open software development for computational biology and bioinformatics. *Genome Biol.* **2004**, *5*, R80. [CrossRef]
39. Wagner, J.; Warren, W.; Minx, P.; Podrabsky, J. Austrofundulus limnaeus 1.0 Draft Genome Assembly with Annotation. Available online: http://www.ncbi.nlm.nih.gov/genome/?term=txid52670[orgn] (accessed on 23 June 2023).
40. Love, M.I.; Huber, W.; Anders, S. Moderated estimation of fold change and dispersion for RNA-seq data with DESeq2. *Genome Biol.* **2014**, *15*, 550. [CrossRef]
41. Emms, D.M.; Kelly, S. OrthoFinder: Phylogenetic orthology inference for comparative genomics. *Genome Biol.* **2019**, *20*, 238. [CrossRef]
42. Wu, T.; Hu, E.; Xu, S.; Chen, M.; Guo, P.; Dai, Z.; Feng, T.; Zhou, L.; Tang, W.; Zhan, L.; et al. clusterProfiler 4.0: A universal enrichment tool for interpreting omics data. *Innovation* **2021**, *2*, 100141. [CrossRef] [PubMed]
43. Liska, O.; Bohár, B.; Hidas, A.; Korcsmáros, T.; Papp, B.; Fazekas, D.; Ari, E. TFLink: An integrated gateway to access transcription factor–target gene interactions for multiple species. *Database* **2022**, *2022*, baac083. [CrossRef]
44. Podrabsky, J.E.; Somero, G.N. An inducible 70 kDa-class heat shock protein is constitutively expressed during early development and diapause in the annual killifish *Austrofundulus limnaeus*. *Cell Stress Chaperones* **2007**, *12*, 199–204. [CrossRef]
45. Stricher, F.; Macri, C.; Ruff, M.; Muller, S. HSPA8/HSC70 chaperone protein. *Autophagy* **2013**, *9*, 1937–1954. [CrossRef]
46. Wang, C.; Zhang, X.; Wang, X.; Zhai, Y.; Li, M.; Pan, J.; Bai, Y.; Rong, X.; Zhou, J. Genetic deletion of *hspa8* leads to selective tissue malformations in zebrafish embryonic development. *J. Cell Sci.* **2022**, *135*, jcs259734. [CrossRef]
47. Podrabsky, J.E.; Hand, S.C. Physiological strategies during animal diapause: Lessons from brine shrimp and annual killifish. *J. Exp. Biol.* **2015**, *218*, 1897–1906. [CrossRef] [PubMed]
48. Podrabsky, J.E.; Hand, S.C. The bioenergetics of embryonic diapause in an annual killifish, *Austrofundulus limnaeus*. *J. Exp. Biol.* **1999**, *202*, 2567–2580. [CrossRef]
49. Buttgereit, F.; Brand, M.D. A hierarchy of ATP-consuming processes in mammalian cells. *Biochem. J.* **1995**, *312*, 163–167. [CrossRef] [PubMed]

50. Buttgereit, F.; Brand, M.D.; Muller, M. ConA induced changes in energy metabolism of rat thymocytes. *Biosci. Rep.* **1992**, *12*, 381–385. [CrossRef] [PubMed]
51. Rolfe, D.F.S.; Brown, G.C. Cellular energy utilization and molecular origin of standard metabolic rate in mammals. *Physiol. Rev.* **1997**, *77*, 731–758. [CrossRef]
52. Hardewig, I.; Anchordoguy, T.J.; Crawford, D.L.; Hand, S.C. Profiles of nuclear and mitochondrial encoded mRNAs in developing and quiescent embryos of *Artemia franciscana. Mol. Cell Biochem.* **1996**, *158*, 139–147. [CrossRef]
53. Morimoto, R.I. Regulation of the heat shock transcriptional response: Cross talk between a family of heat shock factors, molecular chaperones, and negative regulators. *Genes Dev.* **1998**, *12*, 3788–3796. [CrossRef]
54. Milton, S.L.; Nayak, G.; Lutz, P.L.; Prentice, H.M. Gene transcription of neuroglobin is upregulated by hypoxia and anoxia in the brain of the anoxia-tolerant turtle Trachemys scripta. *J. Biomed. Sci.* **2006**, *13*, 509–514. [CrossRef]
55. Bundgaard, A.; Ruhr, I.M.; Fago, A.; Galli, G.L. Metabolic adaptations to anoxia and reoxygenation: New lessons from freshwater turtles and Crucian carp. *Curr. Opin. Endocr. Metab. Res.* **2020**, *11*, 55–64. [CrossRef]
56. Neill, G.; Masson, G.R. A stay of execution: ATF4 regulation and potential outcomes for the integrated stress response. *Front. Mol. Neurosci.* **2023**, *16*, 1112253. [CrossRef]
57. Suragani, R.N.V.S.; Zachariah, R.S.; Velazquez, J.G.; Liu, S.; Sun, C.-W.; Townes, T.M.; Chen, J.-J. Heme-regulated eIF2$\alpha$ kinase activated Atf4 signaling pathway in oxidative stress and erythropoiesis. *Blood* **2012**, *119*, 5276–5284. [CrossRef] [PubMed]
58. Crespillo-Casado, A.; Chambers, J.E.; Fischer, P.M.; Marciniak, S.J.; Ron, D. PPP1R15A-mediated dephosphorylation of eIF2$\alpha$ is unaffected by Sephin1 or Guanabenz. *eLife* **2017**, *6*, e26109. [CrossRef] [PubMed]
59. Dharshini, P.L.C.; Vishnupriya, S.; Sakthivel, K.M.; Rasmi, R.R. Oxidative stress responsive transcription factors in cellular signalling transduction mechanisms. *Cell Signal* **2020**, *72*, 109670. [CrossRef]
60. Ubeda, M.; Vallejo, M.; Habener, J.F. CHOP enhancement of gene transcription by interactions with Jun/Fos AP-1 complex proteins. *Mol. Cell Biol.* **1999**, *19*, 7589–7599. [CrossRef]
61. Yadav, S.; Kalra, N.; Ganju, L.; Singh, M. Activator protein-1 (AP-1): A bridge between life and death in lung epithelial (A549) cells under hypoxia. *Mol. Cell Biochem.* **2017**, *436*, 99–110. [CrossRef]
62. Lin, S.-J.; Huang, Y.-C.; Chen, H.-Y.; Fang, J.-Y.; Hsu, S.-Y.; Shih, H.-Y.; Liu, Y.-C.; Cheng, Y.-C. RGS2 suppresses melanoma growth via inhibiting MAPK and AKT signaling pathways. *Anticancer. Res.* **2021**, *41*, 6135–6145. [CrossRef]
63. Cho, J.; Min, H.-Y.; Lee, H.J.; Hyun, S.Y.; Sim, J.Y.; Noh, M.; Hwang, S.J.; Park, S.-H.; Boo, H.-J.; Lee, H.-J.; et al. RGS2-mediated translational control mediates cancer cell dormancy and tumor relapse. *J. Clin. Investig.* **2021**, *131*, e136779. [CrossRef] [PubMed]
64. Nguyen, C.H.; Ming, H.; Zhao, P.; Hugendubler, L.; Gros, R.; Kimball, S.R.; Chidiac, P. Translational control by RGS2. *J. Cell Biol.* **2009**, *186*, 755–765. [CrossRef]
65. Tao, J.; Gu, P.; Lai, H.; Peng, H.; Guo, Z.; Yuan, Y.; Yu, X.; Shen, X.; Liu, J.; Xier, Z.; et al. TXLNG improves insulin resistance in obese subjects in vitro and in vivo by inhibiting ATF4 transcriptional activity. *Mol. Cell Endocrinol.* **2023**, *568–569*, 111928. [CrossRef]
66. Hotokezaka, Y.; Katayama, I.; Nakamura, T. ATM-associated signalling triggers the unfolded protein response and cell death in response to stress. *Commun. Biol.* **2020**, *3*, 378. [CrossRef]
67. Liu, Y.-X.; Wang, J.; Guo, J.; Wu, J.; Lieberman, H.B.; Yin, Y. *DUSP1* is controlled by p53 during the cellular response to oxidative stress. *Mol. Cancer Res.* **2008**, *6*, 624–633. [CrossRef] [PubMed]
68. Wang, X.; Mader, M.M.; Toth, J.E.; Yu, X.; Jin, N.; Campbell, R.M.; Smallwood, J.K.; Christe, M.E.; Chatterjee, A.; Goodson, T.; et al. Complete inhibition of anisomycin and UV radiation but not cytokine induced JNK and p38 activation by an aryl-substituted dihydropyrrolopyrazole quinoline and mixed lineage kinase 7 small interfering RNA. *J. Biol. Chem.* **2005**, *280*, 19298–19305. [CrossRef] [PubMed]
69. Milton, S.L.; J Dirk, L.; F Kara, L.; Prentice, H.M. Adenosine modulates ERK1/2, PI3K/Akt, and p38MAPK activation in the brain of the anoxia-tolerant turtle Trachemys scripta. *J. Cereb. Blood Flow. Metab.* **2008**, *28*, 1469–1477. [CrossRef]
70. Whitaker, R.H.; Cook, J.G. Stress relief techniques: p38 MAPK determines the balance of cell cycle and apoptosis pathways. *Biomolecules* **2021**, *11*, 1444. [CrossRef]
71. Cargnello, M.; Roux, P.P. Activation and function of the MAPKs and their substrates, the MAPK-activated protein kinases. *Microbiol. Mol. Biol.* **2011**, *75*, 50–83. [CrossRef]
72. Gupta, M.; Gupta, S.K.; Hoffman, B.; Liebermann, D.A. Gadd45a and Gadd45b protect hematopoietic cells from UV-induced apoptosis via distinct signaling pathways, including p38 activation and JNK inhibition. *J. Biol. Chem.* **2006**, *281*, 17552–17558. [CrossRef]
73. Joshi, S. Mnk kinase pathway: Cellular functions and biological outcomes. *World J. Biol. Chem.* **2014**, *5*, 321. [CrossRef]
74. Sheikh, S.; Saxena, D.; Tian, X.; Amirshaghaghi, A.; Tsourkas, A.; Brem, S.; Dorsey, J.F. An integrated stress response agent that modulates DR5-dependent TRAIL synergy reduces patient-derived glioma stem cell viability. *Mol. Cancer Res.* **2019**, *17*, 1102–1114. [CrossRef]

75. Culpepper, K.M.; Podrabsky, J.E. Cell cycle regulation during development and dormancy in embryos of the annual killifish *Austrofundulus limnaeus*. *Cell Cycle* **2012**, *11*, 1697–1704. [CrossRef]
76. Meller, C.L.; Podrabsky, J.E. Avoidance of apoptosis in embryonic cells of the annual killifish *Austrofundulus limnaeus* exposed to anoxia. *PLoS ONE* **2013**, *8*, e75837. [CrossRef]
77. Lesnefsky, E.J.; Chen, Q.; Tandler, B.; Hoppel, C.L. Mitochondrial dysfunction and myocardial ischemia-reperfusion: Implications for novel therapies. *Annu. Rev. Pharmacol. Toxicol.* **2017**, *57*, 535–565. [CrossRef] [PubMed]
78. Hochrainer, K.; Jackman, K.; Anrather, J.; Iadecola, C. Reperfusion rather than ischemia drives the formation of ubiquitin aggregates after middle cerebral artery occlusion. *Stroke* **2012**, *43*, 2229–2235. [CrossRef] [PubMed]
79. Braunersreuther, V.; Jaquet, V. Reactive oxygen species in myocardial reperfusion injury: From physiopathology to therapeutic approaches. *Curr. Pharm. Biotechnol.* **2012**, *13*, 97–114. [CrossRef] [PubMed]
80. Kalogeris, T.; Baines, C.P.; Krenz, M.; Korthuis, R.J. Cell biology of ischemia/reperfusion injury. *Int. Rev. Cell Mol. Biol.* **2012**, *298*, 229–317. [CrossRef]
81. Zhang, Y.; Takemori, H.; Wang, C.; Fu, J.; Xu, M.; Xiong, L.; Li, N.; Wen, X. Role of salt inducible kinase 1 in high glucose-induced lipid accumulation in HepG2 cells and metformin intervention. *Life Sci.* **2017**, *173*, 107–115. [CrossRef]
82. Keenan, S.W.; Hill, C.A.; Kandoth, C.; Buck, L.T.; Warren, D.E. Transcriptomic responses of the heart and brain to anoxia in the Western Painted Turtle. *PLoS ONE* **2015**, *10*, e0131669. [CrossRef]
83. Grober, J.; Lucas, S.; Sörhede-Winzell, M.; Zaghini, I.; Mairal, A.; Contreras, J.-A.; Besnard, P.; Holm, C.; Langin, D. Hormone-sensitive lipase is a cholesterol esterase of the intestinal mucosa. *J. Biol. Chem.* **2003**, *278*, 6510–6515. [CrossRef]
84. Huang, Y.; Mahley, R.W. Apolipoprotein E: Structure and function in lipid metabolism, neurobiology, and Alzheimer's diseases. *Neurobiol. Dis.* **2014**, *72*, 3–12. [CrossRef]
85. Khalil, Y.A.; Rabès, J.-P.; Boileau, C.; Varret, M. APOE gene variants in primary dyslipidemia. *Atherosclerosis* **2021**, *328*, 11–22. [CrossRef]
86. Nye, C.K.; Hanson, R.W.; Kalhan, S.C. Glyceroneogenesis is the dominant pathway for triglyceride glycerol synthesis in vivo in the rat. *J. Biol. Chem.* **2008**, *283*, 27565–27574. [CrossRef]
87. Chatterjee, S.; Choi, A.J.; Frankel, G. A systematic review of Sec24 cargo interactome. *Traffic* **2021**, *22*, 412–424. [CrossRef] [PubMed]
88. Linders, P.T.; Van Der Horst, C.; Ter Beest, M.; Van Den Bogaart, G. Stx5-mediated ER-Golgi transport in mammals and yeast. *Cells* **2019**, *8*, 780. [CrossRef]
89. López-Hernández, T.; Haucke, V.; Maritzen, T. Endocytosis in the adaptation to cellular stress. *Cell Stress* **2020**, *4*, 230–247. [CrossRef]
90. Park, J.; Kim, J.; Park, H.; Kim, T.; Lee, S. ESCRT-III: A versatile membrane remodeling machinery and its implications in cellular processes and diseases. *Anim. Cells Syst.* **2024**, *28*, 367–380. [CrossRef]
91. Kumar, A.V.; Mills, J.; Lapierre, L.R. Selective autophagy receptor p62/SQSTM1, a pivotal player in stress and aging. *Front. Cell Dev. Biol.* **2022**, *10*, 793328. [CrossRef]
92. Hu, X.; Wang, L.; Wang, Y.; Ji, J.; Li, J.; Wang, Z.; Li, C.; Zhang, Y.; Zhang, Z.-R. RNF126-mediated reubiquitination is required for proteasomal degradation of p97-extracted membrane proteins. *Mol. Cell* **2020**, *79*, 320–331.e329. [CrossRef] [PubMed]
93. Loh, X.-Y.; Sun, Q.-Y.; Ding, L.-W.; Mayakonda, A.; Venkatachalam, N.; Yeo, M.-S.; Silva, T.C.; Xiao, J.-F.; Doan, N.B.; Said, J.W.; et al. RNA-binding protein *ZFP36L1* suppresses hypoxia and cell-cycle signaling. *Cancer Res.* **2020**, *80*, 219–233. [CrossRef] [PubMed]
94. She, Z.-Y.; Yang, W.-X. SOX family transcription factors involved in diverse cellular events during development. *Eur. J. Cell Biol.* **2015**, *94*, 547–563. [CrossRef]
95. Svingen, T.; Tonissen, K.F. Hox transcription factors and their elusive mammalian gene targets. *Heredity* **2006**, *97*, 88–96. [CrossRef]
96. Robert-Moreno, À.; Naranjo, S.; De La Calle-Mustienes, E.; Gómez-Skarmeta, J.L.; Alsina, B. Characterization of new otic enhancers of the Pou3f4 gene reveal distinct signaling pathway regulation and spatio-temporal patterns. *PLoS ONE* **2010**, *5*, e15907. [CrossRef]
97. Bell, S.M.; Schreiner, C.M.; Waclaw, R.R.; Campbell, K.; Potter, S.S.; Scott, W.J. *Sp8* is crucial for limb outgrowth and neuropore closure. *Proc. Natl. Acad. Sci. USA* **2003**, *100*, 12195–12200. [CrossRef] [PubMed]
98. Maier, C.J.; Maier, R.H.; Rid, R.; Trost, A.; Hundsberger, H.; Eger, A.; Hintner, H.; Bauer, J.W.; Onder, K. PIM-1 kinase interacts with the DNA binding domain of the vitamin D receptor: A further kinase implicated in 1,25-(OH)2D3 signaling. *BMC Mol. Biol.* **2012**, *13*, 18. [CrossRef]
99. Gu, J.J.; Wang, Z.; Reeves, R.; Magnuson, N.S. PIM1 phosphorylates and negatively regulates ASK1-mediated apoptosis. *Oncogene* **2009**, *28*, 4261–4271. [CrossRef] [PubMed]
100. König, B.; Rauer, C.; Rosenbaum, S.; Brandsch, C.; Eder, K.; Stangl, G.I. Fasting upregulates PPARα target genes in brain and influences pituitary hormone expression in a PPARα dependent manner. *PPAR Res.* **2009**, *2009*, 1–9. [CrossRef] [PubMed]

101. Mandard, S.; Müller, M.; Kersten, S. Peroxisome proliferator-activated receptor a target genes. *Cell. Mol. Life Sci.* **2004**, *61*, 393–416. [CrossRef]
102. Kim, D.; Ha, S.K.; Gonzalez, F.J. CBFA2T3 is PPARA sensitive and attenuates fasting-induced lipid accumulation in mouse liver. *Cells* **2024**, *13*, 831. [CrossRef]

*biomolecules*

MDPI

*Review*

# Computational Approaches to Cancer Cell Dormancy: From Detection to Dynamic Modelling

**Lucas G. N. Spink, Shi Pan, Minyoung Kim , Belis Yener, Borbála Bánfalvi and Maria Secrier ***

UCL Genetics Institute, Department of Genetics, Evolution and Environment, University College London, Gower Street, London WC1E 6BT, UK; lucas.spink.22@ucl.ac.uk (L.G.N.S.); shi.pan@ucl.ac.uk (S.P.); minyoung.kim.23@ucl.ac.uk (M.K.); belis.yener.23@ucl.ac.uk (B.Y.); b.banfalvi@qmul.ac.uk (B.B.)

* Correspondence: m.secrier@ucl.ac.uk

**Abstract**

Cancer cell dormancy is a clinically consequential yet computationally under-defined phenomenon characterised by reversible growth arrest and delayed disease recurrence. Although advances in single-cell and multi-omic profiling have improved detection of dormant and persister populations, their molecular identity and dynamical behaviour remain difficult to resolve. In this review, we examine how computational methods have been applied to infer dormant cell identity, heterogeneity, microenvironmental regulation, state transitions, and reactivation dynamics. We highlight how single-cell transcriptomics, lineage tracing, spatial profiling, and integrative multi-omic analyses reveal substantial context-dependent variability, undermining the notion of a universal dormancy signature. We further discuss emerging mathematical and statistical frameworks to model the awakening from dormancy, alongside approaches linking dormancy-associated features to clinical outcomes. Recurring challenges include fragmented operational definitions, rare-state detection, cross-study incompatibility, and the use of snapshot data to interrogate inherently temporal processes. We argue that progress will depend on computational frameworks that treat dormancy as a dynamic, multi-scale systems problem rather than a static cell-type classification task.

**Keywords:** cancer cell dormancy; computational modelling; cell state transitions; tumour reactivation; drug-tolerant persister cells; disseminated tumour cells

## 1. Introduction

Cellular dormancy is an evolutionarily ancient survival strategy. Across many domains of life, cells have evolved the capacity to enter reversible growth arrest in response to environmental adversity, enabling them to endure otherwise lethal conditions and resume proliferation when favourable conditions return [1,2]. In yeast, nutrient limitation triggers entry into the stationary phase, a quiescent state regulated by conserved nutrient-sensing pathways, including TOR, PKA, and AMPK [3–6]. In the nematode *Caenorhabditis elegans*, larvae enter the stress-resistant dauer larval state under unfavourable conditions, a process regulated by insulin/IGF-1 signalling (IIS) and AMPK pathways, which converge on the transcription factor DAF-16, as well as the PKA pathway [7–9]. Some killifish species live in ponds that desiccate during dry seasons; to survive these periods, killifish embryos undergo diapause, a prolonged developmental arrest that is thought to be mediated by the IIS, AMPK, and p53 pathways, enabling survival through seasonal desiccation [10–14]. In mammals, quiescence is regulated by the p53 and mTOR pathways, integrated with the IIS and AMPK signalling networks [15–18]. These pathways share not only functional logic but also specific molecular components: the AMPK pathway is significant in quiescence regulation across all multicellular eukaryotes examined, while TOR, IIS, and p53 are conserved across overlapping but distinct subsets of species. Similarly, the phenotypic hallmarks of quiescent cells, including arrested growth, decreased cell size, reduced biosynthesis, increased autophagy, and resistance to environmental stresses, are remarkably consistent across organisms [14,19].

This evolutionary conservation is directly relevant to cancer biology. Rather than representing a cancer-specific process, cellular cancer dormancy (not to be confused with population-level forms of dormancy, such as angiogenic or immunogenic dormancy [20]) is best understood as the pathological co-option of these conserved cellular stress-response and developmental quiescence programmes [2]. Cancer cells that enter dormancy exploit pre-existing survival mechanisms that normally enable cells to withstand environmental stress, nutrient limitation, or unfavourable growth conditions [21–23]. In doing so, they become refractory to therapies targeting actively proliferating cells and can persist undetected for extended periods before potentially reawakening to seed disease recurrence, often years or decades after apparently successful treatment [24–26]. Understanding cancer dormancy is therefore of considerable clinical importance, as dormant cancer cells represent a reservoir for relapse and metastasis [27], yet they remain poorly characterised and largely invisible to current therapeutic and diagnostic strategies.

In cancer, cellular dormancy refers to treatment-refractory cancer cells that have entered reversible G0/G1 arrest, with varying levels of metabolic activity [25,27–31]. While the general characteristics of dormant cancer cells are reasonably well established in the literature, the terminology used to describe them remains inconsistent and often overlapping, making precise discussion challenging [32]. Two manifestations of cancer dormancy have been widely characterised in the literature (Figure 1a):

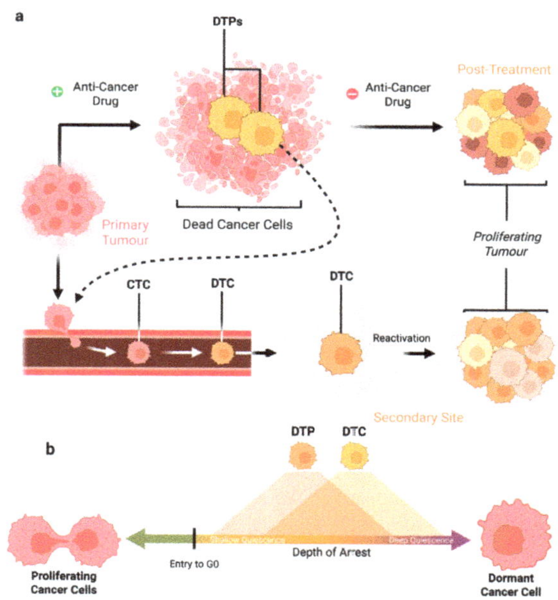


**Figure 1.** Mechanisms of drug-tolerant persister (DTP) and disseminated tumour cell (DTC) formation, and a proposed spectrum of dormant cancer cell states across varying depths of quiescence. (a) Schematic representation of the origins and fates of two different manifestations of cancer dormancy,

DTPs and DTCs. DTPs form from anti-cancer treatment that pushes the cancer cells into a dormant state that allows them to survive therapy. After treatment is removed, DTPs re-enter the cell cycle and begin to proliferate again, forming new tumours. DTCs are circulating tumour cells (CTCs) that have become dormant and then embedded in a secondary site, where they remain dormant for a period before beginning to proliferate and form a metastatic tumour. It is important to note that DTPs can also leave the primary tumour to form metastases in a similar fashion. Arrows indicate transitions between proliferative and dormant states. (**b**) Conceptual model of a dormancy continuum, illustrating a spectrum of cell states with increasing depth of quiescence and significant overlap between dormant subtypes, rather than discrete categories. Created in BioRender. Spink, L. (2026). https://BioRender.com/f05zqdp (accessed on 17 April 2026).

**Drug-Tolerant Persisters (DTPs)** describe a subpopulation of cancer cells that survive cancer treatment that would otherwise be lethal through non-genetic mechanisms. Instead of acquiring resistance mutations, DTPs enter reversible G0 or slowly cycling (long G1) states, enabling them to withstand the drug challenge via epigenetic changes or metabolic rewiring. These cells are thought to form a reservoir of resistance, as they can revert to proliferation after selective pressure from cancer treatment is removed (Figure 1a) [33,34].

**Disseminated Tumour Cells (DTCs)** are cancer cells that enter the blood as circulating tumour cells (CTCs) and then embed themselves in organs and tissues [35]. These DTCs will typically have entered reversible G0/G1 arrest and, once embedded, will remain clinically silent for long periods before "reawakening" to form metastases (Figure 1a). Similar to DTPs, DTCs are also thought to be a key driver of cancer recurrence [36].

Alongside these, evidence of other forms of dormancy in the context of therapeutic resistance is beginning to emerge in multiple cancer types:

**Polyploid Giant Cancer Cells (PGCC)** are a subpopulation of cancer cells characterised by giant or multiple nuclei and polyploidy [37]. PGCCs are thought to result from endoreduplication, cell fusion, cytokinesis failure, and mitotic slippage induced by various stimuli, including therapeutic drugs, radiation treatment, hypoxic environments, and human oncoviruses and cytomegaloviruses. Recent work has also linked their emergence to so-called "treacherous apoptosis", in which cells initiate but fail to complete apoptotic programmes, instead surviving and contributing to tumour heterogeneity and repopulation, suggesting that cell death pathways themselves may paradoxically promote the formation of therapy-resistant, dormant-like states [38,39]. When environmental conditions are unfavourable, PGCCs enter a slow-cycling/quiescent state, becoming drug-resistant. PGCCs were long considered terminally senescent and incapable of further division; however, pioneering work by Walen and Rajaraman and colleagues in the early 2000s demonstrated that polyploid cells can undergo amitotic division to produce viable, rapidly proliferating progeny with tumorigenic potential [40]. When conditions are favourable again, PGCCs exit their dormant state. They can begin to produce progeny via asymmetric non-canonical divisions (e.g., budding, splitting, bursting, neosis, and amitosis), a process dubbed de-polyploidisation. The resulting daughter cells exhibit cancer stem-like properties, high invasion/migration, and chemoradiotherapy resistance and are capable of initiating tumours in vivo [37,41–43]. Moreover, PGCC-derived daughter cells have been repeatedly identified as major drivers of tumour recurrence and distant metastasis [44–47].

**Stably Resistant Cancer Cells (SRCCs)** are a *NOTCH+* subpopulation of dormant cancer cells described in lung cancer exhibiting stable, innate resistance to chemotherapy, targeted therapy, and immunotherapy. SRCCs' cellular and molecular characteristics, such as epithelial state, elevated reactive oxygen species, and anti-apoptotic and anti-ferroptotic survival, differentiate them from other resistant populations, such as cancer stem cells, epithelial–mesenchymal transition cells, and DTPs. Furthermore, inhibition of the non-canonical *NOTCH1* pathway resensitises these cells to cisplatin-induced cell death in vitro and in vivo [48].



https://doi.org/10.3390/biom16050633

While several forms of cancer cell dormancy have been proposed, these definitions overlap. For example, a DTC that has been exposed to therapy could simultaneously meet the definition of a DTP and DTC. Similarly, a DTP exposed to sufficiently high levels of environmental stress could take on characteristics associated with PGCCs. These overlaps have prompted some to ask whether these subpopulations truly represent biologically distinct cellular types or whether cancer dormancy may be better represented by a spectrum of states along a gradient of cell cycle arrest depth (Figure 1b) [29]. While we lack exact and discrete definitions for the different forms of dormancy in cancer, several biological programmes have been linked to the entry to, maintenance of, and exit from this state.

*Known Mechanisms of Cancer Dormancy and the Need for Computational Approaches*

The molecular mechanisms underlying cancer dormancy can be broadly organised into three functional categories: induction of dormancy, maintenance of this state, and reawakening into proliferation. Induction of dormancy is thought to be triggered by a range of factors, including intrinsic cellular stress and microenvironmental cues, and is enacted through cell cycle arrest, TGF-β/BMP signalling, and downregulation of oncogenic pathways [24,49–52]. A well-established hallmark of dormancy is the balance between ERK and p38 signalling. Loss of integrin-mediated extracellular matrix signalling, hypoxia, nutrient stress, and cancer therapy are thought to promote a shift from high ERK/low p38 activity towards low ERK/high p38 activity, associated with entry into quiescence, a transition regulated in part by upstream mediators such as uPAR [51–56].

Maintenance of dormancy is primarily thought to be mediated by continued p27/p21 expression, low CDK activity, and sustained low ERK/high p38 signalling, which keep cells in G0 [30,52,54]. Autophagy, metabolic changes, and reduced biosynthesis are also thought to be modified to support long-term survival with minimal accumulation of damage [51,57]. In parallel, stress-response pathways such as the unfolded protein response contribute to dormancy maintenance, in part through activation of the antioxidant regulator NRF2, which mitigates oxidative stress associated with altered metabolism and has been implicated in defining distinct persister subpopulations [58,59]. At the transcriptional level, factors such as NR2F1, Dec2, and NDRG1 further reinforce this state by coordinating gene programmes associated with quiescence, stemness, and stress adaptation [60–63]. The niche in which the dormant cancer cells reside is also thought to be important to maintenance, with evidence that specific extracellular matrix (ECM) composition, integrin signalling, and perivascular/bone-marrow niches actively help maintain quiescence [24,64]. These pathways form an integrated network rather than independent modules, collectively stabilising reversible growth arrest while preserving long-term cellular viability.

The mechanisms surrounding dormant cancer cells reawakening are more nebulous, but reawakening is thought to be partly mediated by the reversal of the aforementioned programmes (e.g., restoration of the ERK/p38 balance); however, exact triggers for reawakening remain unclear [30]. One example of a potential pro-reawakening trigger is ECM remodelling and chronic inflammation, which provide pro-proliferative cues that could potentially trigger re-entry to the cell cycle, restarting tumour proliferation [65].

While previous experimental work in the field has revealed some pathways and general principles of cancer dormancy, as comprehensively reviewed elsewhere [27,49,66–68], these findings have yet to converge on a unified mechanistic framework, and much of this process remains poorly understood. Recent advances in single-cell and multi-omic profiling technologies, and the computational techniques to analyse their output, have shown promise in helping us better understand cancer dormancy. In this review, we begin by discussing a series of key questions surrounding cancer dormancy and how computational techniques have been used to address them, particularly by enabling the identification,

 https://doi.org/10.3390/biom16050633

stratification, and dynamic modelling of rare dormant cell populations that are otherwise inaccessible to bulk experimental approaches. We then go on to discuss the main challenges for computational analysis of cancer dormancy as well as potential future solutions to these challenges. Given the overlapping definitions and lack of consensus within the field, throughout this review, we use the term "dormant cancer cells" to broadly refer to cancer cells that have, through any mechanism, entered reversible G0/G1 arrest, rendering them refractory to standard cancer treatment. Where a study focuses on a specific subtype, we use the specific term (e.g., DTC, DTP, etc.). This usage is not intended to imply that these subtypes are biologically equivalent but reflects the current state of the field in which the precise differences between the dormant cancer cell subtypes remain unresolved.

## 2. From Multi-Omics Data to Mathematical Models: Computational Studies of Dormancy

Dormant cancer cells are widely considered to be rare in tumour populations [50]. As such, in bulk sequencing datasets, dormant cancer cells are averaged out and are essentially invisible. Advances in single-cell resolution and spatially resolved technologies, including single-cell RNA sequencing, single-cell multi-omics, and spatial transcriptomics, have enabled the detection and characterisation of dormant cancer cells within heterogeneous tumour populations. In doing so, these technologies also revealed the extent of the challenge that dormant cancer cells represent: they are rare in most datasets, difficult to distinguish from other cell cycle arrest states (e.g., quiescence, senescence, etc.), and their regulatory programmes are context-dependent. In the following sections, we examine how computational approaches have been applied to four central questions in cancer dormancy research: inference of dormant cancer cell identity, microenvironmental regulation of dormancy, transitions into and out of the dormant state, and the influence of cancer dormancy on clinical outcomes (Figure 2).

### 2.1. Computational Inference of Dormant Cell Identity

One of the first major challenges in computational research on cancer dormancy is reliably identifying dormant cancer cells. All computational studies of dormancy must first define which cells are dormant, and in the absence of a consensus molecular marker, this definition varies across studies. Some studies use label retention as a proxy for quiescence, while others use survival under sustained drug treatment or growth arrest under therapy. Given this definitional challenge, many computational studies have attempted to identify markers for dormant cancer cells. Here, we discuss a series of papers that apply computational techniques across different modalities to characterise dormant cancer cells.

A natural starting point is to ask whether dormant cancer cells express distinct genes that could serve as molecular markers. To answer this question, Ren et al. [69] used the retention of the membrane dye DiD as a functional readout of proliferative history (quiescent cells retain the membrane dye, while actively dividing cells dilute it) to distinguish proliferative from dormant metastatic breast cancer cells in the lungs and bone marrow of a genetically engineered mouse model. The proliferative and dormant cells were then sequenced using single-cell RNA sequencing (scRNA-seq), and differential gene expression was performed, identifying a group of genes, including *Cfh*, *Gas6*, *Mme*, and *Ogn*, that were all upregulated in dormant breast cancer cells compared to the proliferative breast cancer cells. While these genes correlated with disease-free survival in breast cancer patients, attempts to validate these results experimentally with single-gene perturbation experiments in murine models showed that no individual gene affected dormancy, indicating that dormancy is likely not controlled by a single, universal master regulator but reflects a coordinated programme-level change which can be difficult to capture through standard

 https://doi.org/10.3390/biom16050633

differential analysis, particularly if some of the changes are short-lived. Moreover, the genes identified by Ren et al. are likely specific to metastatic breast cancer in this specific experimental context. Indeed, across the broader literature, the wide range of dormancy-associated markers that have been reported reinforces the view that many dormancy programmes are context-dependent, even if certain core gene programmes are shared.

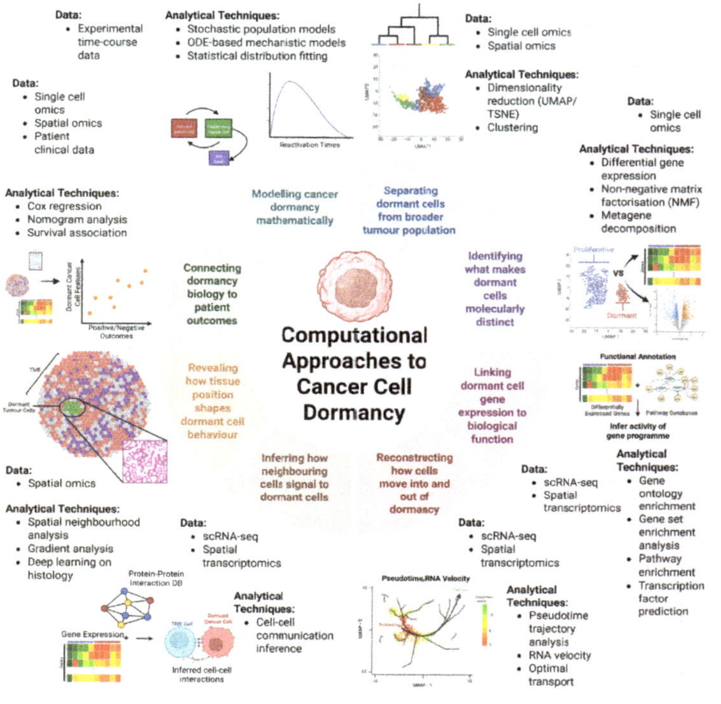


**Figure 2.** Overview of computational approaches to studying cancer cell dormancy. Eight key analytical strategies for interrogating cancer cell dormancy are arranged around a central theme, each accompanied by a schematic illustration, the data modalities it draws on, and the specific techniques it encompasses. Starting from the top and moving clockwise: (1) dimensionality reduction and clustering of single cell or spatial omics data separates dormant cells from the broader tumour population and identifies subtypes within them; (2) differential gene expression analysis, transcription factor prediction, and metagene decomposition across single cell omic data identify what makes dormant cells molecularly distinct; (3) gene ontology, gene set enrichment, and pathway enrichment analyses link dormant cell gene expression changes to biological function; (4) pseudotime trajectory analysis and RNA velocity applied to single-cell and spatial transcriptomic data reconstruct how cells transition into and out of dormancy; (5) cell–cell communication inference from single-cell expression data, combined with protein–protein interaction databases, infers how neighbouring cells in the tumour microenvironment signal to dormant cells; (6) spatial neighbourhood analysis, gradient analysis, and deep learning on histopathology applied to spatial transcriptomic data reveal how tissue position shapes dormant cell behaviour; (7) Cox regression, nomogram analysis, and survival association applied to single-cell/spatial omic and clinical data connect dormancy biology to patient outcomes; (8) stochastic population models, ODE-based mechanistic models, and statistical distribution fitting applied to experimental time-course data model cancer dormancy dynamics mathematically, including reactivation timing. No single data type or analytical method is sufficient to capture the complexity of cancer cell dormancy, and progress in the field relies on the complementary application of multiple approaches across different data modalities. Created in BioRender. Spink, L. (2026) https://BioRender.com/00foaaz (accessed on 17 April 2026).

 https://doi.org/10.3390/biom16050633

While marker-based isolation followed by transcriptomic profiling identifies candidate genes, combining lineage tracing with multi-modal profiling can reveal additional molecular layers of dormant cell identity that transcriptomics alone cannot capture. For example, Oren et al. [70] developed Watermelon, a lentiviral barcoding system that combines lineage tracing with scRNA-seq, and applied it to lung, breast, melanoma and colorectal cancer cell lines treated with clinically relevant kinase inhibitors (osimertinib, lapatinib or dabrafenib) to track the emergence of persistent cells. These DTPs were separated into two phenotypes, cycling and non-cycling persisters, which arose from distinct cell lineages. Subsequent differential expression analysis of scRNA-sequencing and LC-MS/MS metabolic profiling showed that the rare subset of cells capable of re-entering the cell cycle under constitutive drug treatment (cycling persisters) presented nuclear factor erythroid 2-related factor 2 (*NRF2*)-driven antioxidant programmes and a shift towards fatty acid oxidation, which were absent in the non-cycling persisters. The authors demonstrated that these signatures are partially or fully reproducible across multiple lines and in patient tumour samples, indicating that cycling persisters share a reproducible molecular identity defined by antioxidant and metabolic reprogramming and identifying these programmes as potential markers to distinguish cycling from non-cycling persister populations.

While Oren et al.'s combination of transcriptomic and metabolomic profiling with lineage tracing identified *NRF2*-driven antioxidant programmes and metabolic reprogramming as defining features of cycling persisters, these signatures only operate at the transcriptomic and metabolomic levels. Rosano et al. [71] extended this multimodal approach to the epigenetic layer. In their TRADITIOM (**TR**acking **A**daptation, **D**ormancy and Awakening with multi**om**ics) study, the authors developed a long-term in vitro lineage-tracing system that incorporates bulk RNA-seq and scRNA-seq, whole-genome sequencing, and super-SILAC (stable isotope labelling by amino acids in cell culture) histone mass spectrometry [72] to map the adaptive trajectories of $ER^+$ breast cancer cell lines into and out of dormancy under long-term endocrine therapies. By combining clonal barcode tracking with transcriptomic profiling over extended treatment periods, this work demonstrated that endocrine therapies can induce a non-genetic transition into dormancy in a stochastic subset of cells and that individual dormant lineages can later reawaken through divergent epigenetic reprogramming without recurrent genetic mutations. This approach revealed that dormant cell identity in this system operates primarily at the chromatin level and that epigenetic profiling captures aspects of dormancy invisible to transcriptomics or genomics alone. Together, Oren et al. and Rosano et al. illustrate that dormant cancer cell identity spans multiple molecular layers (transcriptomic, metabolic, epigenetic) and that no single modality can capture the full picture. They also illustrate a key tension in the field: whether dormant cell identity is defined by stable, pre-existing molecular programmes, as Oren et al.'s reproducible *NRF2* signatures suggest, or by variable epigenetic states that differ across individual dormant lineages, as Rosano et al. found. These conclusions are not incompatible but rather reflect differences in cancer type and treatment regimen; they have significant implications for whether dormancy can be predicted from the pre-treatment state or only observed retrospectively.

Taken together, these studies show that while computational approaches can identify and stratify dormant and persister populations at single-cell resolution, they do not converge on a stable or universal dormancy signature. Instead, they reveal substantial lineage- and context-specific heterogeneity, with different molecular programmes supporting similar phenotypic outcomes. This reinforces the view that dormancy may not represent a single, fixed cell state, but rather a flexible, adaptive state whose identity is contingent on prior history and environmental conditions.


https://doi.org/10.3390/biom16050633

### 2.2. Computational Analysis of Microenvironmental and Spatial Regulation of Cancer Dormancy

Another key question in the field is how the microenvironment and spatial context may influence cancer dormancy. The idea that the microenvironment was involved in cancer was first put forward by Steven Paget with his "seed and soil" hypothesis in 1889, where he suggested that the spread of tumour cells across the body to form metastases was mediated by an interaction between cancer cells (seed) and the organs/tissues they embedded in (soil) [73]. Paget's hypothesis has proved especially relevant to dormancy, where niche-derived signals have been implicated in both the induction and the disruption of quiescent states. Recent advances in single-cell and spatial profiling technologies, combined with increasingly sophisticated computational approaches, have enabled researchers to interrogate these interactions at high resolution. Here, we discuss a series of studies that have applied such approaches to characterise the microenvironmental regulation of dormancy at increasing scales of complexity, from individual niche interactions to pan-cancer network analyses.

Initial computational studies examining how the microenvironment may regulate cancer dormancy primarily utilised scRNA-seq data alongside various computational analyses. While these studies consistently found that the tumour microenvironment (TME) plays a role in cancer cell dormancy, the manner in which the TME affects cancer cell dormancy differed. For example, Khoo et al. [74] sought to understand how the endosteal bone microenvironment affected myeloma cell dormancy. They experimentally identified dormant and activated myeloma cells in murine bone marrow using DiD label retention as a readout of proliferation and then sequenced them using scRNA-seq. The authors then analysed the scRNA-seq data with differential gene expression analysis, gene ontology enrichment, and transcription factor prediction. They also applied non-negative matrix factorisation (NMF) clustering, which decomposes gene expression data into a small number of interpretable gene expression patterns and shows how strongly each pattern is present in each sample [75]. Their analysis showed that, despite myeloma cells being of lymphoid origin, the dormant myeloma cells had a transcriptional profile enriched for immune-related genes and myeloid differentiation genes. Experimental validation using coçulture and intravital imaging showed that the myeloid transcriptional reprogramming was induced by direct contact with osteoblasts. Inhibition of *AXL*, one of the highly expressed genes found in dormant myeloma cells, reduced dormant cancer cells in their mouse model. In patients with monoclonal gammopathy of uncertain significance, enrichment of *AXL* and genes coregulated with *AXL* expression was associated with better long-term survival compared to controls, consistent with a greater proportion of cancer cells being in a quiescent state. Furthermore, the expression of the myeloid transcriptional signature in patients with multiple myeloma was associated with a twofold increase in overall survival. This study showed that the environment/niche can significantly affect dormant cancer cells and that computational tools can effectively model these interactions.

While Khoo et al. identified specific cell–cell interactions driving cancer cell dormancy-associated reprogramming, Janghorban et al. [76] found evidence of broader TME remodelling. Janghorban et al. generated GFP+ tumours in GFP mice, which were then transplanted into the fat pads of immunogenic mice. After the tumours reached 7–9 mm in diameter, the mice were treated with chemotherapy (FGFR inhibitor) for 14 days. Mice were sacrificed, and their tumours and surrounding microenvironments were sequenced with scRNA-seq at 3 time points: before treatment, after 14 days of chemotherapy treatment and 14 days off treatment, and 1–4 months after treatment, representing primary, dormant, and recurrent populations of cancer cells. Some mice were also treated with α-Jag1-blocking antibodies or with α-IgG2b. The authors then performed differential gene expression, trajectory analysis, and gene ontology analyses, which showed that cancer

 https://doi.org/10.3390/biom16050633

dormancy was associated with significant remodelling of the myeloid and lymphoid compartments. Among newly recruited myeloid and lymphoid cells during dormancy, the majority exhibited immunosuppressive activity. They further showed that Jagged-1/Notch signalling played a key role in dormancy, as treatment with anti-Jag1 antibodies delayed tumour recurrence. Ultimately, this study demonstrated that dormancy was associated with broader changes in TME composition compared to the specific interactions described in Khoo et al. (albeit in a different system).

These studies raise an important question about the scale at which the TME regulates dormancy. In Khoo et al.'s myeloma study, reprogramming effected by the TME appears to be driven by a specific cell–cell interaction between osteoblasts and dormant cancer cells in the bone marrow niche. In Janghorban et al.'s mouse breast cancer model of minimal residual disease, dormancy is associated with broader remodelling of the entire TME. Whether this difference in interactions truly represents a distinct regulatory mechanism or reflects differences in tissue context between bone marrow and mouse mammary fat pad remains unclear.

While the previous two studies used scRNA-seq and standard bioinformatics analyses to investigate how the microenvironment contributes to cancer dormancy, scRNA-seq is blind to the spatial context of these microenvironments. Recent advances in spatial transcriptomics now allow researchers to explore both the transcriptomic and spatial context of dormancy. Rubinstein et al. [77] utilised spatial transcriptomics to explore how the spatial organisation of the tumour microenvironment affects drug-induced dormancy in BRAF-mutant melanoma. The authors used patient-derived xenograft models treated with dabrafenib/trametinib to simulate treatment with samples collected at various time points. Computationally, the authors applied copy number alteration inference and pathway enrichment to profile the molecular characteristics of the dormant cells. They also used pseudotime trajectory analysis, which orders cells along an inferred temporal axis based on transcriptomic similarity [78], and RNA velocity, which estimates future cell states by comparing spliced and unspliced RNA ratios [79], to reconstruct cell state transitions. Their results showed that after treatment dormant cancer cells were characterised by increased oxidative phosphorylation, decreased proliferation, and increased invasive capacity. Strikingly, the authors also found central-peripheral patterns in tumours, where populations at the tumour-stromal boundary were subject to distinct signalling from those in the intra-tumoural regions and, as a result, exhibited distinct transcriptional changes. Crucially, these central-to-peripheral gradients would have been invisible in dissociated scRNA-seq experiments, as tissue dissociation removes the spatial context. Furthermore, deep learning applied to histopathology slides demonstrated that major transcriptional states, treatment-sensitive versus resistant lineages, and the persister phenotype could be distinguished directly from H&E images, despite being indistinguishable by conventional pathological review, with imaging features tracking disease progression through treatment and correlating strongly with RNA expression clusters and, more weakly, with CNV-defined lineages. This raises the possibility that dormancy-associated spatial patterns may eventually be detectable from standard clinical imaging.

Beyond spatial organisation, an increasingly common complementary approach in single-cell and spatial transcriptomics is ligand–receptor-based inference of cell–cell communication (Figure 2). Although widely used to characterise tumour microenvironment signalling, its direct application to cancer dormancy remains limited. Given the central role of niche-derived signalling, this represents a promising direction for identifying microenvironmental regulators of dormancy, with emerging work in related contexts beginning to explore this space [80].


https://doi.org/10.3390/biom16050633

Together, these studies illustrate how computational approaches at increasing molecular resolution have reshaped our understanding of the microenvironmental regulation of dormancy. From niche-specific transcriptional reprogramming in myeloma to immune ecosystem remodelling in breast cancer to spatially resolved metabolic gradients in melanoma, a consistent theme emerges: the microenvironment does not merely provide a passive sanctuary for dormant cells but actively shapes their transcriptional identity, immune interactions, and metabolic state.

### 2.3. Computational Inference of Dormancy State Transitions

Beyond identifying cell states and interactions with the microenvironment, another poorly understood aspect of cancer cell dormancy is the mechanism(s) underpinning entry into and exit from cancer cell dormancy. Specifically, two distinct questions remain unanswered: what are the molecular mechanisms driving entry, maintenance, and exit from the dormant state, and whether the dormant state itself represents a single uniform condition, a graded spectrum of quiescence depth [81], or a set of biologically distinct subtypes. These are clinically relevant questions, as the factors that govern these transitions are prime targets for treatments aimed at reducing the risk of relapse and metastasis. As these transitions are difficult to observe directly in patients and unfold on time scales that are difficult to capture experimentally, computational inference from high-resolution molecular data offers one of the few available routes to study them effectively. Despite its clinical importance, studies directly investigating how this transition occurs computationally remain limited; however, some work has begun to move in this direction.

Fernandez et al. [82] combined scRNA-seq and time-lapse imaging of cell cycle dynamics to examine the range of quiescent/senescent phenotypes in MCF10A mammary epithelial cells following chemotherapy (etoposide). Cells were exposed to increasing doses of etoposide, and *CDK2* activity was used to classify fast-cycling, slow-cycling (quiescent), and arrested (senescent) cells, which were then profiled by scRNA-seq on day 6 after drug washout. The authors then used a combination of dimensionality reduction, clustering, differential gene expression, and pseudotime trajectory analysis to analyse the scRNA-seq data. Strikingly, instead of forming transcriptionally distinct islands, the pseudotime analysis applied by the authors showed that etoposide-treated cells formed a continuous trajectory of cells that gradually tended towards deeper arrest. Gene set enrichment analysis (GSEA) showed progressive downregulation of cell-cycle, RNA-processing, and protein-translation programmes along the pseudotime axes, supporting the idea that cells treated with therapy enter arrest gradually rather than in a switch-like fashion. From this continuum, the study identified four transcriptionally distinct senescent subtypes, termed "senotypes", arising from two cellular arrest routes: cells that arrested gradually via a standard mitosis-to-G0 transition and cells that arrested via mitotic slip. While this study was performed in non-cancerous cell lines, these findings have implications for drug tolerance, persistence after treatment, and, indirectly, for cancer dormancy. Namely, that rather than drug-induced adaptation representing a binary transition into a uniform dormant state, drug-induced arrest may represent a graded progression from a shallow quiescent state to a deeper, more senescence-like arrest and that these states may represent overlapping rather than discrete categories.

Understanding the mechanisms that regulate the re-entry of dormant cancer cells into the cell cycle is critical for developing treatments that aim to reduce recurrence rates after successful cancer treatment. Hu et al. [83] aimed to elucidate this tumour-intrinsic mechanism using indolent mouse and human lung adenocarcinoma cell lines and patient-derived samples, specifically examining tumour-intrinsic regulators of cancer dormancy. The authors performed in vivo CRISPR–Cas9 screens targeting 220 immune-related genes,


https://doi.org/10.3390/biom16050633

combined with sgRNA enrichment sequencing, scRNA-seq, GSEA, chromatin immunoprecipitation sequencing (ChIP–seq), bisulfite sequencing, and functional perturbation experiments. The results implicate the cGAS-STING pathway as an important regulator of dormant cancer cell re-entry to the cell cycle in this model, with STING expression elevated in *SOX2*+ early metastatic progenitors but downregulated in *SOX9*+ macrometastatic cells in both the mouse models and patient-derived data. CHIP-seq showed that STING is epigenetically silenced in metastatic progenitors re-entering the cell cycle via promoter and enhancer hypermethylation, whereas metastatic cells that fail to re-enter the cell cycle and instead re-enter dormancy via TGFβ signalling repress STING through chromatin remodelling. Pharmacological activation of STING reduced dormant metastatic burden in vivo, suggesting that modulation of this pathway may represent a potential strategy to limit lung cancer relapse [83]. Complementing these findings, Rosano et al. [71] further implicates epigenetic reprogramming as a major regulator of dormancy plasticity. As introduced in Section 2.1, they employed long-term, multimodal lineage tracing to track endocrine-treated breast cancer cells as they transitioned in and out of dormancy. The data suggest that entry into dormancy occurs stochastically across lineages through non-genetic mechanisms rather than clonal genetic selection and that reawakening is characterised by divergent epigenetic reprogramming, with different lineages finding distinct chromatin-level routes out of dormancy.

Together, the findings from Rosano et al. and Hu et al. reveal a key unresolved question in the field: when dormant cancer cells reawaken, do they converge on a shared molecular mechanism or awaken via divergent pathways? Hu et al.'s results suggest convergence on epigenetic STING silencing via promoter hypermethylation, consistently associated with reawakening, implying an epigenetic bottleneck to reawakening. Conversely, Rosano et al. support divergent reawakening pathways, with evidence that awakening occurs through distinct epigenetic modifications. This distinction is clinically relevant because, if awakening converges on a single shared mechanism, then, in principle, targeting this mechanism could prevent reawakening across cancers. However, if reawakening is divergent, then no single intervention would be able to prevent reawakening, indicating a need for combination strategies or some form of testing to identify the type of reawakening a specific cancer type utilises. It should be noted that Hu et al.'s and Rosano et al.'s findings come from different cancer types (lung adenocarcinoma and ER+ breast cancer, respectively), different treatments (no treatment and endocrine therapy, respectively), and different experimental time scales. Thus, these divergent results may indicate the context-dependence of cancer cell dormancy reawakening mechanisms.

The studies discussed above showcase how advances in experimental modelling of dormancy, coupled with multi-omics computational analyses, can help elucidate state transitions in cancer dormancy. Beyond standard bioinformatics, cell state transition modelling is an emerging frontier in the field that is powered by increasingly high-throughput, spatially and temporally resolved datasets [81].

### 2.4. Computational Inference of Survival Dynamics and Dormancy Awakening Hazards

Beyond identifying the factors that control entry into and exit from dormancy as targets for therapy, another clinically relevant approach to understanding cancer dormancy is to examine how dormancy affects patient outcomes. Researchers have approached this via two methodologically distinct strategies: forward mathematical modelling and empirical clinical association. Forward mathematical modelling aims to build mechanistic/statistical models from first principles to, for example, predict how dormancy affects treatment efficacy and reactivation timings. In contrast, empirical clinical association studies will work backwards from patient data to see if different features of dormancy correlate with better


https://doi.org/10.3390/biom16050633

or worse patient outcomes. Both strategies are ultimately aimed at gaining a better understanding of how to treat patients where cancer dormancy may be causing cancer recurrence more effectively. Here, we discuss how a combination of mathematical modelling and computational analysis can help better understand the influence of cancer dormancy on patient outcomes.

Two studies have recently used mathematical modelling to better understand dormant cancer cell reactivation timings. Blath et al. [84] developed a stochastic individual-based population model to investigate how short-term cancer dormancy affects therapy success by modelling tumour cell populations undergoing therapy. The authors constructed an idealised mathematical model with active and dormant cancer cell compartments and two classes of drug agents, incorporating spontaneous and drug-induced state switching, competition, and drug degradation. Importantly, this model linked phenotypic plasticity (the ability of cancer cells to become dormant) to therapy evasion. Their model showed that even a very small fraction of dormant cancer cells (~0.5%) is enough to prevent tumour eradication under single-drug therapy. They later show that treatment is most effective when targeting dormant cancer cells with a drug that either "wakes" the cell, making it sensitive to traditional chemotherapy, or directly kills the dormant cancer. While the authors acknowledge that this is a toy model that simplifies many aspects of cancer dormancy, it is one of the few examples of mathematical modelling of cellular dormancy to date and among the limited efforts to explicitly incorporate treatment dynamics into the model. As our understanding of the underlying mechanisms of cancer dormancy increases, mathematical modelling will be an increasingly useful tool for understanding how biological knowledge can be translated into effective treatments that overcome dormancy.

Sfakianakis et al. [85] used a dataset from Cameron et al. [86] to investigate which of eight predetermined statistical distributions (Normal, Logistic, Gamma, Weibull, Log-Normal, Pareto, Gompertz, and Inverse Gaussian) best captured the dormancy reactivation dynamics in murine metastatic melanoma. The dataset from Cameron et al. was produced using fluorescently labelled B16F10 murine melanoma cells that were injected into mice and then sacrificed at six time points over 14 days, with the relative frequencies of solitary cells and small, medium, and large cell clusters recorded to characterise dormancy and reactivation dynamics. The authors constructed a mechanistic ODE-based model that included circulating tumour cell death, extravasation, stochastic dormancy duration (sampled from eight distributions), reactivation, proliferation, and tissue-level death, with parameters fitted using particle swarm optimisation. When evaluated against the Cameron et al. dataset, the Pareto and inverse Gaussian distributions provided the best fits, supporting the idea that heavy-tailed statistical representations of dormancy better predict the dynamics of metastatic dormant cell reactivation. However, it is important to note that the model relies on several strong simplifying assumptions: a constant death rate and proliferation rate, an immediate switch from dormancy to full proliferation, and the assumption that cells are independent of one another, which limits biological interpretability. Nevertheless, the study represents a strong step forward towards formalising reactivation dynamics mathematically.

To better understand the consequences of cancer cell dormancy on patient prognosis, some studies have focused on linking measurable characteristics of cancer dormancy (e.g., transcriptomics and the presence of DTCs) to patient outcomes. Tian et al. [87] investigated tumour dormancy in neuroblastoma using a retrospective transcriptomic dataset, performing consensus clustering on a panel of 19 literature-derived dormancy-associated genes to stratify dormant cells into two clusters. The cluster with the higher dormancy signature score showed a significantly lower survival probability over time. The authors then used univariate and multivariate Cox regression analyses to identify

 https://doi.org/10.3390/biom16050633

dormancy genes associated with prognostic outcomes, resulting in a 6-gene dormancy-associated prognostic signature (*CDKN2A*, *BHLHB3*, *CDKN2B*, *MAPK14*/p38, *CDKN13*, and *BMP7*) that stratified patients into high- and low-risk groups. Interestingly, several of these genes have been previously implicated in cancer cell dormancy, such as the p38 gene family and *BMP7*, or in cell cycle control (*CDKN2A*, *CDKN2B*, and *CDKN1B*). A nomogram analysis incorporating the six-gene dormancy signature alongside sex, age, stage and N-myc status was used to further split neuroblastoma patients into high-, medium-, and low-risk groups. While this study was mainly correlative and retrospective, it highlights that cancer dormancy biology plays an important role in relapse that can be computationally interrogated to predict patient outcomes.

Tjensvoll et al. [88] used data from an operable non-metastatic breast cancer cohort of 191 patients with a median 15.3-year follow-up to show that the presence of DTCs in bone marrow (BM) before or after surgery can be used to predict late recurrence. Bone marrow samples obtained before and after surgery were analysed for the presence of DTCs using a multimarker mRNA quantitative reverse transcription PCR assay, alongside other established markers such as the mitotic activity index (MAI) and lymph node status. Survival analysis showed that patients with pre-operative DTCs recurred much earlier compared to the DTC-negative patients. A high MAI count was also associated with shorter overall survival. Univariable and multivariable Cox regression analyses showed that the highest risk of recurrence was observed in patients who were DTC-positive before and after surgery, with hazard ratios of 6.93 and 8.34, respectively. Furthermore, the presence of DTCs before surgery was the only predictor of late recurrence. This paper supports the idea that DTCs could serve as a biomarker to identify patients at risk of recurrence due to cancer dormancy who may benefit from extended therapy or more frequent monitoring by healthcare professionals.

While transcriptomic signatures can, in principle, be used as clinical tools, as demonstrated by clinical assays such as Oncotype DX in breast cancer [89], Tian et al.'s dormancy-associated signature remains preliminary, as it is derived from a retrospective cohort using a literature-derived gene panel for cancer dormancy. Given the lack of a clear molecular definition for cancer dormancy in the literature, without prospective validation and a standardised definition for the form of cancer cell dormancy being predicted, the signature remains correlative rather than predictive. However, Tian et al.'s results represent a step towards predicting cancer dormancy and recurrence. Tjensvoll et al.'s more direct approach, utilising the physical presence of DTCs in bone marrow, also represents a promising direction for the prediction of recurrence as a result of cancer cell dormancy; however, the invasive bone marrow aspiration needed to check for DTCs is invasive and impractical for routine screening, limiting its scalability.

Taken together, these studies demonstrate two distinct computational strategies for understanding how cancer dormancy may influence patient outcomes: mathematical modelling of reactivation dynamics and empirical associations between dormancy features and clinical outcomes. It is worth noting that both strategies remain largely nascent: mathematical models rely on strong simplifying assumptions about dormancy, while association studies are limited by their retrospective, correlative designs. Currently, these strategies operate in parallel: the mathematical models are not parameterised by the molecular data used in association studies, and the association studies do not incorporate the dynamic frameworks provided by the models. Furthermore, currently, neither strategy accounts for the biological complexity established earlier in Section 2. As this field develops, the mechanistic insights discussed in previous sections (e.g., transcriptional programmes, microenvironmental interactions, and epigenetic modifications) will need to be incorporated into these predictive models to improve these computational strategies.

 https://doi.org/10.3390/biom16050633

## 3. Current Challenges and Future Perspectives: Computational Analysis of Cancer Dormancy

In Section 2, we reviewed the main types of computational inference used to investigate cancer dormancy. This revealed significant methodological diversity but also highlighted key recurring limitations, including reliance on scRNA-seq, fragmented definitions of cancer dormancy, and difficulty combining datasets for meta-analysis. It is also worth noting that the studies in Section 2 are dominated by a small number of cancer types, with breast cancer in particular dominating the scRNA-seq literature. Cancer dormancy remains computationally underexplored in other cancer types, so it is not clear whether the molecular programmes and analytical frameworks discussed in Section 2 apply to the lesser-studied cancer types. In this section, we begin by discussing the main challenges in computational cancer dormancy research, then outline proposed solutions and future directions for the field. Figure 3 provides a schematic overview linking each challenge to corresponding computational strategies.

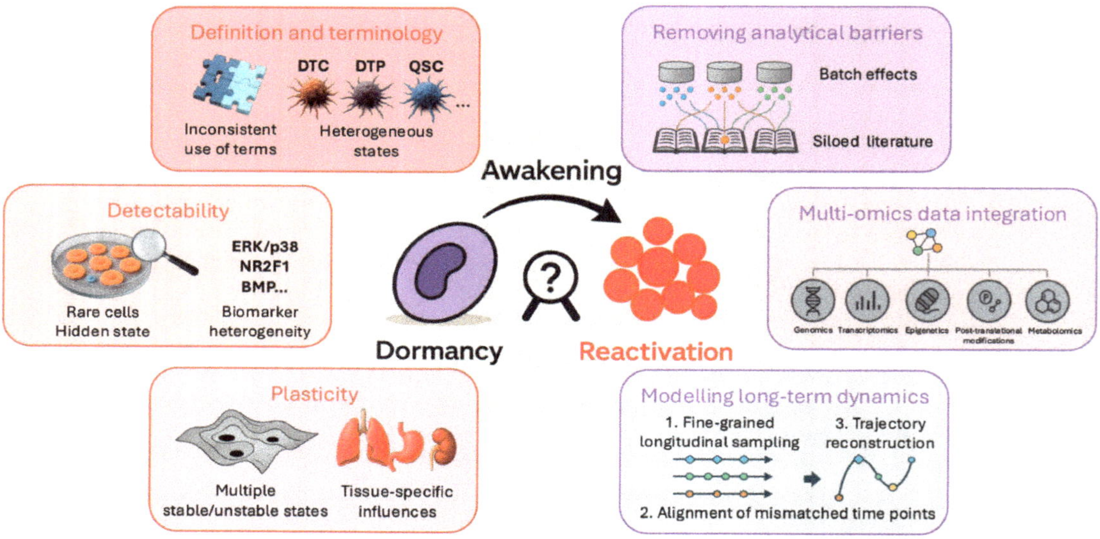


**Figure 3.** Computational challenges and future directions in modelling cancer cell dormancy. Key analytical barriers (left panels) and proposed solutions (right panels) for advancing computational studies of cancer dormancy. Current limitations include fragmented and inconsistent terminology (e.g., DTC, DTP, QSC - quiescent stem cells), difficulty in detecting rare states, the absence of universal biomarkers, and plasticity in dormancy-state switching, also influenced by the tissue context. Moving forward, improving the characterisation and understanding of this complex process will require addressing analytical barriers such as batch effect and multi-study integration, multi-omic profiling (genomics, transcriptomics, epigenomics, proteomics, metabolomics) of dormant cells, and fine-grained longitudinal sampling combined with dynamic modelling frameworks to reconstruct entry, maintenance, and reactivation trajectories. The visual elements in this figure have been generated with the help of Microsoft Copilot version bizchat.20260416.29.2.

### 3.1. The Ground Truth Problem

Despite extensive work characterising dormant cancer cells, there is still no agreed-upon gold standard for computationally identifying cancer dormancy [90,91]. This likely reflects the sheer diversity of cell types encompassed by the term "dormant cancer cells." Although all forms of cancer dormancy share longer-lived and reversible G0/G1 cell cycle arrest, the mechanisms governing entry into, maintenance of, and exit from dormancy are

 https://doi.org/10.3390/biom16050633

not necessarily shared [92]. A range of factors, including cancer type, prior chemotherapy exposure, the specific agents used, and even the niche in which the cells reside, are likely to influence how dormancy is regulated. Therefore, it is not surprising that the field has struggled to identify a universal marker of cancer dormancy; even well-established subtypes such as DTCs and DTPs lack well-defined markers [29,93]. Despite their growing clinical relevance, PGCCs remain computationally understudied compared to DTPs and DTCs, with the few existing examples limited to standard differential expression and pathway enrichment [94]. Extending more advanced computational approaches to other emerging forms of dormancy remains an important future direction.

This issue is further amplified by inconsistent methodological approaches across studies, in which different studies may apply distinct "dormancy signatures" or stratification strategies for dormant cells, hindering the alignment of findings. Consequently, the field may be collectively examining overlapping but not identical dormancy-associated cell states rather than a single discrete identity. Thus, we believe that the lack of consensus markers for cancer dormancy as an umbrella term and even its many proposed subtypes represents one of the largest challenges facing those investigating dormancy computationally.

The absence of standardised marker definitions makes dormancy particularly susceptible to propagation of inconsistent signatures across computational workflows. The lack of consensus markers also becomes apparent when attempting automated literature mining. Computational approaches that can systematically aggregate and reconcile heterogeneous literature may therefore provide a complementary strategy to address this fragmentation. To examine whether recent AI-assisted tools could help mitigate fragmentation in biomarker identification, we compared dormancy-associated genes curated manually by two independent reviewers with those identified by an AI agent using structured prompts. To address this challenge, we developed LitGraph, an AI-assisted literature mining framework designed to systematically extract and structure dormancy-associated knowledge from the literature. LitGraph analyses papers using the Claude large language model and builds knowledge graphs summarising these findings via GraphRAG (see Methods). Conceptually, the LitGraph pipeline operates in three stages: (i) targeted retrieval of domain-specific literature using structured queries, (ii) context-aware extraction of candidate biomarkers and functional roles, and (iii) integration of extracted entities into a graph-based framework that enables relationship tracing across studies. This design allows the system to move beyond isolated gene lists toward a network-level representation of dormancy biology.

Unlike conventional literature mining approaches based on keyword co-occurrence, frequency analysis, or static ontologies, LitGraph is designed to capture structured biological relationships (e.g., gene–function–context triplets) and preserve their contextual meaning across studies. In the context of cancer dormancy, where biomarker definitions are highly fragmented and context-dependent, this enables systematic comparison of reported regulators and helps distinguish consistently supported mechanisms from context-specific or weakly supported claims.

In practice, applying LitGraph to the problem of fragmented dormancy biomarkers allowed us to recover a core set of recurrent regulators, including NR2F1, p38 MAPK, TGFβ2, BMP7, and canonical CDK inhibitors, suggesting that highly cited core mechanisms are consistently retrievable (Tables S1–S5). At the same time, areas of conceptual ambiguity became apparent, where qualitative differences in interpretation were evident. The AI agent frequently conflated experimental readouts used to identify dormant cells (e.g., proliferation or label-retention markers) with mechanistic regulators of dormancy, misattributed claims to incorrect references, or cited papers whose conclusions did not fully support the statement made. Notably, the agent did not generate fictitious citations; rather, inaccuracies arose primarily from contextual misinterpretation or imprecise attribution. Fol-

 https://doi.org/10.3390/biom16050633

lowing manual curation, approximately 75% of AI-suggested entries corresponded to real papers from which relevant information could be appropriately extracted, but systematic validation was required.

In practice, although AI-assisted curation accelerated the initial aggregation of candidate markers and occasionally surfaced studies not captured in the first manual pass, the time required for verification reduced the overall efficiency gain. This exercise underscores the broader challenge facing the field: without standardised definitions and clearly annotated reference datasets, both human and automated approaches risk propagating ambiguity. At present, AI language models are best viewed as complementary hypothesis-generation systems rather than authoritative sources. A hybrid workflow, with initial manual grounding followed by AI-assisted expansion and expert validation, may therefore offer a pragmatic interim solution, while the development of a community-maintained, continuously updated reference database of experimentally supported dormancy markers would provide a more durable foundation for computational analyses.

### 3.2. Barriers to Meta-Analysis

Another factor complicating the computational analysis of dormancy is the difficulty in comparing results and/or combining multiple datasets for meta-analysis. Even within a single data modality like scRNA-seq, cross-study comparison and analysis are limited by batch effects, differences in preprocessing, heterogeneous experimental conditions, and a lack of a standardised definition of what counts as a dormant cancer cell. These issues are exacerbated when attempting to compare studies with more complex designs. For example, it is very difficult to compare or integrate results from longitudinal studies because most longitudinal studies sample at different time points and use different treatment schedules. The result is that the majority of the existing dormancy literature is essentially siloed, so individual studies generate their own clusters, signatures, or models, making it very difficult to validate between studies or combine data to generate a larger picture. Despite these barriers, some studies have attempted cross-dataset integration by stepping up to a level of abstraction to the network scale. For example, Uzuner et al. [95] aggregated 21 dormancy-cancer comparisons from 10 bulk transcriptomic datasets across seven cancer types and then mapped the differentially expressed genes onto protein–protein interaction and gene-regulatory networks to extract modules enriched in dormancy-associated genes. This approach is appealing because it somewhat sidesteps the integration problem. Instead of forcing heterogeneous datasets together for analysis, this approach asks which interactions are recurrently implicated across studies. While a promising direction, it is important to note that this approach inherits significant limitations from its inputs: it operates on bulk rather than single-cell data, depends on the completeness of existing interactome resources, and still relies on study-specific definitions of dormant and proliferative states. Network-level meta-analysis therefore offers a useful complementary lens rather than a full solution, and extending this strategy to single-cell and multi-omic datasets remains an open challenge. Beyond practical limitations, the conceptual representation of cancer dormancy also complicates computational analysis.

### 3.3. Cancer Dormancy as a Continuum or a Set of Discrete States

There is emerging evidence that rather than falling into discrete, well-defined categories, dormant cancer cells may occupy a spectrum of arrest depth, ranging from shallow, easily reversible quiescence through to deeper senescence-like states [29,96]. Under this model, subtypes such as DTPs and DTCs could represent overlapping ranges along this quiescence-depth axis. However, other subtypes are defined by features that cannot be reduced to arrest depth alone. PGCCs, for example, are characterised by polyploidy and


https://doi.org/10.3390/biom16050633

multinucleation, properties that distinguish them morphologically and mechanistically from other dormant subtypes. This suggests that dormancy heterogeneity may operate across multiple axes simultaneously: a continuum of arrest depth within some subtypes, but discrete morphological and gene regulation distinctions between others. In support of a continuum within dormancy subtypes, Fernandez et al.'s pseudotime analysis (discussed in Section 2.3) showed that drug-induced arrest formed a continuous trajectory from shallow quiescence to deeper senescence-like states rather than discrete transcriptional clusters, though this was observed in non-cancerous MCF10A cells. Whether this within-subtype continuity extends to cancer cells and whether different dormancy subtypes (e.g., DTPs vs. DTCs vs. PGCCs) represent discrete categories or points on a broader shared spectrum remain open questions. The answer to this question has significant consequences for how we study cancer cell dormancy computationally.

If cancer cells can progressively deepen their arrest along a quiescence-to-senescence axis, then clustering-based analyses, commonly used to computationally distinguish dormant cancer cells from proliferative cancer cells, impose artificial boundaries on what may actually be a continuous landscape. Furthermore, if we assume that cancer dormancy is a spectrum, a cell occupying an intermediate state may be mistakenly grouped with fully dormant or fully proliferative cells, thereby clouding downstream biological interpretations. Thus, when studying cancer cell dormancy, researchers must consider how their assumptions about the taxonomic structure of dormant cancer cells affect their chosen analytical techniques. If one assumes discrete cell states, then standard clustering with classifier-based approaches is appropriate, as they assign cells to defined categories with clear boundaries. If one assumes a continuum of cell states, then pseudotime [97] and optimal transport [98] approaches are more suitable, as they preserve continuous ordering and do not impose artificial boundaries on cell states. If the reality is some sort of intermediate between the two, then neither technique alone is sufficient. For example, researchers could apply both clustering and trajectory-based approaches to the same dataset and evaluate which better captures the underlying structure, or apply pseudotime analysis within computationally defined subpopulations to test for continuous variation within otherwise discrete groupings.

Ultimately, more research is needed to understand the landscape of cancer dormancy, particularly to determine whether current categorical frameworks are sufficient or whether novel analytical approaches are required to represent this complexity faithfully.

### 3.4. Beyond Transcriptomics: The Need for Multi-Omic Integration and Modelling

The majority of computational studies interrogating cancer dormancy have relied on scRNA-seq as the primary modality; however, it is increasingly clear that cancer dormancy's biology likely spans multiple molecular layers that cannot be captured by scRNA-seq alone. For example, the balance between ERK/p38 phosphorylation has been shown to be a hallmark of dormant cancer cells, but these post-translational changes are invisible to scRNA-seq [51–56]. Epigenetic changes are another example, with multiple studies described in Section 2 showing that epigenetic changes are involved in cancer dormancy [71,83]. Given that cancer dormancy likely operates across multiple molecular scales, we suggest that, moving forward, multi-omic studies combining multiple modalities at single-cell resolution are needed, alongside appropriate computational integration methods. Methods already exist for combining and analysing different single-cell modalities, for example, Multi-Omic Factor Analysis v2 (MOFA+) [99] or graph-linked unified embedding [100].


https://doi.org/10.3390/biom16050633

### 3.5. Foundation Models and Transfer Learning for Learning Rare Cell States

As previously discussed, identifying dormant cancer cells remains a major obstacle to computational research on cancer dormancy. Furthermore, dormant cancer cells are rare in most snapshot studies, making it difficult to develop supervised classifiers, as models will either overfit to dataset-specific artefacts or collapse cancer dormancy into broader quiescent cell states. Foundation models pre-trained on large amounts of scRNA-seq data could help address this limitation. The general-purpose representations learned by pre-trained foundation models can serve as a base for fine-tuning with relatively few annotated examples of dormant cancer cells. A few such models now exist, for example, Geneformer, a foundational model pre-trained on approximately 30 million scRNA transcriptomes [101]; scGPT, a foundational model that is based on a generative pretrained transformer on approximately 33 million scRNA transcriptomes [102]; scFoundation, a foundation model pre-trained on 50 million human scRNA-seq profiles [103]; or scBERT, a pre-trained deep neural network-based model focused on cell type annotation [104]. Importantly, the biggest challenge here will not be identifying dormancy but delineating normal dormancy from cancer dormancy, which may require signatures from different modalities, such as epigenetic or metabolic signatures that current foundation models do not capture.

### 3.6. Temporal Dynamics: Continuous-Time Models

The majority of the computational analyses described in Section 2 were applied to snapshot data, but cancer dormancy is defined by temporal behaviour (entry, dwell time, and exit). For example, a cancer cell that may appear transcriptionally quiescent in scRNA-seq data from a single moment in time could be transiently pausing before re-entering the cell cycle to proliferate again or remain stably dormant for decades. To this end, longitudinal studies examining dormancy over time, which have been greatly enabled through lineage tracing or similar experiments, are crucial for understanding cancer dormancy. Integrating multiple longitudinal studies would, in principle, allow us to capture a broader view of the temporal aspects of dormancy; however, as previously explained, it is difficult to integrate multiple longitudinal studies due to mismatched sampling time points. A potential solution to this problem is to use continuous-time models to sidestep the mismatched sampling time issue. For example, a latent ODE framework, as proposed by Rubanova et al., which learns a shared dynamical model defined by a neural ODE that can be evaluated at any time point, irrespective of when the sample was actually collected [105]. In principle, this would allow one to merge data from longitudinal studies with mismatched time points to inform a single model of cancer cell dormancy dynamics. However, it is important to note that there are some fundamental challenges. For example, because scRNA-seq is destructive, individual cell trajectories cannot be observed directly. Therefore, to apply a latent ODE here, one would need to either use population-level summaries (e.g., collapsing scRNA-seq data into an aggregate representation for each time point) at each time point or utilise lineage tracing to link related cells across time points. Another drawback to consider is that longitudinal datasets typically sample only a few time points, which may not be sufficient for learning complex non-linear dynamics. One would likely need to integrate multiple studies and impose strong biological priors.

Taken together, the challenges outlined in this section highlight a central disconnect in computational cancer dormancy research: the biological phenomenon is dynamic, heterogeneous, and multi-layered, whereas many current analytical frameworks remain static, transcriptome-centric, and dataset-specific. Overcoming these limitations will require not only technical innovation but also greater conceptual alignment regarding how dormancy is defined and modelled across studies. As datasets become increasingly longitudinal, spatially resolved, and multi-modal, the opportunity arises to move beyond isolated signatures


https://doi.org/10.3390/biom16050633

toward integrated representations of dormancy as a dynamic systems process. Establishing such frameworks will be critical for ensuring that computational advances translate into biologically coherent and clinically actionable insights (Figure 3).

While this review has, for the most part, focused on computational approaches to investigate cancer cell dormancy, it is important to note that these computational approaches are not possible without rigorous experimental research, and the most effective experimental and computational methods are used iteratively. Several of the studies discussed in Section 2 demonstrate this synergy. For example, Khoo et al. used scRNA-seq to identify *AXL* as a marker of dormant cancer cells in myeloma and then validated these results experimentally with coculture and intravital imaging in vivo. Similarly, Hu et al. used computational analyses to interpret results from a CRISPR screen and then validated those analyses pharmacologically, ultimately showing that STING activation reduced the micrometastatic burden. These examples demonstrate just how effective the closed loop is; computational analyses allow us to generate testable hypotheses, whose results can then be used to further refine computational analyses.

A clinically relevant extension of this iterative cycle is the translation of computationally identified dormancy regulators into therapeutic targets. The studies reviewed in Section 2 collectively identified several examples of candidate therapeutic targets. Alongside ongoing experimental efforts to translate such candidates into cancer treatments [106–108], computational techniques such as network-based drug target prioritisation [109], drug sensitivity prediction [110,111], and drug repurposing approaches [112] could, in the future, help bridge the gap between molecular insights derived from computational studies and cancer therapies.

## 4. Materials and Methods

To investigate fragmentation in cancer dormancy biomarker identification, we compared manual literature curation with an automated AI-assisted pipeline developed in-house (LitGraph). The objective was to evaluate concordance, divergence, and systematic error patterns between human and AI-driven biomarker extraction workflows. The study consisted of three components:

- Manual biomarker curation by two independent reviewers.
- AI-assisted literature mining using the LitGraph framework.
- Manual reviewing and structured comparison of outputs by a third reviewer.

### 4.1. Manual Literature Curation

Two independent reviewers manually curated molecular regulators associated with cancer cell dormancy/quiescence from peer-reviewed literature. Inclusion criteria required experimental evidence supporting a mechanistic role in entry into, maintenance of, or exit from cellular dormancy. Experimental readouts used solely to identify dormant cells (e.g., label-retention dyes or proliferation markers such as Ki67) were distinguished from mechanistic regulators.

### 4.2. AI-Assisted Literature Mining Using LitGraph

AI-assisted curation was performed using LitGraph, an automated literature analysis pipeline developed by the authors and publicly available at https://github.com/RunningStone/LitGraph (accessed on 26 September 2025).

LitGraph is designed to automatically query literature databases, retrieve and process publications, analyse textual content using a large language model, and construct structured knowledge graphs summarising extracted relationships.


https://doi.org/10.3390/biom16050633

In this study, LitGraph was configured to retrieve literature related to cancer dormancy and quiescence published between 1995 and 2024 using predefined keyword queries. Retrieved documents were processed and analysed using the Claude large language model (Anthropic) for structured extraction of:

- Candidate biomarkers
- Claimed functional roles
- Associated references
- Supporting evidence statements

Extracted entities and relationships were organised using a GraphRAG (Graph-based Retrieval-Augmented Generation) framework. This enabled:

- Construction of a structured knowledge graph linking biomarkers to mechanisms and references
- Relationship tracing between signalling pathways, transcription factors, and dormancy states
- Context-aware retrieval of supporting claims.

The AI-generated biomarker tables (Tables S1–S3) were exported in a structured format to enable direct comparison with the manually curated dataset (Table S4), and a final synthesis table was generated (Table S5).

*4.3. Verification and Error Classification*

All AI-derived entries were manually verified against the cited primary literature. Each entry was classified into one of the following categories:

- Correct citation and accurate contextual extraction
- Real paper cited with incorrect bibliographic details
- Real paper cited but claim not supported by source
- Conceptual conflation (e.g., experimental identification marker misclassified as mechanistic regulator)
- Non-verifiable claim

No fictitious (non-existent) publications were identified during review. Approximately 75% of AI-suggested entries corresponded to real papers from which relevant dormancy-related information could be extracted following adjudication. An annotated comparison table detailing validation outcomes is provided in Table S5.

## 5. Conclusions

Cancer dormancy reflects a form of cell state plasticity with profound clinical consequences. Although increasingly accessible through modern profiling technologies, it remains resistant to simple classification, suggesting that dormancy is better understood as a dynamic and context-dependent process rather than a fixed cellular identity. Computational approaches have begun to reshape how dormancy is conceptualised: not merely as a rare quiescent population, but as a systems-level phenomenon governed by state transitions across molecular and temporal scales. The challenge ahead is to translate these conceptual advances into coherent modelling frameworks and clinically interpretable predictors. Achieving this will be essential if dormancy biology is to move beyond descriptive characterisation toward strategies that anticipate and prevent cancer recurrence.

 https://doi.org/10.3390/biom16050633

**Supplementary Materials:** The following supporting information can be downloaded at https://www.mdpi.com/article/10.3390/biom16050633/s1. Table S1: Structured biomarker annotation table derived by LitGraph from curated literature and contextual validation. Table S2: Contextual interpretation table for expert quality control and semantic validation. Table S3: Literature cross-reference table linking extracted entries to original publications. Table S4: Cancer dormancy biomarkers retrieved by human experts through manual curation of the literature. Table S5: Comparative evaluation of manually and AI-curated cancer dormancy biomarkers.

**Author Contributions:** Conceptualisation, M.S. and L.G.N.S.; formal analysis, S.P. and L.G.N.S.; investigation, L.G.N.S., S.P., M.K., B.Y. and B.B.; writing—original draft preparation, L.G.N.S.; writing—review and editing, M.S.; visualisation, L.G.N.S. and M.S.; supervision, M.S.; funding acquisition, M.S. All authors have read and agreed to the published version of the manuscript.

**Funding:** M.S. and S.P. were supported by a UKRI Future Leaders Fellowship (MR/Y034031/1).

**Institutional Review Board Statement:** Not applicable.

**Informed Consent Statement:** Not applicable.

**Data Availability Statement:** Curated biomarkers with contextual synthesis, literature mappings, and annotated comparisons are provided in Table S1. No new datasets were generated. All referenced data originate from previously published studies cited in the manuscript. The LitGraph framework used for biomarker retrieval in this study is publicly available at the following repository: https://github.com/RunningStone/LitGraph (accessed on 26 September 2025).

**Acknowledgments:** During the preparation of this manuscript, the authors used Generative AI with the LitGraph pipeline to analyse retrieved literature and extract structured biomarker information using the Claude Opus 4.1 large language model. AI-generated outputs were manually validated against primary sources prior to inclusion in comparative analysis. Microsoft Copilot version bizchat.20260416.29.2 was used to generate the visual elements in Figure 2, but the conceptual design and content were generated by the authors. AI was not used to generate experimental data. All scientific interpretation and final conclusions were performed by the authors. The authors have reviewed and edited the output and take full responsibility for the content of this publication.

**Conflicts of Interest:** The authors declare no conflicts of interest. The funders had no role in the design of the study; in the collection, analyses, or interpretation of data; in the writing of the manuscript; or in the decision to publish the results.

## Abbreviations

The following abbreviations are used in this manuscript:

| | |
|---|---|
| AI | Artificial Intelligence |
| AMPK | AMP-Activated Protein Kinase |
| CDK | Cyclin-Dependent Kinase |
| ChIP-seq | Chromatin Immunoprecipitation Sequencing |
| CRISPR | Clustered Regularly Interspaced Short Palindromic Repeats |
| CTC | Circulating Tumour Cell |
| DTC | Disseminated Tumour Cell |
| DTP | Drug-Tolerant Persister |
| ECM | Extracellular Matrix |
| ERK | Extracellular Signal-Regulated Kinase |
| GSEA | Gene Set Enrichment Analysis |
| GraphRAG | Graph-based Retrieval-Augmented Generation |
| IIS | Insulin/Insulin-like Growth Factor Signalling |
| LC-MS/MS | Liquid Chromatography–Tandem Mass Spectrometry |
| MOFA | Multi-Omic Factor Analysis |
| NMF | Non-negative Matrix Factorisation |
| NRF2 | Nuclear Factor Erythroid 2–Related Factor 2 |

 https://doi.org/10.3390/biom16050633

| | |
|---|---|
| ODE | Ordinary Differential Equation |
| PGCC | Polyploid Giant Cancer Cell |
| scRNA-seq | Single-Cell RNA Sequencing |
| sgRNA | Single Guide RNA |
| SRCC | Stably Resistant Cancer Cell |
| super-SILAC histone mass spectrometry | Super-Stable-isotope labelling by amino acids in cell culture histone mass spectrometry |
| TOR | Target of Rapamycin |

## References


1. Nath, R.D.; Chow, E.S.; Wang, H.; Schwarz, E.M.; Sternberg, P.W. *C. elegans* Stress-Induced Sleep Emerges from the Collective Action of Multiple Neuropeptides. *Curr. Biol.* **2016**, *26*, 2446–2455. [CrossRef]
2. Gomatou, G.; Syrigos, N.; Vathiotis, I.A.; Kotteas, E.A. Tumor Dormancy: Implications for Invasion and Metastasis. *Int. J. Mol. Sci.* **2021**, *22*, 4862. [CrossRef] [PubMed]
3. Galdieri, L.; Mehrotra, S.; Yu, S.; Vancura, A. Transcriptional Regulation in Yeast during Diauxic Shift and Stationary Phase. *OMICS J. Integr. Biol.* **2010**, *14*, 629–638. [CrossRef] [PubMed]
4. Smets, B.; Ghillebert, R.; De Snijder, P.; Binda, M.; Swinnen, E.; De Virgilio, C.; Winderickx, J. Life in the Midst of Scarcity: Adaptations to Nutrient Availability in *Saccharomyces cerevisiae*. *Curr. Genet.* **2010**, *56*, 1–32. [CrossRef]
5. De Virgilio, C. The Essence of Yeast Quiescence. *FEMS Microbiol. Rev.* **2012**, *36*, 306–339. [CrossRef]
6. Dhawan, J.; Laxman, S. Decoding the Stem Cell Quiescence Cycle–Lessons from Yeast for Regenerative Biology. *J. Cell Sci.* **2015**, *128*, 4467–4474. [CrossRef]
7. Kenyon, C.J. The Genetics of Ageing. *Nature* **2010**, *464*, 504–512. [CrossRef]
8. Roux, A.E.; Langhans, K.; Huynh, W.; Kenyon, C. Reversible Age-Related Phenotypes Induced during Larval Quiescence in *C. elegans*. *Cell Metab.* **2016**, *23*, 1113–1126. [CrossRef]
9. Nelson, M.D.; Trojanowski, N.F.; George-Raizen, J.B.; Smith, C.J.; Yu, C.-C.; Fang-Yen, C.; Raizen, D.M. The Neuropeptide NLP-22 Regulates a Sleep-like State in Caenorhabditis Elegans. *Nat. Commun.* **2013**, *4*, 2846. [CrossRef]
10. Hu, C.-K.; Brunet, A. The African Turquoise Killifish: A Research Organism to Study Vertebrate Aging and Diapause. *Aging Cell* **2018**, *17*, e12757. [CrossRef] [PubMed]
11. Romney, A.L.T.; Davis, E.M.; Corona, M.M.; Wagner, J.T.; Podrabsky, J.E. Temperature-Dependent Vitamin D Signaling Regulates Developmental Trajectory Associated with Diapause in an Annual Killifish. *Proc. Natl. Acad. Sci. USA* **2018**, *115*, 12763–12768. [CrossRef] [PubMed]
12. Hu, C.-K.; Wang, W.; Brind'Amour, J.; Singh, P.P.; Reeves, G.A.; Lorincz, M.C.; Alvarado, A.S.; Brunet, A. Vertebrate Diapause Preserves Organisms Long Term through Polycomb Complex Members. *Science* **2020**, *367*, 870–874. [CrossRef]
13. Meller, C.L.; Meller, R.; Simon, R.P.; Culpepper, K.M.; Podrabsky, J.E. Cell Cycle Arrest Associated with Anoxia-Induced Quiescence, Anoxic Preconditioning, and Embryonic Diapause in Embryos of the Annual Killifish Austrofundulus Limnaeus. *J. Comp. Physiol. B* **2012**, *182*, 909–920. [CrossRef]
14. MacRae, T.H. Gene Expression, Metabolic Regulation and Stress Tolerance during Diapause. *Cell. Mol. Life Sci.* **2010**, *67*, 2405–2424. [CrossRef]
15. Dulic, V. Senescence Regulation by mTOR. In *Cell Senescence*; Galluzzi, L., Vitale, I., Kepp, O., Kroemer, G., Eds.; Methods in Molecular Biology; Humana Press: Totowa, NJ, USA, 2013; Volume 965, pp. 15–35.
16. Loayza-Puch, F.; Drost, J.; Rooijers, K.; Lopes, R.; Elkon, R.; Agami, R. P53 Induces Transcriptional and Translational Programs to Suppress Cell Proliferation and Growth. *Genome Biol.* **2013**, *14*, R32. [CrossRef]
17. Fujimaki, K.; Li, R.; Chen, H.; Della Croce, K.; Zhang, H.H.; Xing, J.; Bai, F.; Yao, G. Graded Regulation of Cellular Quiescence Depth between Proliferation and Senescence by a Lysosomal Dimmer Switch. *Proc. Natl. Acad. Sci. USA* **2019**, *116*, 22624–22634. [CrossRef]
18. O'Neill, L.A.J.; Hardie, D.G. Metabolism of Inflammation Limited by AMPK and Pseudo-Starvation. *Nature* **2013**, *493*, 346–355. [CrossRef] [PubMed]
19. Van Velthoven, C.T.J.; Rando, T.A. Stem Cell Quiescence: Dynamism, Restraint, and Cellular Idling. *Cell Stem Cell* **2019**, *24*, 213–225. [CrossRef] [PubMed]
20. Wiecek, A.J.; Jacobson, D.H.; Lason, W.; Secrier, M. Pan-Cancer Survey of Tumor Mass Dormancy and Underlying Mutational Processes. *Front. Cell Dev. Biol.* **2021**, *9*, 698659. [CrossRef]
21. Marescal, O.; Cheeseman, I.M. Cellular Mechanisms and Regulation of Quiescence. *Dev. Cell* **2020**, *55*, 259–271. [CrossRef]
22. Pranzini, E.; Raugei, G.; Taddei, M.L. Metabolic Features of Tumor Dormancy: Possible Therapeutic Strategies. *Cancers* **2022**, *14*, 547. [CrossRef]



https://doi.org/10.3390/biom16050633

23. Braun, S.; Kilchert, C.; Bulut-Karslioglu, A.; Ruault, M.; Taddei, A.; Rabbani, F.; Włoch-Salamon, D. Nuclear Dynamics in Quiescent Cells: Conserved Mechanisms from Yeasts to Mammals. *Biomolecules* **2026**, *16*, 203. [CrossRef] [PubMed]
24. Sosa, M.S.; Bragado, P.; Aguirre-Ghiso, J.A. Mechanisms of Disseminated Cancer Cell Dormancy: An Awakening Field. *Nat. Rev. Cancer* **2014**, *14*, 611–622. [CrossRef]
25. Butturini, E.; Carcereri de Prati, A.; Boriero, D.; Mariotto, S. Tumor Dormancy and Interplay with Hypoxic Tumor Microenvironment. *Int. J. Mol. Sci.* **2019**, *20*, 4305. [CrossRef]
26. Endo, H.; Inoue, M. Dormancy in Cancer. *Cancer Sci.* **2019**, *110*, 474–480. [CrossRef]
27. Phan, T.G.; Croucher, P.I. The Dormant Cancer Cell Life Cycle. *Nat. Rev. Cancer* **2020**, *20*, 398–411. [CrossRef]
28. Francescangeli, F.; De Angelis, M.L.; Rossi, R.; Cuccu, A.; Giuliani, A.; De Maria, R.; Zeuner, A. Dormancy, Stemness, and Therapy Resistance: Interconnected Players in Cancer Evolution. *Cancer Metastasis Rev.* **2023**, *42*, 197–215. [CrossRef]
29. Truskowski, K.; Amend, S.R.; Pienta, K.J. Dormant Cancer Cells: Programmed Quiescence, Senescence, or Both? *Cancer Metastasis Rev.* **2023**, *42*, 37–47. [CrossRef] [PubMed]
30. Weston, W.A.; Barr, A.R. A Cell Cycle Centric View of Tumour Dormancy. *Br. J. Cancer* **2023**, *129*, 1535–1545. [CrossRef]
31. Sabit, H.; Abdel-Ghany, S.; Albrahim, Y.; Wadan, A.-H.S.; Rashwan, S.; Arneth, R.; Arneth, B. Bridging the Gap in Breast Cancer Dormancy: Models, Mechanisms, and Translational Challenges. *Pharmaceuticals* **2025**, *18*, 961. [CrossRef] [PubMed]
32. Vallette, F.M.; Olivier, C.; Lézot, F.; Oliver, L.; Cochonneau, D.; Lalier, L.; Cartron, P.-F.; Heymann, D. Dormant, Quiescent, Tolerant and Persister Cells: Four Synonyms for the Same Target in Cancer. *Biochem. Pharmacol.* **2019**, *162*, 169–176. [CrossRef]
33. Cabanos, H.F.; Hata, A.N. Emerging Insights into Targeted Therapy-Tolerant Persister Cells in Cancer. *Cancers* **2021**, *13*, 2666. [CrossRef]
34. Mikubo, M.; Inoue, Y.; Liu, G.; Tsao, M.-S. Mechanism of Drug Tolerant Persister Cancer Cells: The Landscape and Clinical Implication for Therapy. *J. Thorac. Oncol.* **2021**, *16*, 1798–1809. [CrossRef] [PubMed]
35. Ring, A.; Spataro, M.; Wicki, A.; Aceto, N. Clinical and Biological Aspects of Disseminated Tumor Cells and Dormancy in Breast Cancer. *Front. Cell Dev. Biol.* **2022**, *10*, 929893. [CrossRef] [PubMed]
36. Risson, E.; Nobre, A.R.; Maguer-Satta, V.; Aguirre-Ghiso, J.A. The Current Paradigm and Challenges Ahead for the Dormancy of Disseminated Tumor Cells. *Nat. Cancer* **2020**, *1*, 672–680. [CrossRef] [PubMed]
37. Liu, P.; Wang, L.; Yu, H. Polyploid Giant Cancer Cells: Origin, Possible Pathways of Formation, Characteristics, and Mechanisms of Regulation. *Front. Cell Dev. Biol.* **2024**, *12*, 1410637. [CrossRef] [PubMed]
38. Mirzayans, R. Changing the Landscape of Solid Tumor Therapy from Apoptosis-Promoting to Apoptosis-Inhibiting Strategies. *Curr. Issues Mol. Biol.* **2024**, *46*, 5379–5396. [CrossRef] [PubMed]
39. Dhanasekaran, R. Treacherous Apoptosis- Cancer Cells Sacrifice Themselves at the Altar of Heterogeneity. *Hepatology* **2022**, *76*, 549–550. [CrossRef] [PubMed]
40. Mirzayans, R.; Murray, D. Amitotic Cell Division, Malignancy, and Resistance to Anticancer Agents: A Tribute to Drs. Walen and Rajaraman. *Cancers* **2024**, *16*, 3106. [CrossRef]
41. Liu, J.; Erenpreisa, J.; Sikora, E. Polyploid Giant Cancer Cells: An Emerging New Field of Cancer Biology. *Semin. Cancer Biol.* **2022**, *81*, 1–4. [CrossRef] [PubMed]
42. Liu, J.; Liu, K.; Yan, M.; Xu, J.; Liu, Y.; Zhang, S. Dormant and Activated Polyploid Giant Cancer Cells: From Chemoradiotherapy Resistance to Cancer Progression. *Cancer Lett.* **2026**, *636*, 218137. [CrossRef]
43. Jiao, Y.; Yu, Y.; Zheng, M.; Yan, M.; Wang, J.; Zhang, Y.; Zhang, S. Dormant Cancer Cells and Polyploid Giant Cancer Cells: The Roots of Cancer Recurrence and Metastasis. *Clin. Transl. Med.* **2024**, *14*, e1567. [CrossRef] [PubMed]
44. Mirzayans, R.; Murray, D. Intratumor Heterogeneity and Treatment Resistance of Solid Tumors with a Focus on Polyploid/Senescent Giant Cancer Cells (PGCCs). *Int. J. Mol. Sci.* **2023**, *24*, 11534. [CrossRef]
45. Huang, P.; Wu, R.; Yang, Z.; Li, Y.; Fei, F.; Yu, Y. Polyploid Giant Cancer Cells and Tumor Budding: Translation from Basic Research to Clinical Application. *Front. Oncol.* **2025**, *15*, 1611920. [CrossRef] [PubMed]
46. Pan, S.; Liu, H.; Xiao, H.; Yuan, L.; Zhang, K.; Yin, H.; Han, Y.; Wu, R.; Xia, T.; Zhang, Z.; et al. ITGB6 Promotes Tumor Recurrence and Metastasis by Mediating the Resistance of Daughter Cells of PGCCs to Anoikis. *Sci. Rep.* **2025**, *15*, 33571. [CrossRef]
47. Zhou, X.; Zhou, M.; Zheng, M.; Tian, S.; Yang, X.; Ning, Y.; Li, Y.; Zhang, S. Polyploid Giant Cancer Cells and Cancer Progression. *Front. Cell Dev. Biol.* **2022**, *10*, 1017588. [CrossRef] [PubMed]
48. Huang, H.; Chai, Y.; Wu, X.; Wang, S.; Li, B.; Wang, Y.; Zheng, W.; Wu, Y.; Meng, D.; Wang, H.; et al. Non-Canonical NOTCH1 Signaling Regulates Ferroptosis Vulnerability in Dormant Lung Cancer Cells with Stable Resistance. *Cell Death Dis.* **2025**, *17*, 1. [CrossRef]
49. Aguirre-Ghiso, J.A. Models, Mechanisms and Clinical Evidence for Cancer Dormancy. *Nat. Rev. Cancer* **2007**, *7*, 834–846. [CrossRef]
50. Yeh, A.C.; Ramaswamy, S. Mechanisms of Cancer Cell Dormancy—Another Hallmark of Cancer? *Cancer Res.* **2015**, *75*, 5014–5022. [CrossRef] [PubMed]


 https://doi.org/10.3390/biom16050633

51. Collignon, E. Unveiling the Role of Cellular Dormancy in Cancer Progression and Recurrence. *Curr. Opin. Oncol.* **2024**, *36*, 74 [CrossRef]
52. Tufail, M.; Jiang, C.-H.; Li, N. Tumor Dormancy and Relapse: Understanding the Molecular Mechanisms of Cancer Recurrence *Mil. Med. Res.* **2025**, *12*, 7. [CrossRef]
53. Sosa, M.S.; Avivar-Valderas, A.; Bragado, P.; Wen, H.-C.; Aguirre-Ghiso, J.A. ERK1/2 and P38α/β Signaling in Tumor Cell Quiescence: Opportunities to Control Dormant Residual Disease. *Clin. Cancer Res.* **2011**, *17*, 5850–5857. [CrossRef]
54. Recasens, A.; Munoz, L. Targeting Cancer Cell Dormancy. *Trends Pharmacol. Sci.* **2019**, *40*, 128–141. [CrossRef]
55. Guereño, M.; Delgado Pastore, M.; Lugones, A.C.; Cercato, M.; Todaro, L.; Urtreger, A.; Peters, M.G. Glypican-3 (GPC3) Inhibits Metastasis Development Promoting Dormancy in Breast Cancer Cells by P38 MAPK Pathway Activation. *Eur. J. Cell Biol.* **2020** *99*, 151096. [CrossRef]
56. Rutecki, S.; Leśniewska-Bocianowska, A.; Chmielewska, K.; Matuszewska, J.; Naumowicz, E.; Uruski, P.; Radziemski, A. Mikuła-Pietrasik, J.; Tykarski, A.; Książek, K. Serum Starvation-Based Method of Ovarian Cancer Cell Dormancy Induction and Termination in Vitro. *Biol. Methods Protoc.* **2023**, *8*, bpad029. [CrossRef] [PubMed]
57. Min, H.-Y.; Lee, H.-Y. Cellular Dormancy in Cancer: Mechanisms and Potential Targeting Strategies. *Cancer Res. Treat.* **2023**, *55*, 720–736. [CrossRef] [PubMed]
58. Harris, C.J.; Stewart, G.R.; Foston, A.; Green, A.C. Cancer Cell Dormancy in the Bone Microenvironment. *Curr. Osteoporos. Rep.* **2025**, *23*, 46. [CrossRef] [PubMed]
59. Fox, D.B.; Garcia, N.M.G.; McKinney, B.J.; Lupo, R.; Noteware, L.C.; Newcomb, R.; Liu, J.; Locasale, J.W.; Hirschey, M.D.; Alvarez, J.V. NRF2 Activation Promotes the Recurrence of Dormant Tumour Cells through Regulation of Redox and Nucleotide Metabolism. *Nat. Metab.* **2020**, *2*, 318–334. [CrossRef] [PubMed]
60. Sosa, M.S.; Parikh, F.; Maia, A.G.; Estrada, Y.; Bosch, A.; Bragado, P.; Ekpin, E.; George, A.; Zheng, Y.; Lam, H.-M.; et al. NR2F1 Controls Tumour Cell Dormancy via SOX9- and RARβ-Driven Quiescence Programmes. *Nat. Commun.* **2015**, *6*, 6170. [CrossRef]
61. Yang, X.; Wu, J.; Li, M.; Zhang, W.; Gao, X.; Wang, H.; Yu, X.; Pang, X.; Zhang, M.; Liang, X.; et al. Inhibition of DEC2 Is Necessary for Exiting Cell Dormancy in Salivary Adenoid Cystic Carcinoma. *J. Exp. Clin. Cancer Res.* **2021**, *40*, 169. [CrossRef]
62. Merlot, A.M.; Porter, G.M.; Sahni, S.; Lim, E.G.; Peres, P.; Richardson, D.R. The Metastasis Suppressor, NDRG1, Differentially Modulates the Endoplasmic Reticulum Stress Response. *Biochim. Biophys. Acta (BBA)-Mol. Basis Dis.* **2019**, *1865*, 2094–2110. [CrossRef]
63. Tamamouna, V.; Pavlou, E.; Neophytou, C.M.; Papageorgis, P.; Costeas, P. Regulation of Metastatic Tumor Dormancy and Emerging Opportunities for Therapeutic Intervention. *Int. J. Mol. Sci.* **2022**, *23*, 13931. [CrossRef]
64. Ohta, Y.; Fujii, M.; Takahashi, S.; Takano, A.; Nanki, K.; Matano, M.; Hanyu, H.; Saito, M.; Shimokawa, M.; Nishikori, S.; et al. Cell–Matrix Interface Regulates Dormancy in Human Colon Cancer Stem Cells. *Nature* **2022**, *608*, 784–794. [CrossRef]
65. Park, S.-Y.; Nam, J.-S. The Force Awakens: Metastatic Dormant Cancer Cells. *Exp. Mol. Med.* **2020**, *52*, 569–581. [CrossRef] [PubMed]
66. Agudo, J.; Aguirre-Ghiso, J.A.; Bhatia, M.; Chodosh, L.A.; Correia, A.L.; Klein, C.A. Targeting Cancer Cell Dormancy. *Nat. Rev. Cancer* **2024**, *24*, 97–104. [CrossRef] [PubMed]
67. Wang, Y.; Wang, L.; Wei, Y.; Wei, C.; Yang, H.; Chen, Q.; Zhang, R.; Shen, H. Advances in the Molecular Regulation Mechanism of Tumor Dormancy and Its Therapeutic Strategy. *Discov. Oncol.* **2024**, *15*, 184. [CrossRef] [PubMed]
68. Cho, J. Understanding Tumor Dormancy: From Experimental Models to Mechanisms and Therapeutic Strategies. *Biomol. Ther.* **2025**, *33*, 770–784. [CrossRef] [PubMed]
69. Ren, Q.; Khoo, W.H.; Corr, A.P.; Phan, T.G.; Croucher, P.I.; Stewart, S.A. Gene Expression Predicts Dormant Metastatic Breast Cancer Cell Phenotype. *Breast Cancer Res.* **2022**, *24*, 10. [CrossRef] [PubMed]
70. Oren, Y.; Tsabar, M.; Cuoco, M.S.; Amir-Zilberstein, L.; Cabanos, H.F.; Hütter, J.-C.; Hu, B.; Thakore, P.I.; Tabaka, M.; Fulco, C.P.; et al. Cycling Cancer Persister Cells Arise from Lineages with Distinct Programs. *Nature* **2021**, *596*, 576–582. [CrossRef]
71. Rosano, D.; Sofyali, E.; Dhiman, H.; Ghirardi, C.; Ivanoiu, D.; Heide, T.; Vingiani, A.; Bertolotti, A.; Pruneri, G.; Canale, E.; et al. Long-Term Multimodal Recording Reveals Epigenetic Adaptation Routes in Dormant Breast Cancer Cells. *Cancer Discov.* **2024**, *14*, 866–889. [CrossRef] [PubMed]
72. Noberini, R.; Longhi, E.; Bonaldi, T. A Super-SILAC Approach for Profiling Histone Posttranslational Modifications. *Methods Mol. Biol.* **2023**, *2603*, 87–102. [CrossRef]
73. Maman, S.; Witz, I.P. A History of Exploring Cancer in Context. *Nat. Rev. Cancer* **2018**, *18*, 359–376. [CrossRef] [PubMed]
74. Khoo, W.H.; Ledergor, G.; Weiner, A.; Roden, D.L.; Terry, R.L.; McDonald, M.M.; Chai, R.C.; De Veirman, K.; Owen, K.L.; Opperman, K.S.; et al. A Niche-Dependent Myeloid Transcriptome Signature Defines Dormant Myeloma Cells. *Blood* **2019**, *134*, 30–43. [CrossRef]
75. Brunet, J.-P.; Tamayo, P.; Golub, T.R.; Mesirov, J.P. Metagenes and Molecular Pattern Discovery Using Matrix Factorization. *Proc. Natl. Acad. Sci. USA* **2004**, *101*, 4164–4169. [CrossRef] [PubMed]

 https://doi.org/10.3390/biom16050633

76. Janghorban, M.; Yang, Y.; Zhao, N.; Hamor, C.; Nguyen, T.M.; Zhang, X.H.-F.; Rosen, J.M. Single-Cell Analysis Unveils the Role of the Tumor Immune Microenvironment and Notch Signaling in Dormant Minimal Residual Disease. *Cancer Res.* **2022**, *82*, 885–899. [CrossRef]
77. Rubinstein, J.C.; Domanskyi, S.; Sheridan, T.B.; Sanderson, B.; Park, S.; Kaster, J.; Li, H.; Anczukow, O.; Herlyn, M.; Chuang, J.H. Spatiotemporal Profiling Defines Persistence and Resistance Dynamics During Targeted Treatment of Melanoma. *Cancer Res.* **2025**, *85*, 987–1002. [CrossRef]
78. Reid, J.E.; Wernisch, L. Pseudotime Estimation: Deconfounding Single Cell Time Series. *Bioinformatics* **2016**, *32*, 2973–2980. [CrossRef] [PubMed]
79. La Manno, G.; Soldatov, R.; Zeisel, A.; Braun, E.; Hochgerner, H.; Petukhov, V.; Lidschreiber, K.; Kastriti, M.E.; Lönnerberg, P.; Furlan, A.; et al. RNA Velocity of Single Cells. *Nature* **2018**, *560*, 494–498. [CrossRef]
80. Celik, C.; Withnell, E.; Pan, S.; Chu, T.; Labbadia, J.; Secrier, M. Balancing Tumour Proliferation and Sustained Cell Cycle Arrest through Proteostasis Remodelling Drives Immune Niche Compartmentalisation in Breast Cancer. *bioRxiv* **2025**. [CrossRef]
81. Wiecek, A.J.; Cutty, S.J.; Kornai, D.; Parreno-Centeno, M.; Gourmet, L.E.; Tagliazucchi, G.M.; Jacobson, D.H.; Zhang, P.; Xiong, L.; Bond, G.L.; et al. Genomic Hallmarks and Therapeutic Implications of G0 Cell Cycle Arrest in Cancer. *Genome Biol.* **2023**, *24*, 128. [CrossRef]
82. Fernandez, B.; Passanisi, V.J.; Ashraf, H.M.; Spencer, S.L. Single-Cell RNA Sequencing Reveals a Quiescence-Senescence Continuum and Distinct Senotypes Following Chemotherapy. *Nat. Commun.* **2025**, *17*, 169. [CrossRef] [PubMed]
83. Hu, J.; Sánchez-Rivera, F.J.; Wang, Z.; Johnson, G.N.; Ho, Y.; Ganesh, K.; Umeda, S.; Gan, S.; Mujal, A.M.; Delconte, R.B.; et al. STING Inhibits the Reactivation of Dormant Metastasis in Lung Adenocarcinoma. *Nature* **2023**, *616*, 806–813. [CrossRef] [PubMed]
84. Blath, J.; Kraut, A.; Paul, T.; Tóbiás, A. A Stochastic Population Model for the Impact of Cancer Cell Dormancy on Therapy Success. *J. Theor. Biol.* **2025**, *597*, 111995. [CrossRef] [PubMed]
85. Sfakianakis, N.; Katsaounis, D.; Chaplain, M.A.J. Modelling Cancer Cell Dormancy in Mice: Statistical Distributions Predict Reactivation Times and Survival Dynamics. *bioRxiv* **2025**. [CrossRef]
86. Cameron, M.D.; Schmidt, E.E.; Kerkvliet, N.; Nadkarni, K.V.; Morris, V.L.; Groom, A.C.; Chambers, A.F.; MacDonald, I.C. Temporal Progression of Metastasis in Lung: Cell Survival, Dormancy, and Location Dependence of Metastatic Inefficiency. *Cancer Res.* **2000**, *60*, 2541–2546. [PubMed]
87. Tian, X.; Cao, F.; Li, X.; He, Y.; Xia, Y.; Chen, L.; Zhao, Q. Tumor Dormancy Is Closely Related to Prognosis Prediction and Tumor Immunity in Neuroblastoma. *Transl. Pediatr.* **2023**, *12*, 445–461. [CrossRef]
88. Tjensvoll, K.; Nordgård, O.; Skjæveland, M.; Oltedal, S.; Janssen, E.A.M.; Gilje, B. Detection of Disseminated Tumor Cells in Bone Marrow Predict Late Recurrences in Operable Breast Cancer Patients. *BMC Cancer* **2019**, *19*, 1131. [CrossRef]
89. Mamounas, E.P.; Tang, G.; Fisher, B.; Paik, S.; Shak, S.; Costantino, J.P.; Watson, D.; Geyer, C.E.; Wickerham, D.L.; Wolmark, N. Association Between the 21-Gene Recurrence Score Assay and Risk of Locoregional Recurrence in Node-Negative, Estrogen Receptor–Positive Breast Cancer: Results From NSABP B-14 and NSABP B-20. *J. Clin. Oncol.* **2010**, *28*, 1677–1683. [CrossRef]
90. Kim, M.J. Tracing Quiescent Cancer Cells In Vivo. *Cancers* **2024**, *16*, 3822. [CrossRef]
91. Kim, M.J. Diagnosis and Therapeutic Targeting of Quiescent Cancer Cells: Road to Conquer Cancer Recurrence. *BMB Rep.* **2025**, *58*, 277–287. [CrossRef] [PubMed]
92. Lim, A.R.; Ghajar, C.M. Thorny Ground, Rocky Soil: Tissue-Specific Mechanisms of Tumor Dormancy and Relapse. *Semin. Cancer Biol.* **2022**, *78*, 104–123. [CrossRef] [PubMed]
93. Li, H.; Xu, W.; Cheng, W.; Yu, G.; Tang, D. Drug-Tolerant Persister Cell in Cancer: Reversibility, Microenvironmental Interplay, and Therapeutic Strategies. *Front. Pharmacol.* **2025**, *16*, 1612089. [CrossRef]
94. Zhou, M.; Ma, Y.; Chiang, C.-C.; Rock, E.C.; Butler, S.C.; Anne, R.; Yatsenko, S.; Gong, Y.; Chen, Y.-C. Single-Cell Morphological and Transcriptome Analysis Unveil Inhibitors of Polyploid Giant Breast Cancer Cells in Vitro. *Commun. Biol.* **2023**, *6*, 1301. [CrossRef]
95. Uzuner, D.; Akkoç, Y.; Peker, N.; Pir, P.; Gözüaçık, D.; Çakır, T. Transcriptional Landscape of Cellular Networks Reveal Interactions Driving the Dormancy Mechanisms in Cancer. *Sci. Rep.* **2021**, *11*, 15806. [CrossRef] [PubMed]
96. Fujimaki, K.; Yao, G. Cell Dormancy Plasticity: Quiescence Deepens into Senescence through a Dimmer Switch. *Physiol. Genom.* **2020**, *52*, 558–562. [CrossRef] [PubMed]
97. Tirosh, I.; Suva, M.L. Cancer Cell States: Lessons from Ten Years of Single-Cell RNA-Sequencing of Human Tumors. *Cancer Cell* **2024**, *42*, 1497–1506. [CrossRef]
98. Bunne, C.; Schiebinger, G.; Krause, A.; Regev, A.; Cuturi, M. Optimal Transport for Single-Cell and Spatial Omics. *Nat. Rev. Methods Primers* **2024**, *4*, 58. [CrossRef]
99. Argelaguet, R.; Arnol, D.; Bredikhin, D.; Deloro, Y.; Velten, B.; Marioni, J.C.; Stegle, O. MOFA+: A Statistical Framework for Comprehensive Integration of Multi-Modal Single-Cell Data. *Genome Biol.* **2020**, *21*, 111. [CrossRef]


 https://doi.org/10.3390/biom16050633

100. Cao, Z.-J.; Gao, G. Multi-Omics Single-Cell Data Integration and Regulatory Inference with Graph-Linked Embedding. *Nat. Biotechnol.* **2022**, *40*, 1458–1466. [CrossRef] [PubMed]
101. Theodoris, C.V.; Xiao, L.; Chopra, A.; Chaffin, M.D.; Al Sayed, Z.R.; Hill, M.C.; Mantineo, H.; Brydon, E.M.; Zeng, Z.; Liu, X.S.; et al. Transfer Learning Enables Predictions in Network Biology. *Nature* **2023**, *618*, 616–624. [CrossRef]
102. Cui, H.; Wang, C.; Maan, H.; Pang, K.; Luo, F.; Duan, N.; Wang, B. scGPT: Toward Building a Foundation Model for Single-Cell Multi-Omics Using Generative AI. *Nat. Methods* **2024**, *21*, 1470–1480. [CrossRef]
103. Hao, M.; Gong, J.; Zeng, X.; Liu, C.; Guo, Y.; Cheng, X.; Wang, T.; Ma, J.; Zhang, X.; Song, L. Large-Scale Foundation Model on Single-Cell Transcriptomics. *Nat. Methods* **2024**, *21*, 1481–1491. [CrossRef]
104. Yang, F.; Wang, W.; Wang, F.; Fang, Y.; Tang, D.; Huang, J.; Lu, H.; Yao, J. scBERT as a Large-Scale Pretrained Deep Language Model for Cell Type Annotation of Single-Cell RNA-Seq Data. *Nat. Mach. Intell.* **2022**, *4*, 852–866. [CrossRef]
105. Rubanova, Y.; Chen, R.T.Q.; Duvenaud, D. Latent ODEs for Irregularly-Sampled Time Series. *arXiv* **2019**, arXiv:1907.03907. [CrossRef]
106. Chen, M.; Mainardi, S.; Lieftink, C.; Velds, A.; De Rink, I.; Yang, C.; Kuiken, H.J.; Morris, B.; Edwards, F.; Jochems, F.; et al. Targeting of Vulnerabilities of Drug-Tolerant Persisters Identified through Functional Genetics Delays Tumor Relapse. *Cell Rep. Med.* **2024**, *5*, 101471. [CrossRef]
107. De Conti, G.; Dias, M.H.; Bernards, R. Fighting Drug Resistance through the Targeting of Drug-Tolerant Persister Cells. *Cancers* **2021**, *13*, 1118. [CrossRef]
108. Chen, Y.-C.; Gowda, K.; Amin, S.; Schell, T.D.; Sharma, A.K.; Robertson, G.P. Pharmacological Agents Targeting Drug-Tolerant Persister Cells in Cancer. *Pharmacol. Res.* **2024**, *203*, 107163. [CrossRef]
109. Koutsandreas, T.; Tsafou, K.; Horn, H.; Barrett, I.; Petsalaki, E. Network-Based Approaches for Drug Target Identification. *Annu. Rev. Biomed. Data Sci.* **2025**, *8*, 423–446. [CrossRef]
110. Chen, Y.; Juan, L.; Lv, X.; Shi, L. Bioinformatics Research on Drug Sensitivity Prediction. *Front. Pharmacol.* **2021**, *12*, 799712. [CrossRef]
111. Dawood, M.; Vu, Q.D.; Young, L.S.; Branson, K.; Jones, L.; Rajpoot, N.; Minhas, F.u.A.A. Cancer Drug Sensitivity Prediction from Routine Histology Images. *npj Precis. Oncol.* **2024**, *8*, 5. [CrossRef]
112. Kulkarni, V.S.; Alagarsamy, V.; Solomon, V.R.; Jose, P.A.; Murugesan, S. Drug Repurposing: An Effective Tool in Modern Drug Discovery. *Russ. J. Bioorganic Chem.* **2023**, *49*, 157–166. [CrossRef]




 https://doi.org/10.3390/biom16050633

biomolecules

MDPI

*Review*

# Nuclear Dynamics in Quiescent Cells: Conserved Mechanisms from Yeasts to Mammals

**Sigurd Braun [1,*], Cornelia Kilchert [2,*], Aydan Bulut-Karslioglu [3], Myriam Ruault [4], Angela Taddei [4], Fatemeh Rabbani [5] and Dominika Włoch-Salamon [5]**

[1] Institute of Genetics, Justus Liebig University Giessen, 35392 Giessen, Germany
[2] Institute of Biochemistry, Justus Liebig University Giessen, 35392 Giessen, Germany
[3] Max Planck Institute for Molecular Genetics, 14195 Berlin, Germany
[4] UMR 3664 Nuclear Dynamics, CNRS, Institut Curie, Université PSL, Sorbonne University, 75248 Paris, France
[5] Institute of Environmental Sciences, Faculty of Biology, Jagiellonian University, 30-387 Kraków, Poland
* Correspondence: sigurd.braun@gen.bio.uni-giessen.de (S.B.); cornelia.kilchert@chemie.bio.uni-giessen.de (C.K.)

**Abstract**

Quiescence is a reversible, non-proliferative cellular state that enables survival under nutrient limitation while preserving the capacity to resume growth. Rather than representing a passive default, quiescence is an actively regulated program conserved from unicellular eukaryotes to metazoans. This review focuses on the nuclear mechanisms underlying quiescence entry, maintenance, and exit, with primary emphasis on mechanistic insights from yeast models while highlighting conserved principles in multicellular systems. Across species, quiescence is characterized by global transcriptional repression, chromatin compaction, and the extensive reorganization of nuclear architecture, coordinated by nutrient-sensing pathways centered on TOR/mTOR signaling. We discuss how transcriptional reprogramming is achieved through redistribution of RNA polymerases, dynamic transcription factor activities, and large-scale remodeling of histone modifications, alongside repressive chromatin formation. In parallel, post-transcriptional mechanisms—including intron retention, alternative polyadenylation, and accumulation of non-coding RNAs—fine-tune gene expression while limiting biosynthetic output. We further examine how changes in nuclear organization, such as nucleolar condensation, condensin-mediated chromosome rearrangements, and telomere hyperclusters, support long-term viability and genome stability. Collectively, this review highlights nuclear dynamics as an integrative regulatory layer that links metabolic state to cellular identity, adaptability, and long-term survival, with broad implications for development, stem cell function, and disease.

**Keywords:** cellular quiescence; nuclear organization; chromatin remodeling; transcriptional reprogramming; post-transcriptional regulation; epigenetic regulation

## 1. Introduction

Quiescence, the reversible arrest of cell proliferation, is the predominant state of life. Unlike terminally differentiated and senescent cells, quiescent cells (Q cells) maintain the full capacity to return from G0 and reenter the cell cycle (Figure 1). Unicellular organisms benefit from quiescence by surviving periods of growth limitation and resume active proliferation when favorable conditions return. Quiescence is also essential for various aspects of metazoan biology including reproduction, immunological memory, and tissue regeneration [1,2]. For example, unfertilized mammalian oocytes are generated during fetal development but can remain dormant for decades until release during ovulation. Even after fertilization, embryonic development can be temporally suspended: in some organisms, diapause, a form of developmental dormancy, is triggered to increase the chances of embryo survival under starvation or other stressors. In multicellular organisms, quiescent tissue-resident stem cells serve as reservoirs for tissue homeostasis and regeneration. At the same time, Q cells in tumors and fungal biofilms—often termed persister cells—also contribute to therapy resistance and disease recurrence, underscoring the clinical relevance of understanding quiescent cell biology.

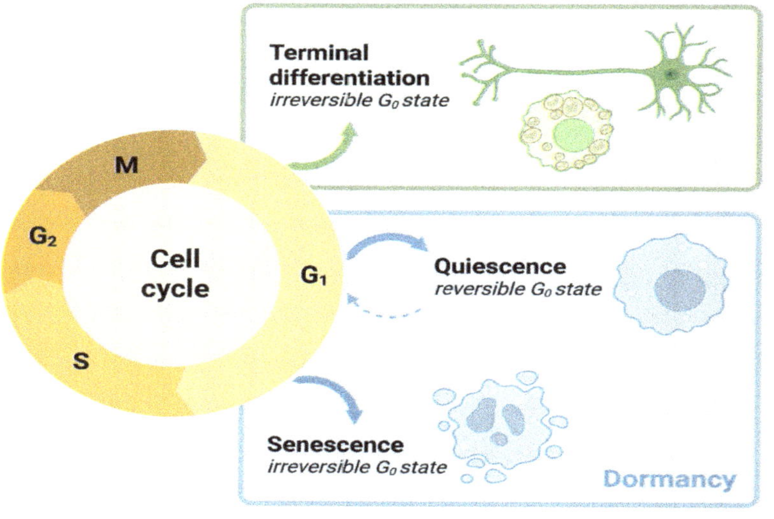


**Figure 1. Definitions of different non-proliferative cellular states.** Non-proliferative cells can exist in several distinct states that differ in their reversibility and developmental context. However, terminology varies considerably across fields: the term "dormancy," for instance, has been applied inconsistently to describe reversibly arrested cells, irreversibly arrested cells, or both, depending on the system and context. To avoid ambiguity, we primarily use the more stringently defined terms "quiescence" and "senescence" throughout this review. Quiescence refers to a reversible cell cycle arrest in which cells retain the capacity to resume division upon appropriate stimulation. Examples include starved cells, adult stem cells, immunological memory cells, dormant cancer cells (often termed cancer stem-like cells or persister cells), and diapaused embryos. Senescence is characterized as an irreversible cell cycle arrest that can arise as a consequence of aging, DNA damage, or other forms of cellular stress. Terminal differentiation, in contrast, describes cells that exit the cell cycle as part of a developmental program, acquiring specialized functions that typically preclude further division. Created in BioRender. Kilchert, C. (2026) https://BioRender.com/85t3r35 (accessed on 16 January 2026).

Quiescence is believed to be an early evolutionary phenomenon that is universal across life. Periods of hardship and nutrient scarcity are inevitable companions of life—when conditions are favorable, life explodes, and intense competition for nutrients, and their depletion, are the consequence. It begs no question what an immense advantage the ability to withstand poor conditions must have posed from life's inception. In the course of evolution, quiescence has been non-uniformly adapted to suit the specialized purpose of diverse biological contexts [1]. However, despite the diversity and complexity of eukaryotic cellular arrest models, cellular processes associated with entry into, maintenance, and exit from the quiescent state are often shared. Across eukaryotes, quiescence is characterized by reduced cell size, chromatin compaction, and global downregulation of transcription and


https://doi.org/10.3390/biom16020203

translation, reflecting a profound metabolic shift in which cells suppress growth-associated biosynthesis while selectively activating catabolic pathways and antioxidant defenses that support long-term survival [3]. Across organisms, these changes are driven by starvation-responsive signaling pathways, prominently including inactivation of the highly conserved Target of Rapamycin (TOR) kinase complexes or its metazoan homolog Mechanistic Target of Rapamycin mTOR (for key terms and abbreviations, see Table 1). TOR and mTOR integrate information from multiple nutrient-sensing pathways and promote cell growth, proliferation, and survival when conditions are favorable; their inactivation initiates a conserved "hold and protect" program that gradually develops into quiescence [4,5] (Figure 2). The central nature of TOR signaling in quiescence establishment is underscored by the fact that pharmacological TOR inhibition, for example with rapamycin, recapitulates most hallmarks of quiescence even in complex systems, and has been used experimentally as a model of diapause. For example, partial inhibition of mTOR in mouse blastocysts and embryonic stem cells induces a reversible paused state that preserves pluripotency, reduces transcription and translation, alters histone modifications such as H4K16ac and H3K36me2, and allows full developmental reactivation upon release from inhibition [6]. Importantly, TOR signaling is not restricted to its well-characterized cytoplasmic functions in translational control and metabolism. In addition to regulating the nucleocytoplasmic distribution of downstream signaling effectors, both TORC1 and TORC2 possess active nuclear pools that shuttle between the cytoplasm and nucleus and associate directly with chromatin, positioning TOR as a direct regulator of gene expression and epigenetic state [5] (Figure 2).

Together, these observations establish quiescence as a highly regulated, multilayered program rather than a uniform arrested state. By integrating cytoplasmic control of metabolism and translation with direct nuclear regulation of transcription, chromatin organization, RNA processing, and epigenetic stability, adaptations during quiescence enable cells to survive prolonged periods of unfavorable conditions while retaining the capacity for rapid reactivation when environments improve. This kind of rewiring is found in unicellular organisms as well as metazoan models, but can exhibit substantial heterogeneity across cell types and even among clonal cell populations [3,7–13]. This apparent contradiction between universal molecular pathways and diverse cellular responses raises critical questions: How can conserved molecular mechanisms give rise to such distinct phenotypic outcomes? And what cell type-specific factors shape quiescence dynamics?

**Table 1.** Glossary of key terms and abbreviations.

| Abbreviation | Definition | Explanation | Species Context |
|---|---|---|---|
| G0 | G0 phase | Reversible cell-cycle arrest in which cells stop dividing but retain the ability to re-enter proliferation. | Conserved |
| Q cells | Quiescent cells | Cells in a stable, non-proliferative state with reduced biosynthesis and extensive nuclear remodeling; also referred to as reversible G0 cells. | Conserved |
| NQ cells | Non-quiescent cells | Cells that fail to achieve deep quiescence and exist in an unstable intermediate state, transitioning more rapidly than Q cells into irreversible senescence during prolonged nutrient deprivation | Functionally conserved |
| log cells | Proliferating cells | Exponentially growing cells dividing at the maximal rate that conditions allow; often referred to as "logarithmic" because cell number plotted against time on a semi-logarithmic scale yields a straight line. | Conserved |
| Diapause | Embryonic diapause | Reversible developmental arrest of the mammalian embryo, typically at the blastocyst stage. | Metazoans (mammals) |


https://doi.org/10.3390/biom16020203

**Table 1.** *Cont.*

| Abbreviation | Definition | Explanation | Species Context |
|---|---|---|---|
| TOR | Target of Rapamycin | Nutrient-sensing kinase pathway coordinating growth, metabolism, and quiescence. | Yeast |
| mTOR | Mechanistic Target of Rapamycin | Metazoan TOR pathway integrating nutrient and growth-factor signaling to regulate dormancy and growth. | Metazoans |
| TORC1 | TOR complex 1 | TOR-containing complex promoting ribosome biogenesis and translation; inhibited during quiescence. | Conserved |
| TORC2 | TOR complex 2 | TOR-containing complex regulating stress responses and survival pathways. | Conserved |
| PKA | Protein kinase A | cAMP-dependent kinase promoting growth-associated transcription and antagonizing quiescence. | Conserved |
| DS | Diauxic shift | Metabolic transition in which yeast cells switch from fermentative growth on glucose to respiratory metabolism of alternative carbon sources after glucose depletion. | *S. cerevisiae* |
| MuSC | Muscle stem cell | Quiescent adult stem cells involved in muscle maintenance and repair; also known as satellite cells | Metazoans |
| RNAPII | RNA polymerase II | Enzyme transcribing mRNAs and many non-coding RNAs; redistributes during quiescence. | Conserved |
| RNAPI | RNA polymerase I | Enzyme transcribing ribosomal RNA genes at the rDNA locus; tightly linked to growth state. | Conserved |
| RiBi | Ribosome biogenesis | Coordinate transcriptional program encoding ribosomal proteins, rDNA processing factors, and ribosome assembly components. | *S. cerevisiae* (conserved growth regulon in eukaryotes) |
| RRPE | rRNA processing element | Cis-regulatory promoter motif enriched in ribosome biogenesis genes that mediates coordinated transcriptional repression via binding of Stb3. Related elements (Homo1D and Homo1E motifs) also exist in *S. pombe*. | *S. cerevisiae* |
| RP(G) | Ribosomal protein(s) (gene(s)) | Structural and functional components of ribosomes; their expression is tightly repressed during quiescence. | Conserved |
| Stb3 | Sin3-binding protein 3 | Nutrient-regulated transcriptional repressor of RiBi genes that recruit Rpd3 to promote chromatin compaction and biosynthetic shutdown | *S. cerevisiae* |
| Xbp1 | XhoI site-binding protein 1 | Stress-induced transcriptional repressor that binds promoter regions of growth-associated genes | *S. cerevisiae* |
| HDAC | Histone deacetylase | Enzymes removing acetyl groups from histones, promoting chromatin compaction and repression. | Conserved |
| Rpd3 | Reduced potassium dependency 3 | Class I HDAC broadly retargeted to promoters during budding yeast quiescence. | *S. cerevisiae* |
| Set1C/COMPASS | Set1 complex | H3K4 methyltransferase complex depositing H3K4me3 at promoters. | Conserved |
| SAGA | Spt-Ada-Gcn5 acetyltransferase complex | Transcriptional coactivator complex with histone acetyltransferase and deubiquitylation activities that promotes transcription initiation and stress- and growth-related genes. | Conserved |
| H3K4me3 | H3 lysine 4 trimethylation | Active promoter mark selectively retained or redistributed during quiescence. | Conserved |
| H3K36me3 | H3 lysine 36 trimethylation | Histone modification associated with transcription elongation and gene body marking, largely maintained during quiescence (in yeast). | Conserved |
| Ino80C | Ino80 chromatin remodeling complex | ATP-dependent remodeler evicting H2A.Z and regulating transcription during quiescence. | Conserved |
| H2A.Z | H2A.Z histone variant | Histone variant modulating promoter responsiveness and chromatin boundaries. | Conserved |


https://doi.org/10.3390/biom16020203

**Table 1.** *Cont.*

| Abbreviation | Definition | Explanation | Species Context |
| --- | --- | --- | --- |
| Hho1 | Histone H1 homolog | Linker histone contributing to chromatin compaction without enforcing full silencing. | *S. cerevisiae* |
| Paf1C | Polymerase-associated factor 1 complex | Transcription elongation complex that associates with RNAPII and regulates histone modifications linked to active transcription | Conserved |
| Clr3/Clr6 | Cryptic loci regulators 3/6 | HDACs contributing to heterochromatin formation and quiescence maintenance. | *S. pombe* |
| Clr4 | Clr4 methyltransferase | Histone H3K9 methyltransferase/Suvar 39 h homolog, heterochromatin initiation and spreading. | *S. pombe* |
| H3K9me | H3 lysine 9 methylation | Repressive histone mark defining constitutive and facultative heterochromatin. | Conserved (absent in *S. cerevisiae*) |
| H3K27me3 | H3 lysine 27 trimethylation | Polycomb-associated repressive mark controlling facultative heterochromatin repression. | Metazoans |
| H4K20me3 | H4 lysine 20 trimethylation | Deeply repressive mark associated with chromatin compaction and long-term arrest. | Metazoans |
| PRC2 | Polycomb repressive complex 2 | Complex depositing H3K27me3 to mediate stable transcriptional repression. | Metazoans |
| EZH1 | Enhancer of zeste homolog 1 | PRC2 catalytic subunit prevalent in non-proliferative and Q cells. | Metazoans |
| EZH2 | Enhancer of zeste homolog 2 | PRC2 catalytic subunit associated with proliferation and activation. | Metazoans |
| HP1 | Heterochromatin protein 1 | Chromodomain protein binding H3K9me and promoting heterochromatin spreading. | Conserved (functional analogs in yeast) |
| RNAi | RNA interference | Small RNA–guided silencing pathway coupling transcripts to heterochromatin formation. | *S. pombe*. metazoans |
| Dcr1 | Dicer | RNase III enzyme producing siRNAs; also performs RNAi-independent roles in quiescence survival. | *S. pombe* |
| Ago1 | Argonaute 1 | Small-RNA effector protein targeting transcripts and chromatin. | *S. pombe*, metazoans |
| RITS | RNA-induced transcriptional silencing complex | RNAi effector complex linking siRNAs to chromatin-based repression. | *S. pombe* |
| lncRNA | Long non-coding RNA | >200-nt RNAs regulating chromatin, transcription, and nuclear organization. | Conserved |
| asRNA | Antisense RNA | Non-coding RNA transcribed opposite a gene, often mediating transcriptional interference. | Conserved |
| PAS | Polyadenylation site | RNA cleavage site determining 3′-UTR length and post-transcriptional regulation. | Conserved |
| 3-UTR | 3′-untranslated region | mRNA region controlling stability, localization, and translation efficiency. | Conserved |
| CID | Chromosomal interaction domain | Local self-interacting chromatin unit detected by Hi-C. | Yeast (conceptually conserved) |
| L-CID | Large chromosomal interaction domain | Larger chromatin domains that become more insulated during quiescence. | *S. cerevisiae* |
| Condensin | Condensin complex | SMC complex organizing higher-order chromosome structure via loop formation. | Conserved |


https://doi.org/10.3390/biom16020203

**Table 2.** The nuclear hallmarks and molecular effectors of the quiescent state in unicellular and multicellular organisms described in this review. Superscript 1: yeast (*S. cerevisiae* or *S. pombe*); superscript 2: metazoans or mammalian cells.

| Cellular Process | Hallmarks/Phenotype | Nuclear Mechanism/Factors |
|---|---|---|
| Cell cycle | Stable arrest with ability to re-enter proliferation | TOR [1]/mTOR [2] inhibition, PKA [1] signaling, Myc [2] downregulation |
| Growth and metabolism | Reduced protein synthesis and metabolic activity | Repression of ribosomal RNA and protein biogenesis [1,2], translational control [1,2], intron retention [1,2] |
| Autophagy and survival | Increased autophagy and stress resistance | TOR [1,2] inactivation, nutrient sensing pathways [1,2] |
| Transcriptional regulation | Global transcriptional repression with selective gene activation | RNAPII redistribution [1,2], transcriptional repressors (Xbp1 [1], Stb3 [1]), chromatin remodeling [1,2] |
| Chromatin state | Increased chromatin compaction and heterochromatin formation | HDACs (Rpd3 [1], Clr3 [1], Clr6 [1]), histone methyltransferases (Clr4 [1], PRC2 [2]) |
| Histone modifications | Global hypoacetylation, accumulation of repressive methylation marks | Histone deacetylation [1,2], H3K9 [1,2]/H3K27 [2]/H4K20 [2] methylation pathways |
| Genome organization | Nuclear and 3D genome reorganization | Condensin [1,2], heterochromatin dynamics [1,2], telomere clustering [1], stress-induced gene cluster relocalization [1] |
| RNA abundance | Strong reduction in total mRNA levels | Reduced RNAP II initiation and elongation [1,2], RNA decay modulation [1,2] |
| RNA processing | Widespread intron retention | Splicing repression [1,2], spliceosome retention [1], TOR signaling [1] |
| Post-transcriptional regulation | 3′-UTR lengthening, altered RNA stability | Polyadenylation site selection bias [1,2], reduced cleavage [1] |
| Transcript storage | Accumulation of RNAs in condensates | RNA-Protein phase separation [1,2] |
| Non-coding RNA levels | Increased lncRNAs and other ncRNAs | Reduced nuclear exosome activity [1,2], lncRNA-mediated recruitment of chromatin modifiers [2] |
| Stress and metabolic genes | Relative enrichment of stress-response and nutrient scavenging transcripts | Ino80C [1], Paf1C [1], histone turnover [1], replication-independent chromatin remodeling [1] |
| Re-proliferation capacity | Transcriptional reactivation upon nutrient readdition | Histone acetyltransferases (SAGA [1], NuA4 [1]), TOR reactivation [1] |

**Genetic screens have revealed the central importance of nuclear processes.** Screens using yeast deletion collections or temperature-sensitive mutants have confirmed the central importance of TOR signaling, repression of protein biosynthesis, and an increase in autophagy in establishing quiescence [3], while also identifying numerous genes that are essential for survival during G0 or for re-entry into the cell cycle [14,15]. Notably, genes governing nuclear dynamics were frequently recovered in these screens, reflecting the extensive remodeling of nuclear architecture and chromatin reorganization that accompanies entry into quiescence across organisms [16]. This observation prompted Zahedi et al. to perform a targeted flow cytometry screen of known chromatin regulator deletion mutants in fission yeast [17], monitoring the proportion of cells in G0 as well as cell mortality over extended periods of quiescence to distinguish between genes essential for entering G0 and those required for long-term viability. Strikingly, among genes annotated with the Gene Ontology term "chromatin organization", 68% of the mutants had defects in establishing or stably maintaining quiescence. Similarly high defect rates were observed for genes involved in "chromatin remodeling" (69%), "regulation of transcription" (71%),


https://doi.org/10.3390/biom16020203

or "DNA repair" (76%), underscoring the importance of nuclear processes during cellular quiescence [17,18].

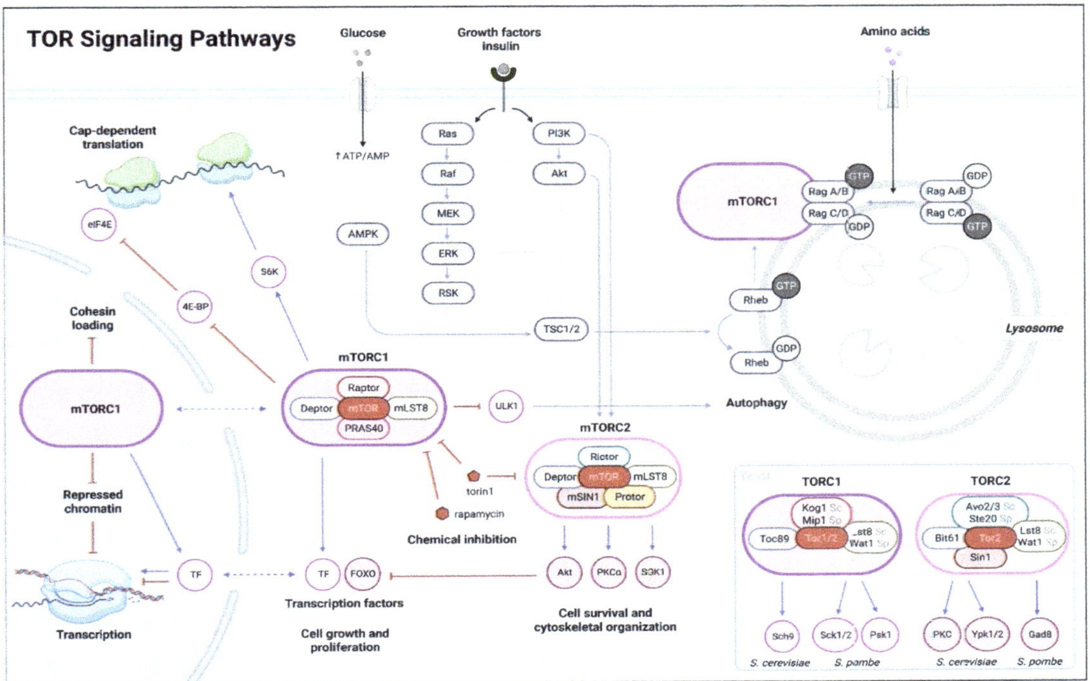


**Figure 2. mTOR signaling integrates information on nutrient availability to regulate cell growth and proliferation.** Signaling from different nutrient-sensing pathways converges on the mechanistic target of rapamycin (mTOR) kinase, which assembles into two different complexes, mTORC1 and mTORC2, to regulate central cellular processes related to growth and survival. In the presence of amino acids, guanine nucleotide exchange enables the Rag A/B-C/D complex to recruit mTORC1 to the lysosomal surface, where it can be activated by the small GTPase Rheb in its GTP-bound state. High Rheb-GTP levels are maintained in response to growth factor and insulin signaling via protein kinase cascades (including the kinases AKT and ERK) that inhibit the tuberous sclerosis complex (TSC1/2), a GTPase activating protein for Rheb. Activated mTORC1 phosphorylates cytoplasmic substrates and regulates nuclear transcriptional programs to promote growth. Key targets include Unc-51 Like Autophagy Activating Kinase 1 (ULK1; inhibitory phosphorylation suppresses autophagy); eIF4E-binding proteins (4E-BPs; phosphorylation prevents inhibition of cap-dependent translation); and S6 kinase (S6K), which phosphorylates diverse substrates including ribosomal protein S6 and transcription factors. In the nucleus, mTORC1 signaling maintains chromatin in an active state permissive for transcription. mTORC1 signaling is further modulated by cellular energy status via AMP-activated protein kinase (AMPK), which is activated when ATP levels drop. Growth factor signaling through phosphoinositide-3 kinase (PI3K) and AKT also activates mTORC2 at the plasma membrane, which regulates cell survival and cytoskeletal organisation. mTORC2-dependent phosphorylation retains the key transcription factor FOXO in the cytoplasm, and its regulation of acetyl-CoA pools modulates histone acetylation levels. Inhibition of mTOR signaling—either by nutrient deprivation or chemical inhibitors such as rapamycin or Torin1—triggers entry into quiescence. Created in BioRender. Braun, S. (2026) https://BioRender.com/ki38916 (accessed on 17 January 2026).

**Scope and organization of this review.** Despite the accumulating evidence for the critical role of nuclear remodeling during quiescence, our understanding of the under-


https://doi.org/10.3390/biom16020203

lying mechanisms remains incomplete. Here, we provide a comprehensive overview of current knowledge regarding nuclear dynamics during quiescence entry, maintenance, and exit, with an emphasis on features conserved across eukaryotes (Table 2). Because most mechanistic data derive from yeast models, we focus primarily on these systems, while highlighting similarities and differences in higher eukaryotes where data are available. Throughout, we emphasize TOR signaling as a master coordinator of nuclear remodeling: strikingly, many complex nuclear rearrangements in Q cells can be triggered by TOR inhibition alone, suggesting TOR translates quiescence-inducing signals into specific architectural and functional adaptations of the nucleus. We link observed changes in nuclear dynamics to their upstream regulation by TOR signaling where such mechanisms are understood, while also pointing out knowledge gaps. For readers interested in the mechanistic details of TOR function and signaling in the nucleus, we refer to a comprehensive review [5]. Topics beyond our scope, including genome stability and DNA damage repair, as well as non-nuclear aspects of quiescence, including signaling pathways, metabolic reprogramming, and the temporal dynamics of quiescence regulation, are covered in recent publications and dedicated reviews [2,8,19–22].

## 2. Established Models for Studying Nuclear Dynamics During Quiescence

### 2.1. Unicellular Eukaryotic Models

Yeast species are powerful models for studying both cellular proliferation and quiescence. *Saccharomyces cerevisiae* (budding yeast) and *Schizosaccharomyces pombe* (fission yeast) have been particularly instrumental in identifying essential conserved mechanisms involved in quiescence establishment. These species diverged 300–350 million years ago, yet they share fundamental cellular processes. Mechanisms conserved between these two evolutionarily distant yeasts are therefore likely to be relevant for metazoans as well. In both organisms, nutrient deprivation leads to a reversible cell cycle arrest and the transition into a quiescent state.

**Budding yeast (*Saccharomyces cerevisiae*).** *S. cerevisiae* has served as a leading eukaryotic model for quiescence research [19]. Early studies showed that cultures starved for carbon, nitrogen, or phosphate accumulate small G1-arrested cells [23], which were later referred to as G0 cells [24–27]. An extensive body of work has led to the development of established guidelines, toolkits and databases that facilitate experimental design and analysis. However, variation in strain background, chronological age, starvation protocols, and criteria to identify Q cells in yeast has resulted in at least 48 possible experimental setups across published studies [28]. One commonly used approach induces quiescence through exhaustion of rich medium, resulting in gradual depletion of the carbon source. Exhaustion or abrupt starvation of other key nutrients (nitrogen, phosphate) also triggers a quiescent state, for which the specific cellular responses can differ. Importantly, the path used to trigger quiescence leads to very different functional outcomes, including gene expression profile, stress resistance, and lifespan [2]. For instance, exponentially growing cells abruptly starved from glucose will rapidly die, while cells from a stationary phase culture will survive in water for several weeks [29]. This is why Q cells resulting from the progressive exhaustion of rich medium are commonly used to study chronological aging [4]. It also emphasizes the importance of providing precise descriptions of experimental methods.

Work from the Werner-Washburne laboratory demonstrated that cells in stationary-phase cultures are heterogeneous and can be separated based on density. Most properties attributed to deep stationary-phase cells are, in fact, features of the dense fraction. This dense population corresponds to Q cells (or deep Q cells), whereas the less dense fraction is referred to as NQ cells (for non-Q). NQ cells more rapidly transition into a senescent


https://doi.org/10.3390/biom16020203

state and lose the ability to reenter the cell cycle during prolonged dormancy. Q cells begin to differentiate as glucose levels decline [30,31], prior to the metabolic transition from fermentation to respiration known as the diauxic shift (DS). They are predominantly small G1-arrested daughter cells, whereas NQ cells are more heterogeneous in both size and cell-cycle phase. Q cells are also characterized by a thickened cell wall that enhances thermotolerance [29]. Notably, the relative abundance of Q and NQ cells varies across genetic backgrounds, contributing to discrepancies in the literature. Strains with respiration defects, such as the commonly used BY background, produce lower yields of Q cells than W303 strains [32]. Auxotrophies—disruptions in key metabolic pathways introduced to facilitate genetic manipulation—also limit the ability of cells to enter the Q-cell state. It is therefore recommended to minimize auxotrophies and to culture cells under conditions that favor respiration [2].

**Fission yeast (*Schizosaccharomyces pombe*).** *S. pombe* is predominantly haploid and contains three chromosomes, with a genome size comparable to that of *S. cerevisiae*. Although these two yeast diverged early and differ substantially in genome organization, chromosome architecture, and regulatory pathways, they retain many conserved cellular processes; nonetheless, *S. pombe* offers complementary advantages as a quiescence model [16]. Quiescence in *S. pombe* is most commonly induced by nitrogen starvation. Starved cells undergo two more rounds of cell division without cellular growth and then arrest in G1, facing two possible outcomes: in the presence of opposite mating types, cells initiate a mating program that ultimately produces resilient spores. In contrast, heterothallic strains—which are fixed in one mating type and self-infertile—progressively enter the G0 state, in which they can remain viable for weeks or months in the presence of a carbon source [33]. Alternative conditions, such as glucose elimination or growth to saturation, cause G2 and G1 arrest, respectively, which resemble nitrogen starvation-induced quiescence but fail to support long-term cell survival [16]. Unlike *S. cerevisiae*, *S. pombe* does not have enzymes for the glyoxylate cycle required for ethanol metabolism and therefore does not undergo the DS. An important experimental advantage of nitrogen starvation is that G0 entry and exit occur synchronously within the cell population [16,34].

### 2.2. Multicellular Models

Temporary entry into quiescent states in response to developmental or environmental cues is widespread in multicellular organisms, which are composed of complex tissues containing diverse cell types that must function in coordination. Quiescence programs can be executed by specific lineages, such as neuroblasts in *Drosophila* or the germline in *C. elegans*. Alternatively, quiescence can affect the entire organism, as exemplified by the *C. elegans* dauer diapause. These forms of quiescence have been extensively studied and are the subject of several recent reviews [35–37].

**Adult stem cells.** Many multicellular tissues harbor stem or progenitor cells that sustain tissue integrity under homeostatic conditions and, when required, mediate repair and regeneration [38]. In most cases, a subset of these cells remains in a quiescent state and becomes activated only in response to defined physiological cues. These tissue-resident adult stem cells occupy specialized niches—microenvironments composed of neighboring cells, extracellular matrix components, and soluble factors—that protect them from cellular damage by restricting replication-associated and metabolic activity. Adult stem cells therefore constitute highly informative models for studying quiescence They share characteristic features, including low RNA content and absence of proliferation markers, and frequently coexist with more rapidly cycling progenitor pools within the same tissue compartment. Their behavior is tightly regulated by niche-derived signals, and disruption of these cues, for example during ageing, results in impaired stem cell activation and tissue



https://doi.org/10.3390/biom16020203

decline. Across systems such as muscle stem cells (MuSC), hematopoietic stem cells (HSC), hair follicle cells (HFSC), and neural stem cells (NSC), quiescence is associated with a distinct epigenetic landscape. Among these, HSCs represent a paradigmatic example of adult stem cell quiescence. In adult mammals, HSCs reside in specialized bone-marrow niches, where a subset of long-term HSCs remains deeply quiescent for extended periods, dividing only rarely [39,40]. Lineage-tracing and label-retention studies indicate that HSCs contribute little to steady-state hematopoiesis, which is instead largely sustained by more actively cycling progenitors. However, in response to stresses such as inflammation or irradiation, HSCs reversibly re-enter the cell cycle and regenerate the hematopoietic system, underscoring quiescence as a protective yet readily deployable state [41–43]. This functional separation between quiescent stem cells and active progenitors mirrors principles observed across adult stem cell systems, reinforcing their value as general models for quiescence. For a detailed overview of the regulation of adult stem cell quiescence, the reader is referred to a recent comprehensive review [21].

**Dormant mouse embryo**. Systemic quiescence in mammals is rare, occurring only in unfertilized oocytes and during early embryonic development. While adult stem cells maintain quiescence to support tissue homeostasis, repair, and regeneration, embryos employ dormancy as a reproductive strategy—either to optimize the timing of birth or to survive environmental challenges—a phenomenon known as embryonic diapause [44]. In the context of embryonic diapause, the period of dormancy can last from days to months, depending on the species [45]. In most cases, the embryo pauses at the blastocyst stage, delaying implantation, and further development. Documented to date in over 130 mammalian species, diapause remains a mystery due to the inaccessibility of embryos from most wildlife species. Mouse models have been instrumental in studying diapause, as it can be reliably induced through ovariectomy or hormone treatment [46,47]. Over the last decade, in vitro models of diapause have been developed and, together with advances in detection methods, have significantly propelled the field forward.

Similar to unicellular organisms, the central cellular growth regulator mTOR is a key factor controlling diapause in mammals [6,48]. Additionally, nutrient depletion, suppression of insulin signaling, Myc activity, and lipid metabolism have been documented to trigger a diapause-like response [6,49,50]. For a detailed comparative analysis of dormancy in adult stem cells and embryos, readers are referred to [51].

While quiescence is used across various tissues in multicellular organisms to enable preservation and repair, for the remainder of the review, we will mainly focus on its use and regulation in the early mammalian embryo. The undifferentiated progenitor cells within the early embryo are in many ways reminiscent of yeast, exhibiting low heterochromatin load and globally high transcriptional activity, and employing similar strategies to adapt to their immediate environment.

## 3. What Drives Transcriptional Reprogramming in Quiescent Cells?

### 3.1. Transcriptional Reprogramming During Metabolic Transitions in Yeast

In log cells, TORC1 directly regulates transcription by all three RNA polymerases (RNAPI, II, and III). It binds promoters of rDNA and tRNA genes to stimulate transcription, while modulating RNAPII-dependent expression of metabolic genes through interactions with transcription factors and coactivators. Moreover, TORC1 promotes transcriptionally permissive states by stimulating histone acetylation, antagonizing histone deacetylases including sirtuins, influencing histone methylation, and coordinating ATP-dependent chromatin remodeling complexes that maintain open promoter architecture [5]. As cells transition into quiescence, cellular processes shift from biomass production to maintenance. This is reflected in profound changes to the transcriptome and, to a lesser extent, the

 https://doi.org/10.3390/biom16020203

proteome. Global transcriptional repression results in a marked decrease in growth-related mRNAs, for example those encoding ribosomal proteins, while the relative expression of genes involved in nutrient uptake and scavenging increases. In *S. cerevisiae*, glucose deprivation results in an approximately 30-fold reduction in mean transcript abundance [52–54]. Similarly, measurements of absolute molecule numbers in fission yeast have revealed a global reduction in RNA levels upon nitrogen starvation, even when accounting for the smaller size of Q cells [55]. Recent studies have identified several mechanisms that likely act in concert to robustly limit mRNA production under these conditions.

In *S. cerevisiae*, quiescence is established over the course of several days, with Q cells exhibiting distinct transcriptional reprogramming. RNAPII distribution undergoes significant changes, characterized by reduced levels of both initiating and elongating forms of RNAPII [53]. Different studies have reported distinct RNAPII distributions, likely reflecting differences in the kinetics of quiescence entry arising from variations in growth conditions, such as aeration and medium pH [28,56]. Monitoring RNAPII over time revealed the evolution of RNAPII distribution during entry into quiescence [57]. Shortly after the DS (24 h of culture), RNAPII decreases in promoter-proximal regions and accumulates near transcription end sites (TES), potentially reflecting regulatory changes in transcription termination [53]. RNAPII then accumulates slowly at upstream-activating sequences of one third of coding genes in late Q cells [57–59]; intergenic enrichment correlated with fast resumption of transcription upon quiescence exit. As mentioned above, this decrease in RNAPII levels on chromatin accompanies a global transcription shut-down [52–54]. Nevertheless, the majority of transcripts remain detectable in mature Q cells, with 5105 genes expressed compared to 6205 in logarithmically growing (log) cells [53]. Notably, approximately 40% of these RNAs are found in an extraction-resistant compartment that is sensitive to protease treatment, indicative of sequestration in RNA-protein (RNP) complexes [53,60]. Consistent with this, two types of RNP granules are observed in Q-cells: processing bodies (P-bodies) and stress granules. While the former assemble after the DS upon PKA inactivation, the latter do not assemble until cells have reached the stationary phase [61]. Although most genes are shut down, approximately 3% of coding genes remain active, with seven detected in more than 80% of the cells in single-cell transcript analyses [57], indicating a very high-level of expression.

Similarly, *S. pombe* exhibits pronounced global repression of transcription during quiescence, with total mRNA levels reduced to approximately 15% of those in log cells [55]. This is accompanied by decreased levels of RNAPII engaged in transcription initiation and elongation [62]. Transcriptional reprogramming following nitrogen deprivation is dynamic, and distinct profiles are observed during the initial response (1–2 h) and the establishment phase (10–24 h) [63]. During establishment, several hundred genes, many of them involved in stress responses, detoxification, and recycling of amino acids and nucleotides, are transiently activated [55,63,64]. Once stable quiescence is achieved, expression of these transiently induced genes is decreased, and 97% of the genome becomes downregulated. Only 16 'core quiescence genes' remain highly expressed even after two weeks of nitrogen starvation [64].

### *3.2. Transcription Factor Dynamics Orchestrate Chromatin Remodeling*

What then drives the extensive transcriptional reprogramming during quiescence entry? In *S. cerevisiae*, differential gene expression is coordinated by a network of transcriptional activators and repressors. Their activity is tightly regulated by integrated signaling networks that include PKA, TORC1, Snf1, and Pho85, which collectively regulate metabolism, stress resistance, translation, autophagy, and the accumulation of energy reserves, such as glycogen, trehalose, and neutral lipids [2,3].



https://doi.org/10.3390/biom16020203

**Global gene regulation through altered activities of transcriptional activators and repressors.** In the presence of glucose, the C2H2-ZF transcription factors Mig1, Mig2, and Mig3 repress hundreds of genes. When glucose is depleted, Mig1 and Mig2 relocalize to the cytoplasm upon phosphorylation by the yeast AMPK Snf1, while Mig3 (and also Mig2) become transcriptionally repressed [2,65,66]. Upon the DS, the C2H2-family transcription factors Msn2 and Msn4 induce expression of a broad set of glucose-repressed genes through binding to specific stress-response elements (STREs) present in their promoter regions [2,67]. Msn2 also contributes to large-scale chromatin reorganization by promoting condensin recruitment to STRE-containing genes, thereby regulating higher-order chromatin structure (see Section 5.2). Msn2/4 activity is negatively regulated by the TORC1 and PKA pathways and positively regulated by the Rim15 protein kinase, which acts downstream of TORC1, through differential phosphorylation that controls nuclear localization of Msn2 and Msn4 [2,68].

Another layer of regulation is imposed by two transcriptional repressors, Xbp1 and Stb3. Xbp1 is transcriptionally induced by Msn2/4 during the DS and other forms of stress and becomes one of the most abundant transcripts in Q cells [2,31,69]. Its accumulation correlates with the frequency and duration of stress, suggesting that Xbp1 serves as a "stress memory" [2]. Xbp1 binds to specific promoter motifs present in over 500 genes, including the G1 cyclin *CLN3* [2,31]. Stb3, in contrast, binds ribosomal RNA processing elements (RRPEs) and represses ribosomal biogenesis (RiBi) gene expression. Its recruitment to ribosomal protein gene (RPG) promoters is inhibited by TOR-dependent phosphorylation, which prevents Stb3 nuclear localization [70,71]. Both repressors are intimately linked to dynamic changes in histone modifications.

**Altered histone modifications and nucleosome density.** Histone acetylation is typically associated with active transcription by modulating nucleosome structure and promoting the recruitment of effector proteins or 'readers'. In addition, lysine methylation, occurring both at N-terminal tails and within core histone domains, is associated with transcriptional activation in budding yeast. Transcriptional repression in Q cells is accompanied by a global increase in histone density and a marked decrease in acetylation at multiple residues at histone H3 and H4 [2,52,54,72]. These chromatin changes are initiated early during the DS, particularly at ribosomal genes, but become fully established after six days in stationary culture. Although histone hypoacetylation occurs genome-wide, gene-specific patterns emerge: growth-related genes exhibit a pronounced decrease in acetylation, whereas stress-response genes upregulated in quiescence gain acetylation marks, indicating a tight correlation between histone acetylation and transcriptional activity [52].

**Rpd3 is a key mediator of quiescence-specific chromatin remodeling.** Stationary cells lacking the conserved class I histone deacetylase Rpd3 (HDAC1/2 in mammals) display histone acetylation levels and nucleosome density similar to proliferating cells or cells in early quiescence [52]. In log cells, Rpd3 primarily localizes to coding regions and is absent from promoters, whereas in Q cells, Rpd3 redistributes to the promoter regions of approximately 50% of genes [52]. Promoter recruitment is mediated by Stb3 and Xbp1, each of them performing unique, non-redundant roles through targeting distinct sets of genes [52] (Figure 3). Xbp1 has been shown to associate in vivo with the Rpd3L complex, likely via its subunit Ume1 [73], suggesting a direct targeting mechanism for histone deacetylation at specific promoters. Notably, while *rpd3Δ* cells do not show a major growth phenotype in log cells, deletion of *RPD3* or *XBP1* impairs the ability to enter quiescence and reduces chronological lifespan [52]. These findings support a critical role for Rpd3-mediated histone deacetylation in enabling transcriptional and chromatin remodeling as cells enter quiescence, although regulation of non-histone substrates cannot be excluded. Beyond its role in transcriptional repression, bulk histone deacetylation may also contribute to metabolic adaptation by generating free acetate for acetyl-CoA synthesis,


https://doi.org/10.3390/biom16020203

thereby supporting energy production and anabolic processes under nutrient-limited conditions [74].

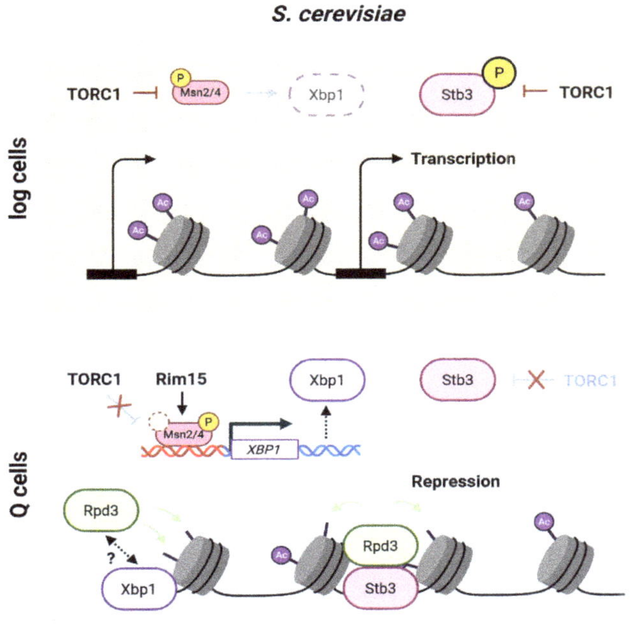


**Figure 3. Mechanisms of transcriptional reprogramming through the concerted actions of repressor proteins.** Histone deacetylation drives quiescence-associated transcriptional repression in *S. cerevisiae*. In log cells (top), high default levels of histone acetylation maintain open promoters and a chromatin structure permissive for transcription, as TORC1 and PKA signaling keep the transcriptional repressors Stb3 phosphorylated and inactive. In Q cells (bottom), TORC1 inactivation activates Stb3, which binds rRNA-processing elements (RRPEs) at ribosome biogenesis genes, and directly recruits the histone deacetylase complex Rpd3L. Concurrently, differential phosphorylation promotes nuclear localization of the stress-regulated transcription factors Msn2 and Msn4, which in return induce *XBP1* expression. Xbp1 associates with Rpd3L in vivo, suggesting direct recruitment to Xbp1-response elements, but the precise molecular mechanism remains unknown (indicated by question mark). Created in BioRender. Kilchert, C. (2026) https://BioRender.com/2hhc1hs (accessed on 17 January 2026).

**Restoration of histone acetylation during quiescence exit.** Upon glucose addition, stationary *S. cerevisiae* cells reinitiate transcription in a few minutes and restore histone acetylation before DNA replication resumes, marking their exit from the quiescent state [54,72,75]. Histone H3 and H4 acetylation is mediated by the SAGA and NuA4 histone acetyltransferase complexes, respectively [75]. In particular, TOR signaling promotes NuA4 recruitment to RPG promoters, ensuring that chromatin remodeling and transcriptional activation are tightly coupled to nutrient availability (for a detailed review, see [3]). The rapid re-initiation of transcription is largely dependent on the chromatin remodeling complex RSC that contributes to maintain NDRs upstream of specific genes and help RNAPII to progress into gene bodies upon release [76]. The intergenic accumulation of RNAPII upstream of specific genes including RPGs was also proposed to contribute to their rapid and robust transcriptional activation within minutes of refeeding [57].


https://doi.org/10.3390/biom16020203

**Transcription factor networks in metazoans.** In metazoan cells, quiescence is likewise governed by interconnected transcription factor networks that regulate thousands of genes and integrate nutrient-, stress-, and niche-derived signals to coordinately repress cell cycle genes while activating programs that support stress resistance, metabolic adaptation, and long-term viability. Rather than relying on single master regulators, quiescent mammalian cells deploy combinatorial TF assemblies—including E2F–RB/DREAM complexes, FOXO factors, Notch effectors, and lineage-specific transcription factors—that act through promoter- and enhancer-associated chromatin remodeling to establish reversible cell cycle arrest. As in yeast, these networks are dynamically modulated at multiple levels, including TF localization, stability, post-translational modification, and chromatin accessibility, enabling rapid and synchronous transitions between quiescent and proliferative states. For example, the autophagy-regulating transcription factors TFEB and FOXO3 are retained in the cytoplasm upon phosphorylation by mTORC1 and PI3K-AKT signaling, respectively, but translocate to the nucleus during quiescence following kinase inactivation. For a detailed overview of mammalian transcription factor networks during quiescence, the reader is referred to a recent comprehensive review [77].

### *3.3. Establishment of Silent Chromatin Through Repressive Epigenetic Modifications*

In quiescent *S. cerevisiae* cells, nearly half of the genome becomes targeted by Rpd3 [52], driving global histone deacetylation and transcriptional repression. While histone hypoacetylation is a common feature of quiescence across eukaryotes, the mechanisms guiding histone deacetylase recruitment differ substantially. Notably, neither Xbp1 nor Stb3 orthologs are found outside the *Saccharomycetaceae* clade, although functional analogs exist (e.g., the stress-responsive transcription factor Atf1 in *S. pombe*), and rRNA processing element (RRPE)-like sequence motifs at ribosomal synthesis genes are conserved (e.g., the Homo1D and Homo1E motifs in *S. pombe*) [78]. In contrast to budding yeast, fission yeast and many other fungi rely more heavily on repressive histone methylation—a mechanism broadly conserved across metazoans but absent in *S. cerevisiae*—to establish silent chromatin states.

**Mechanisms of constitutive heterochromatin establishment in fission yeast.** In *S. pombe*, silent chromatin is formed through locus-specific recruitment of the histone H3 lysine 9 (H3K9) methyltransferase Clr4, which is targeted through two complementary pathways [79]. Direct recruitment is mediated by sequence-specific DNA-binding factors, such as the heterodimeric transcription factor Atf1/Pcr1. RNA-guided recruitment involves the RNA interference (RNAi) machinery: Dcr1 (dicer) generates small interfering RNAs (siRNAs) that are loaded into Ago1, a core component of the RNA-induced transcriptional silencing (RITS) complex. Ago1-bound siRNAs guide RITS to nascent transcripts through base-paring, while the RNA-directed RNA polymerase Rdp1 amplifies siRNA production to reinforce the signal. Once recruited, Clr4 deposits H3K9me, which is recognized by Heterochromatin Protein 1 (HP1) family members (Swi6, Chp2) via their chromodomains. HP1 oligomerization, mediated by its chromoshadow domain, and additional Clr4 recruitment promote spreading of this histone mark across extended genomic regions. H3K9me is also recognized by RITS via its chromodomain protein Chp1, resulting in additional signal amplification and robust heterochromatin assembly. The resulting heterochromatin domains act as platforms for additional repressive factors, such as the class II HDAC Clr3. In log cells, this machinery establishes repressed chromatin at constitutive heterochromatin domains—pericentromeres, the silent mating type locus, rDNA, telomeres, and subtelomeres—as well as at dispersed facultative heterochromatin islands [79,80].

**Heterochromatin remodeling and function during quiescence.** Emerging evidence indicates that heterochromatin factors play important roles in both quiescence establishment and long-term viability during extended nutrient depletion. Multiple heterochromatin

 https://doi.org/10.3390/biom16020203

mutants display defects in entering quiescence, and cells lacking Dcr1, Ago1, or Rdp1 show a progressive decline in viability following nitrogen starvation [17,63,81]. Reduced survival of Q cells is also reported for mutants of the Clr6 HDAC complex—the *S. pombe* homolog of *S. cerevisiae* Rpd3L [17]—and for *clr4Δ* cells, although the severity varies across studies [63,81]. During quiescence entry, the heterochromatin landscape is extensively remodeled. Facultative heterochromatin is acquired at approximately 200 euchromatic loci across the genome, marked by de novo H3K9me deposition and accompanied by quiescence-specific siRNAs bound by Ago1 [63] (Figure 4). Many genes downregulated within 24 h of nitrogen starvation—including transcripts with functions in reproduction, carbohydrate metabolism, and transport—require Clr4 for silencing [63]. Consistent with the need to shut down protein biogenesis, the rDNA locus shows increased H3K9me levels and accounts for half of all siRNA production during quiescence, whereas other heterochromatin regions remain largely unchanged [63,81]. An exception are the subtelomeric regions that show a transient decrease in H3K9me levels, concomitant with the upregulation of several subtelomeric genes [63,82]. Consistently, among all surveyed histone methylation marks, both activating and repressive, H3K9me3 exhibits the most pronounced changes [62], underscoring the extent of heterochromatin remodeling.

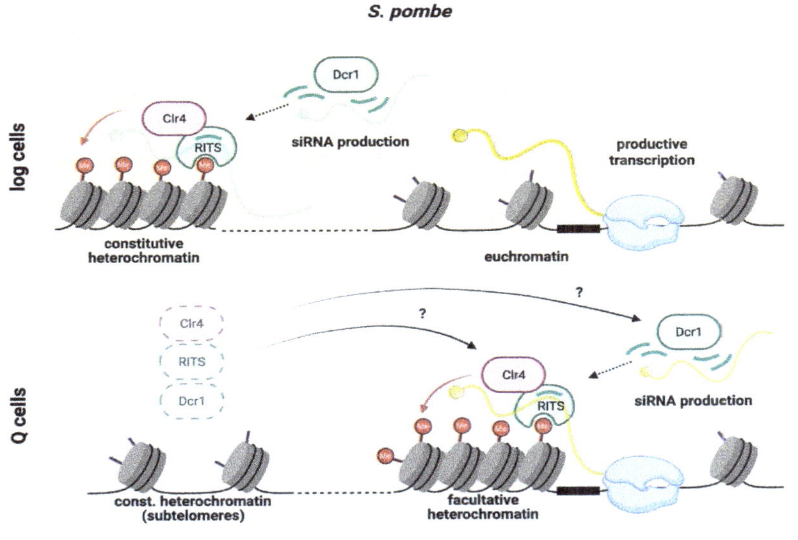


**Figure 4. Repressive histone methylation contributes to transcriptional reprogramming in *S. pombe*.** In log cells (top), silencing factors primarily act on constitutive heterochromatin (left). In Q cells, the RNase III enzyme Dicer (Dcr1) redistributes to euchromatic loci to generate small interfering RNAs (siRNAs), supported by RNA-dependent RNA polymerase (Rdp1). These siRNAs guide the RNA-induced transcriptional silencing complex (RITS) to nascent RNA, where it recruits the H3K9 methyltransferase Clr4, depositing repressive marks to form facultative heterochromatin at the silenced loci. RITS is stabilized on heterochromatin through recognition of H3K9me via its chromodomain, thereby reinforcing heterochromatin assembly. For simplicity, assembly of HP1 proteins is not shown. The initial mechanism targeting Dcr1 and RITS to euchromatic loci remains unknown (indicated by question mark), and Clr4 may also be recruited by RNAi-independent mechanisms. Created in BioRender. Braun, S. (2026) https://BioRender.com/t8cro8m (accessed on 17 January 2026).

**Dcr1 functions beyond RNAi.** While these findings support the notion of a key role for RNAi- and Clr4-dependent heterochromatin during early quiescence, genome-wide H3K9me remodeling at later stages, when nearly 97% of the genome is downregulated [64],

 https://doi.org/10.3390/biom16020203

remains less well explored. Notably, Dcr1 performs an additional, RNAi-independent function at the rDNA locus that is linked to the release of stalled RNAPI and is essential for long-term quiescence survival [81]. Cells lacking Dcr1 accumulate both RNAPI and H3K9me at rDNA repeats at substantially higher levels than in WT during quiescence. Two non-mutually exclusive mechanisms may explain this H3K9me accumulation: First, stalled RNAPI may directly recruit Clr4, in analogy to the recruitment of G9a by RNAPI in mammalian cells [83]. Second, elevated H3K9me levels at rDNA repeats may arise through an indirect mechanism in *dcr1*Δ cells. Specifically, this increase may result from the redistribution of silencing factors following their release from genomic regions that strictly depend on RNAi. This phenomenon has been documented in log cells, where loss of RNAi at one locus leads to enhanced heterochromatin formation at other genomic sites [84–86]. Interestingly, the viability defects of *dcr1*Δ can be rescued by mutations that impair heterochromatin formation (*clr4*Δ, *swi6*Δ) or by defects in RNAPI recruitment or stability. However, while *clr4*Δ suppresses the viability defects of *dcr1*Δ, it does not prevent RNAPI accumulation. This is in contrast to mutations that impair RNAPI recruitment and stability, which reduce both its accumulation and H3K9me levels in *dcr1*Δ cells [81]. This suggests that RNAPI retention alone is insufficient to account for the viability defect. Rather, although heterochromatin normally accumulates at rDNA during quiescence, excessive H3K9me levels—such as those associated with heterochromatin redistribution or RNAPI retention—are deleterious for survival and therefore require tight regulation. Notably, increased RNAPI association and rRNA accumulation are also observed in cells lacking the coilin homolog Mug174, which associates with Cajal bodies and is implicated in RNA splicing and small ribonucleoprotein biogenesis (see Section 5.1). However, although the absence of Mug174 also leads to viability defects during quiescence, its role is genetically independent of Dcr1 and Clr4, and these defects cannot be suppressed by additional deletion of *clr4*. Instead, this phenotype is more likely directly related to the function of Mug174 in splicing, consistent with the observation that splicing mutants also display quiescence defects [87].

**Diverse repressive histone methylation patterns in metazoans.** Similar regulatory principles, including transcriptional repression and histone hypoacetylation, operate in higher eukaryotes during quiescence, yet the accumulation and function of repressive histone marks vary substantially across cell types and physiological contexts. As in fission yeast, H3K9 methylation in metazoans is closely associated with constitutive heterochromatin at pericentromeric repeats and other repetitive elements. However, H3K9 methylation exhibits no uniform pattern during quiescence: several types of Q cells, such as hair follicle stem cells and adult fibroblasts, show reduced H3K9me2/3, whereas others display moderate increases or little detectable change [72]. A similar heterogenous behavior is observed for H3K27me3. In some Q cell types, including fibroblasts, B lymphocytes, and hair follicle stem cells, H3K27me3 levels decrease together with reduced polycomb repressive complex 2 (PRC2) depositing this mark, whereas MuSCs maintain low promoter H3K27me3 in quiescence but gain this mark upon activation [72,88]. These divergent patterns likely reflect differential use of PRC2 paralogs, with EZH1 prevailing in Q cells and EZH2 associated with proliferative states [72,89]. For a detailed overview of polycomb repressive complex function in adult stem cell quiescence, the reader is referred to a dedicated review [90].

In contrast, H40K20me3 displays a more consistent association with quiescence. H4K20me3 is a repressive mark found in metazoans at pericentromeric heterochromatin and other repetitive elements (including transposable elements and subtelomeric regions) but is absent in *S. cerevisiae* and *S. pombe*. During quiescence, H4K20me3 levels robustly increase across constitutive heterochromatin and can extend into facultative domains, contributing to the establishment of a deeply repressed state [72,91]. Primary human fibroblasts


https://doi.org/10.3390/biom16020203

accumulate global levels of H4K20me2 and H4K20me3 during quiescence in association with enhanced chromatin compaction [92], and similar increases occur in serum-deprived mouse NIH 3T3 fibroblasts, where H4K20me3 is enriched at rDNA and transposable elements [93]. Elevated H4K20me3 is likewise observed in MuSCs, whereas loss of this mark induces proliferation and depletion of the MuSC pool [91]. Consistent with a broader role for H4K20me3 in deeply arrested cellular states, drug-tolerant persister cancer cells exhibit marked enrichment of H4K20me3 at promoters of stress-responsive genes, reinforcing transcriptional repression and supporting their low-activity, immune-evading phenotype [94].

**Dynamic remodeling of epigenetic landscapes during mammalian embryonic diapause.** Pluripotent cells in the early embryo, along with embryonic stem cells (ESCs), sustain highly adaptable transcriptional programs that maintain their undifferentiated state through a balance of global permissiveness and localized, polycomb-mediated repression. Entry into quiescence during mouse embryonic diapause triggers a rapid, genome-wide decrease in transcriptional output [6,95], mirroring the suppression of other energy-intensive anabolic processes such as translation. The overall decrease in cellular activity during this state is accompanied by a loss of transcription-associated histone marks, such as H4K16ac and H3K9ac and increased accumulation of heterochromatin at the nuclear lamina in pluripotent epiblast cells [6,96–98]. DNA methylation, which is almost completely erased in the proliferative epiblast, likewise globally increases in the diapaused epiblast [99]. Despite this repressive epigenetic landscape that significantly diverges from the canonical state of proliferative epiblast cells, the diapaused epiblast retains the transcriptional signatures and functionality of naïve pluripotency [6,49]. Although many mechanisms that ensure fidelity of cell identity remain to be discovered, several signaling pathways have been shown to protect the epiblast from premature differentiation or cell death during diapause, including Wnt, LIF/Stat, Nodal/Smad, and Yap/Taz pathways [95,100,101]. In addition, TET DNA demethylases, in cooperation with the transcription factor TFE3, have been shown to specifically target cis-regulatory elements and maintain them in a lowly methylated state despite an overall increase in global DNA methylation [99]. Taken together, multiple regulatory layers from cellular signaling to targeted epigenetic maintenance ensure transcriptional fidelity in the face of increased genomic repression in dormant states.

### 3.4. Persistence and Redistribution of Active Histone Methylation Marks

In contrast to the drastic reduction in histone acetylation in *S. cerevisiae*, methylation marks associated with active transcription (H3K4me3, H36me3, and H3K79me3) largely persist when cells enter quiescence [2,53,54]. Genome-wide profiles of these marks correlate well between Q and log cells, yet they differ at specific loci. In particular, all three methylation marks are strongly increased at stress-induced genes early in quiescence, coinciding with a sharp rise in transcription; in contrast, growth-related genes that are transcriptionally repressed exhibit a 50% decrease in H3K4me3 levels, without a comparable reduction of H3K36me3 or H3K79me3 [53]. As cells exit quiescence, histone methylation shifts from stress-related genes to growth genes, yet this redistribution exhibits slower kinetics than the rapid acetylation bursts and occurs only after DNA replication has resumed [54].

Notably, the histone methyltransferases responsible for H3K4me and H3K36me—Set1 and Set2, respectively—are both downregulated in Q cells, suggesting that these histone marks are deposited during logarithmic growth or early in quiescence [53]. Histone demethylases remain expressed in Q cells but may be less active, and decreased replication-dependent and transcription-coupled nucleosome turnover may further contribute to the persistence of these marks [53]. Thus, rather than serving as direct indicators of


https://doi.org/10.3390/biom16020203

transcriptional activity, these histone methylation marks may help maintain a chromatin environment that remains permissive to transcription [53,54].

Additional evidence from *S. pombe* further supports the view that H3K4me3 becomes selectively redistributed rather than globally erased during quiescence. Although H3K4me3 is broadly retained, its promoter-proximal peak is markedly diminished in Q cells, accompanied by extensive locus-specific remodeling: after 24 h of nitrogen starvation, hundreds of genes show multi-fold increases or decreases in H3K4me3, even though most promoters lose the sharp TSS-centered peak characteristic of log cells [62]. A subset of quiescence-induced genes retains or gains H3K4me3 together with elevated levels of elongating RNAPII, particularly genes encoding metabolic enzymes, membrane transporters, and noncoding RNAs. Notably, H3K4me3 becomes enriched at 'core quiescence' genes, with several of them residing in subtelomeric regions. Their transcriptional induction partially depends on Set1, consistent with genetic evidence that Set1C/COMPASS components are required for efficient quiescence entry [17,62,64]. A comparable pattern is observed in quiescent adult stem cells of higher eukaryotes. However, unlike embryonic stem cells, adult stem cells display limited bivalent H3K4me3/H3K27me3 marks but retain widespread H3K4me3 at promoters, thereby maintaining a transcriptionally poised state that enables rapid activation upon exit from quiescence [38,72]. Together, these findings indicate that, despite global reduction of promoter-centered H3K4me3, the targeted retention and redeployment of this mark contribute to establishing a transcriptionally competent yet selectively reprogrammed chromatin landscape in Q cells.

**H3K79me3: a molecular timer of quiescence duration?** H3K79 methylation is a multifaceted chromatin mark with roles in diverse cellular processes including meiosis and G2/M progression [102]. In *S. cerevisiae*, H3K79me3 remains detectable or even increases in Q cells, while levels of H3K79me1 and H3K79me2 are strongly reduced, suggesting their conversion to the fully methylated state [53]. Intriguingly, this conversion is not observed in NQ cells, which are unable to reenter the cell cycle after prolonged starvation when nutrients are restored, implying a regulatory mechanism involved in maintenance of stable quiescence that remains unidentified [2,53]. Notably, H3K79 methylation is absent in *S. pombe*, indicating this potential timing mechanism is not universally conserved.

### 3.5. Epigenetic Reprogramming of Metabolic Genes Through Altered Histone Dynamics and Nucleosome Remodeling

During quiescence establishment, numerous genes become transiently or persistently upregulated in *S. pombe* [17,55,63,64,82]. Many of these genes are located in subtelomeric regions and encode vacuolar transmembrane transporters essential for recycling amino acids and hexoses during the metabolic shift of Q cells. Multiple genetic studies converge on the conclusion that this transcriptional activation depends on regulated histone exchange and nucleosome remodeling, processes that remain active even when DNA replication—and thus replication-coupled histone turnover—is halted.

**Ino80C mediates the removal of the histone variant H2A.Z.** A major component of this regulatory network is the Ino80C chromatin-remodeling complex (Ino80C), which repositions nucleosomes and catalyzes the removal of the histone variant H2A.Z (Pht1 in *S. pombe*). Loss of Ino80C subunits, such as Iec1 (ino eighty complex subunit 1), compromises survival during quiescence and induces premature, global transcriptional downregulation [17,64]. Accordingly, genes normally upregulated at quiescence onset in WT cells fail to be activated in *iec1Δ* and other Ino80C mutants. In line with this, whereas H2A.Z is normally depleted from subtelomeric regions during quiescence entry, it remains aberrantly enriched in *iec1Δ*, particularly at boundary elements that contain multiple long terminal repeat (LTR) sequences [64]. These findings imply that Ino80-mediated eviction of H2A.Z promotes the

 https://doi.org/10.3390/biom16020203

formation of a transcriptionally competent subtelomeric environment, facilitating activation of metabolic genes required for the establishment of the quiescent state.

This dependency is further modulated by inositol polyphosphate signaling: the inositol kinase Asp1, which influences Ino80C activity, is likewise required for survival under prolonged quiescence [64,103], underscoring the integration of metabolic cues with chromatin remodeling at this transition. Conversely, the H2A.Z-specific histone chaperone Swr1C is dispensable for viability during quiescence, indicating that H2A.Z removal, rather than its deposition, is critical at this stage [17]. Nonetheless, loss of H2A.Z itself also causes reduced viability [64], suggesting a distinct, Ino80C-independent function for this variant.

**Paf1C and histone turnover contribute to epigenetic reprogramming.** The Paf1 complex (Paf1C) promotes histone exchange during vegetative growth and counteracts the accumulation of H3K9 methylation at constitutive and facultative heterochromatin [104]. Q cells lacking Leo1 or other Paf1C components exhibit substantial viability defects [82], highlighting the importance of ongoing, DNA replication-independent histone replacement. In WT cells, heterochromatic H3K9me2/3 levels transiently decline during quiescence entry, particularly at subtelomeric regions; this decrease is attenuated in *leo1Δ*, resulting in elevated H3K9me and reduced expression of transporter genes required for metabolic adaptation [82]. Conversely, other gene clusters, including those involved in DNA repair and chromatin organization, become aberrantly upregulated, reflecting broader disruptions to transcriptional control.

Genetic and pharmacological evidence further positions Leo1 downstream of the TORC2–Gad8$^{Akt}$ pathway: in fission yeast, TORC2 is required to maintain heterochromatin integrity at subtelomeric regions, where it stabilizes repressive chromatin states and limits inappropriate transcription. Through this activity, TORC2 helps preserve epigenetic stability by restraining transcriptional permissiveness [105]. Conversely, TORC2 inhibition by torin induces transcriptional changes that depend on Leo1, while Gad8$^{Akt}$ protein abundance transiently decreases during quiescence entry [106]. These relationships suggest that TORC2 signaling modulates Paf1C-dependent histone turnover to enable epigenetic reprogramming of metabolic pathways, particularly during the early phase of quiescence establishment [5,82].

Together, these findings illustrate that histone variant dynamics and replication-independent histone turnover are central determinants of gene regulation during quiescence. Ino80C-dependent H2A.Z eviction and Paf1C-mediated histone replacement act in concert to remodel subtelomeric chromatin, permit activation of key metabolic genes, and maintain viability during long-term quiescence.

## 4. Post-Transcriptional Mechanisms Shape Quiescence-Specific Gene Expression

### *4.1. Intron Retention Regulates Protein Biogenesis Genes*

Transcriptional adaptations during quiescence are accompanied by significant changes in co-transcriptional RNA processing, particularly splicing [107]. Although *S. cerevisiae* has largely lost splicing capacity during evolution, retaining introns in only 5% of its genes, introns are nevertheless strikingly preserved in ~73% of ribosomal protein genes (RPGs) During amino acid starvation, but not in response to other stress conditions, splicing of RPG mRNAs is selectively repressed, and strains lacking these introns exhibit reduced fitness during prolonged starvation [108–110]. This suggests that regulated intron retention within RPGs functions as a post-transcriptional mechanism to attenuate ribosome biogenesis, thereby limiting energy expenditure as cells enter quiescence. Notably, a subset of excised introns accumulates as a stable, linear form in saturated *S. cerevisiae* cultures or after TORC1 inhibition. These stable introns associate with spliceosomes and may themselves contribute

 https://doi.org/10.3390/biom16020203

to reduced splicing activity under starvation conditions [111]. A subset of stress-responsive genes escapes splicing inhibition and is even more efficiently spliced under starvation conditions, correlating with their induction. These transcripts selectively recruit auxiliary splicing factors, including the U1-associated proteins Nam8 and Mud1, to promote 5′ splice site recognition and intron removal [109].

In *S. pombe*, where splicing is more prevalent (~46% of genes contain introns), transcriptome profiling after 24 h of nitrogen starvation reveals a global increase in intron retention, with RPGs again being particularly affected [55] (Figure 5). This conservation across evolutionarily divergent yeast species indicates that splicing repression of RPGs represents a fundamental regulatory mechanism in quiescence establishment. Consistently, widespread intron retention is also a hallmark of mammalian quiescent adult stem cells, with enrichment among transcripts encoding proteins involved in translation and splicing [112,113]; diapaused embryos likewise exhibit reduced splicing activity [48]. Functionally, intron retention appears to be essential for maintaining quiescence: activation of MuSCs requires quiescence exit, which can be blocked by removing the chromatin-associated factor Dek that promotes intron removal [112].

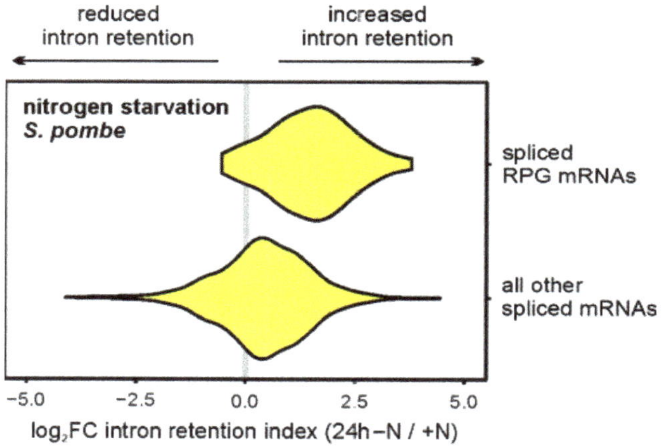


**Figure 5. Increased intron retention regulates ribosome synthesis genes during quiescence.** Comparison of intron retention rates in *S. pombe* during log phase (+N) and after 24 h of nitrogen starvation (−N) for mRNAs encoding ribosomal proteins and all other spliced mRNAs. Intron retention index based on published RNA-seq data (E-MTAB-1154; Ref. [55]) calculated as the ratio of intronic reads over exonic reads for each transcript.

Beyond population-level changes, RPG splicing can also vary among individual cells within populations, creating phenotypic heterogeneity with distinct consequences for adaptation to stress. In *S. cerevisiae*, the RPG Rps22B exhibits a bimodal splicing pattern, with different subpopulations displaying different survival outcomes during starvation: cells with intron retention (low levels of Rps22B) show enhanced survival during prolonged starvation, while cells with efficient intron removal (high levels of Rps22B) resume growth more quickly after transient starvation [114]. Stochastic cell-to-cell variation may represent a bet-hedging strategy that exploits phenotypic diversity to maximize fitness in unpredictable environments. Remarkably, regulated splicing of an intron in the 5′ untranslated region (5′-UTR) is conserved in the homologous *S. pombe* gene, *rps2202*, where it activates a post-transcriptional decay mechanism that is specifically inactivated during nitrogen starvation in homothallic yeast [115,116].

 https://doi.org/10.3390/biom16020203

In mammalian cells, mTORC1 signaling regulates alternative splicing patterns more broadly. Serine/arginine-rich (SR) proteins, which play a central role in splice site selection, are controlled by phosphorylation via the SR protein kinase SRPK2. mTORC1-S6K signaling promotes nuclear translocation of SRPK2, thereby linking mTORC1 activity to splicing regulation [117].

*4.2. 3′-UTR Lengthening Expands the Repertoire of Post-Transcriptional Regulation*

Quiescence is characterized by widespread shifts in polyadenylation site (PAS) selection. While proliferating cells preferentially use proximal PAS and thereby generate transcripts with short 3′-untranslated regions (3′-UTRs), non-proliferating cells (including both differentiated and Q cells) more commonly use distal PAS, resulting in global 3′-UTR lengthening [118,119]. This shift toward distal PAS usage appears to be a conserved feature of Q cells across eukaryotes, and has been documented in *S. pombe* and *S. cerevisiae* grown under nutrient-poor conditions [120], in *S. pombe* after 24 h and 7 d of nitrogen starvation [121], or upon rapamycin treatment [122].

**Molecular mechanisms driving 3′-UTR extension.** Kinetic coupling between transcription and 3′-end processing renders PAS selection highly sensitive to both the transcription elongation rate and the availability of the 3′-end processing machinery. Slow transcription and abundant 3′-end processing factors favour proximal PAS usage and 3′-UTR shortening, whereas faster elongation or reduced processing factor availability shifts selection towards distal sites [123,124]. Accordingly, chromatin changes that modulate elongation kinetics, together with the reduced expression of RNA cleavage and polyadenylation factors in Q cells, are expected to shift the equilibrium towards distal PAS usage and 3′-UTR lengthening [55,63,113]. Emerging evidence indicates that nutrient-sensing pathways help coordinate these shifts in PAS usage. In *S. cerevisiae*, rapamycin-induced alternative PAS selection requires the methyltransferases Set1 and Set2, linking TOR inhibition to transcription-dependent histone modifications (see above; [122]). In mammalian cells, hyperactivation of mTORC1 also drives 3′-UTR shortening, selectively enhancing translation of genes involved in ubiquitin-mediated proteolysis and facilitating targeted degradation of cell-cycle regulators without altering global mRNA abundance [125].

**Regulatory consequences of 3′-UTR extension.** Extended 3′-UTRs typically harbor additional binding sites for RNA-binding proteins (RBPs) and microRNAs, thereby increasing the potential for post-transcriptional control. In differentiated cells, this expanded regulatory landscape generally attenuates gene expression by enhancing opportunities for transcript destabilization or translational repression. Whether this paradigm also applies to Q cells, however, remains unclear. RNA stability measurements in a dermal fibroblast model of quiescence revealed stabilization, rather than destabilization, of transcripts with extended 3′-UTRs [113], suggesting that Q cells may exploit 3′-UTR lengthening to achieve regulatory outcomes distinct from those observed in differentiated tissues. This observation aligns with recent analyses in *S. cerevisiae* showing that the impact of 3′-UTR isoforms on mRNA half-life is highly condition-dependent, with transcripts becoming progressively more stable when cells divide more slowly [126].

**Suppression by microRNAs.** In metazoans, microRNAs (miRNAs) provide an important layer of post-transcriptional control acting at the level of 3′-UTRs. These small RNAs function exclusively as repressors, either suppressing translation or inducing RNA cleavage in a sequence-specific manner after incorporation into RNA-induced silencing complexes (RISC) [127]. Interestingly, hypoxia-inducible factor 1$\alpha$ (HIF-1 $\alpha$), a transcription factor involved in the generation and maintenance of quiescent cancer stem cells developing in hypoxic niches [128], appears to additionally dampen miRNA-dependent regulation. HIF-1$\alpha$ sequesters the miRNA processing component Dgcr8 and prevents its


https://doi.org/10.3390/biom16020203

incorporation into microprocessors, thereby limiting miRNA production in the cancer stem cell niche [129]. However, evidence from multiple other mammalian quiescence models suggests that miRNAs actively regulate quiescent states. In MuSCs, for example, the quiescence-specific miRNA-489 maintains quiescence by repressing Dek along with other targets [130]. Similarly, entry into embryonic diapause involves extensive reprogramming of miRNA expression [48,131,132]. The upregulation of diapause-associated miRNAs is partly controlled by the nutrient-responsive transcription factors Tfe3/TfeB, which act downstream of mTOR signaling. Upon mTOR inhibition, these factors partially translocate to the nucleus and promote miRNA expression [48]. While miRNAs are dispensable for normal blastocyst development, they are critical for initiating diapause, evidenced by the inability of embryos and embryonic stem cells lacking Dgcr8 to enter dormancy [48,133]. One key miRNA, let-7—originally identified as a regulator of developmental timing in *C. elegans*—is upregulated during diapause and plays a role in modulating embryo implantation [131,134]. Let-7 inhibits *Myc*, mTOR signaling, and polyamine biosynthesis, thereby promoting a diapause-like state [134]. Given that embryonic diapause is predominantly regulated by maternal cues, it has been proposed that endometrial extracellular vesicles—key mediators of communication between the uterus and the embryo—deliver let-7 miRNA to the blastocyst to help establish the dormant state [134].

### *4.3. Regulatory Non-Coding RNAs Facilitate Rewiring of Gene Expression*

A striking and conserved feature across diverse quiescence models is the widespread upregulation of non-coding RNAs (ncRNAs). Transcriptome-wide analyses in human somatic quiescence have identified numerous upregulated long non-coding RNAs (lncRNAs), and elevated ncRNA expression is similarly observed in both budding and fission yeast [55,56] (Figure 6A). This conservation across evolutionarily distant organisms suggests that ncRNA accumulation represents a fundamental feature of Q cells, rather than a species-specific adaptation.

**What drives ncRNA accumulation in quiescence?** Beyond transcriptional changes, reduced RNA decay rates are thought to contribute to the distinct lncRNA profiles observed in Q cells. In proliferating cells, nuclear RNA stability depends largely on susceptibility to degradation by the nuclear exosome, a conserved exonucleolytic complex central to RNA surveillance that is guided to its substrates by a modular system of targeting factors [135]. In Q cells, levels of many exosome substrates are substantially increased (Figure 6B). Three non-mutually exclusive mechanisms have been proposed to explain this phenomenon: (i) reduced exosome expression or activity (in fission yeast, exosome mRNA levels decrease to ~50% [63]); (ii) altered exosome targeting, either through inactivation of specific targeting pathways or activation of quiescence-specific mechanisms (e.g., HIF1$\alpha$-dependent retargeting of human microprocessor protein Dgcr8 to the exosome [129,136]); or (iii) substrate competition, where retargeting of the exosome to a broader substrate pool, including incompletely processed mRNAs with retained introns or extended 3′-ends (see above), creates kinetic competition that stabilizes ncRNAs [137,138]. By itself, exosome depletion in mouse embryonic stem cells induces nuclear structural changes that resemble senescent cells, including chromatin compaction [139].

Although the regulatory potential of quiescence-associated lncRNAs is clear, only few have been functionally characterized, sometimes in contexts unrelated to quiescence. LncRNAs, or the act of their transcription, can regulate cellular functions through various mechanisms, or a combination of them, including:

**Recruitment of epigenetic modifiers:** Some lncRNAs recruit chromatin-modifying enzymes to specific loci through direct interaction or via RBP intermediaries, establishing epigenetic signatures that regulate quiescence programs (Figure 7). This mechanism is

 https://doi.org/10.3390/biom16020203

conserved from yeast to mammals. In fission yeast, the lncRNA *prt* recruits H3K9 methyltransferase Clr4 to silence the adjacent *pho1* locus under phosphate-replete conditions but is transcriptionally downregulated when cells are starved for phosphate [141]. This recruitment involves a sequence-specific RBP and a scaffold protein mediating the interaction [80]. In metazoans, similar lncRNA-dependent targeting mechanisms regulate chromatin compaction during quiescence. While the histone H4K20 methyltransferase Suv4-20h2 is typically recruited to constitutive heterochromatin in an HP1-dependent manner, mouse Q cells employ an alternative HP1-independent mechanism. Quiescence-induced lncRNAs produced in cis at ribosomal RNA repeats and other repetitive regions recruit Suv4-20h2, potentially through direct interaction with the enzyme, promoting chromatin compaction at repeat regions [93] (Figure 7).

**ncRNA-mRNA interactions:** Other lncRNAs modulate activity of individual target mRNAs through base-pairing interactions, leading to a range of potential outcomes. The lncRNA NR2F1-AS1, for example, is significantly upregulated in human breast cancer stem-like cells, which are known to give rise to metastatic persister cells in the lungs. NR2F2-AS1 functionally contributes to establishing or maintaining the quiescent state—its knock-down leads to increased proliferation [142]. Mechanistically, NR2F1-AS1 binds an inhibitory region within the 5′-UTR of NR2F1 mRNA, facilitating recruitment of the polypyrimidine tract-binding protein PTBP1 and thereby promoting translation of the transcription factor NR2F1. Similarly, the fission yeast lncRNA *aal1* (ageing-associated lncRNA 1), has been suggested to suppress proliferation by interacting with mRNAs that encode ribosomal proteins, limiting their expression and decreasing the cellular ribosome content [143] (Figure 7).

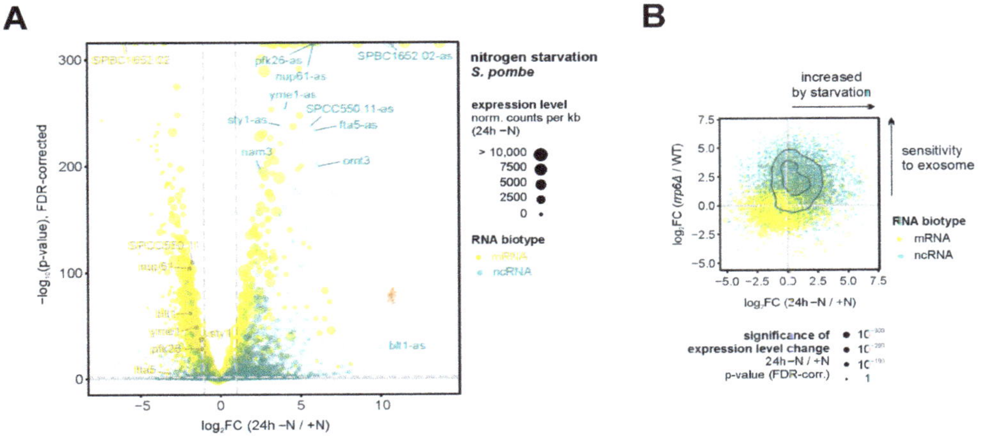


**Figure 6. Accumulation of non-coding RNAs is a conserved feature of quiescent cells.** (**A**). Differential gene expression analysis in *S. pombe* comparing RNA levels during log phase (+N) and after 24 h of nitrogen starvation (−N) for mRNAs (yellow) and ncRNAs (turquoise) (DESeq2). *p*-values adjusted for multiple testing with the Benjamini-Hochberg procedure to control the false discovery rate (FDR). Size of circles indicates absolute expression levels in nitrogen-starved cells (read counts per kilobase). RNA-seq data from E-MTAB-1154 (Ref. [55]; *n* = 2). (**B**). Comparison of RNA expression level changes in nitrogen-starved cells and cells that lack the exonuclease Rrp6, a catalytic subunit of nuclear exosome for mRNAs (yellow) and ncRNAs (turquoise). Size of circles indicates statistical significance of the expression level change (FDR-corrected *p*-values). Contours of the 2d density estimates for both RNA biotypes are shown in yellow and turquoise, respectively. RNA-seq data from E-MTAB-1154 (Ref. [55]; *n* = 2) and GSE148799 (Ref. [140]; *n* = 3).

 https://doi.org/10.3390/biom16020203

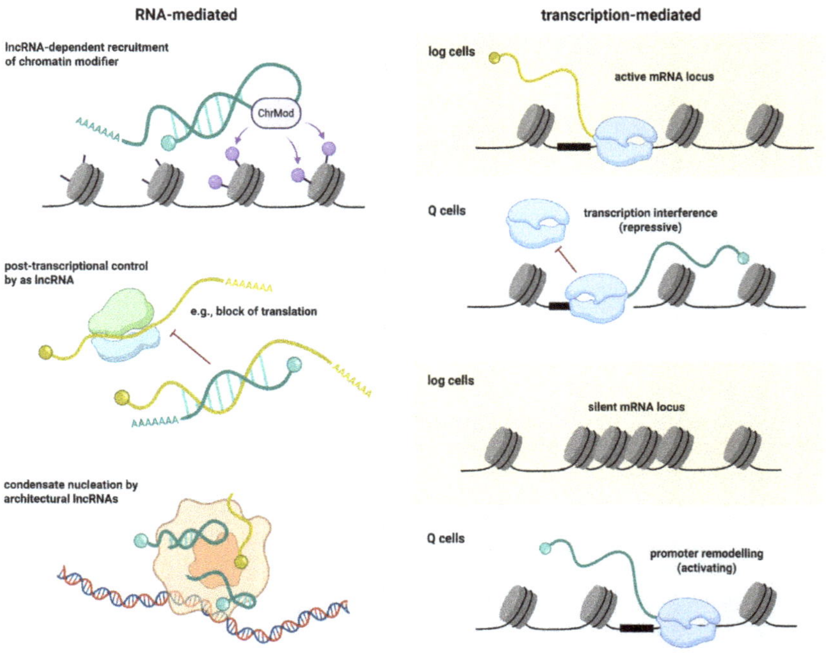


**Figure 7. Multiple mechanisms of lncRNA-dependent regulation operate during quiescence.** lncRNA expression can regulate gene expression through mechanisms mediated by the RNA molecule itself (**left**). These include recruiting epigenetic modifiers to chromatin to alter local histone marks and transcription (**top**); base-pairing with a complementary RNA to alter protein binding and its expression (**middle**); or nucleating biomolecular condensates that control the stability or activity of sequestered mRNAs (**bottom**). Alternatively, the act of transcription can be regulatory independent of the lncRNA product (**right**). This can involve transcriptional interference, where lncRNA synthesis competes with and represses an adjacent gene (**top**), or promoter remodeling, where the passage of RNA polymerase opens an occluded promoter to activate a nearby gene (**bottom**). RNA biotype indicated by colour—mRNAs (yellow) and ncRNAs (turquoise). Created in BioRender. Kilchert, C. (2026) https://BioRender.com/i8yei16 (accessed on 17 January 2026).

**Architectural lncRNAs and condensate nucleation.** Many lncRNAs are large, structurally flexible molecules capable of interacting with multiple proteins through various binding sites. Such multivalent interactions can result in the formation of biomolecular condensates with high densities of RBPs that spatially or temporally regulate RNA-dependent processes [144] (Figure 7). The protein components of nuclear condensates often contain intrinsically disordered low-complexity regions that contribute to the network of multivalent interactions. A notable example of such condensates are paraspeckles—mammalian-specific nuclear structures that require a specific RNA scaffold, NEAT1 lncRNA, for their formation—which sequester specific RBPs and RNAs [145]. In *S. pombe*, certain quiescence-associated lncRNAs, such as *omt3* [55] (Figure 6A), form condensates at the non-coding gene locus. During meiosis (when these lncRNAs are also induced), such condensates facilitate the pairing of homologous chromosomes through a tethering mechanism [146]. The proteins involved in these condensates are components of the RNA 3′-end processing machinery that, in log cells, localize to cleavage bodies—structures thought to buffer nucleoplasmic concentrations of 3′-end formation factors and not known to be associated with lncRNAs [147]. Their behavior in Q cells remains uncharacterized.


https://doi.org/10.3390/biom16020203

**Regulation through lncRNA transcription.** The act of lncRNA transcription—rather than the lncRNA product itself—can strongly influence the expression of adjacent genes through distinct and context-dependent mechanisms. In transcriptional interference, passage of RNAPII through a promoter region can displace transcription factors necessary for initiation, effectively suppressing gene expression [148] (Figure 7). This repressive mechanism operates independently of gene orientation and can occur when the lncRNA gene and the regulated gene are oriented in tandem or when they are in an antisense configuration. Silencing of the *S. pombe pho1* locus by *prt* lncRNA also depends on this mechanism [141]. Transcriptional interference appears to be widely deployed during quiescence: in *S. cerevisiae*, even short periods of glucose or phosphate starvation induce distinct sets of antisense lncRNAs (asRNAs) [149,150]. In *S. pombe*, asRNAs are among the most strongly upregulated ncRNAs in Q cells and often accumulate to high levels [55,151]. This elevated expression typically coincides with downregulation of the corresponding sense gene, illustrating the regulatory impact of transcriptional interference (Figure 6A). Widespread changes in antisense transcription have also been observed in human models of senescence, with a subset of asRNAs showing inverse expression patterns relative to their corresponding sense transcript [152]. Moreover, recent work using synthetic antisense loci in mouse cells demonstrates that antisense transcription can establish stable promoter repression that persists for days, suggesting that transcriptional interference may create a form of expression memory [153]. Conversely, lncRNA transcription can facilitate gene activation through promoter remodeling. When RNAPII traverses compacted chromatin, transcription itself disrupts nucleosome organization, rendering previously inaccessible promoters competent for transcription factor binding. At the *S. pombe fbp1* locus, for example, glucose starvation induces cascading lncRNA transcription, which precedes and enables activation of *fbp1* mRNA transcription [154]. Such pioneering lncRNA transcription, which primes chromatin for productive mRNA synthesis, has been observed at other genes responding to nutritional signals, and may represent a widespread mechanism for transcriptional reprogramming during quiescence entry and exit [155].

## 5. Reorganization of Nuclear Structures and Genome Architecture

### 5.1. Biosynthetic Condensates

**Nucleolar condensation as a conserved marker of quiescence.** The biogenesis of core components of the gene expression machinery is spatially organized in specialized nuclear condensates whose morphology reflects cellular biosynthetic activity. Nucleoli serve as hubs for ribosome maturation, while spliceosome and small nucleolar ribonucleoprotein (snoRNP) production occurs in distinct compartments (see below). During quiescence, nucleoli undergo dramatic reorganization, condensing into compact structures as ribosomal RNA (rRNA) synthesis declines with growth arrest. This condensation is a highly conserved hallmark of Q cells, observed across phylogenetically distant organisms and triggered by different cues: in *S. cerevisiae* following DS [156]; in *S. pombe* cells during nitrogen starvation [33,157]; in *C. elegans* under insulin pathway inhibition [158]; and in mouse embryos during diapause [98]. This reorganization is reversible: upon exit from diapause, nucleoli decondense within 12 h, coinciding with restored biosynthetic activity in the reactivating embryos [96,98]. To date, the molecular composition of condensed nucleoli remains largely unexplored. Ultrastructural analysis of quiescent *S. pombe* nuclei (nitrogen-starved for 24 h) revealed an accumulation of multi-mega Dalton molecular complexes in the condensed nucleolus [157], but their nature and functional role remain uncharacterized.

**Cajal bodies coordinate small RNA biogenesis with growth state.** Cajal bodies are nuclear structures essential for the maturation of small nuclear RNAs (snRNAs) and small nucleolar RNAs (snoRNAs) [145]. These RNAs are highly abundant components of the


https://doi.org/10.3390/biom16020203

gene expression machinery: snRNAs form the catalytic core of the spliceosome, while snoRNAs guide the chemical modification of rRNA required for ribosome biogenesis. Because their production contributes significantly to the biosynthetic load, snRNA and snoRNA biogenesis is tightly coordinated with nutrient availability. In fission yeast, the Cajal body scaffold protein Mug174 (the orthologue of metazoan coilin) is required for viability of Q cells and their subsequent return to proliferation, indicating that long-term survival depends on Cajal body-mediated coordination of small ribonucleoprotein biogenesis with the cellular growth state [87]. The molecular mechanisms underlying this nutrient-responsive regulation are better characterized in mammalian systems, where TOR signaling plays a central role: the survival of motor neurons (SMN) protein, responsible for assembling sn(o)RNPs from sn(o)RNAs and Sm proteins, is a direct mTOR target. Cajal body localization of SMN—which reflects its activity—is regulated by mTOR-dependent phosphorylation, providing a direct link between nutrient sensing, sn(o)RNP biogenesis, and the cellular capacity of RNA splicing [159]. Intriguingly, not all organisms employ Cajal body-based regulation. *S. cerevisiae* lacks orthologues of coilin or SMN, and the site and mechanism of sn(o)RNP assembly in budding yeast remain unresolved. Assembly may occur diffusely within the nucleus, without discrete compartmentalization, raising the question of which alternative mechanisms coordinate small RNA biogenesis with nutrient availability in this organism.

### *5.2. Changes in Local and Higher-Order Chromatin Structures*

Quiescence induces profound reorganization of genome architecture at both the local chromatin level and the genome-wide scale, resulting in dramatic genome compaction and nuclear volume decrease across species [2]. In *S. cerevisiae*, these changes arise through two complementary mechanisms involving histone tail–dependent chromatin fiber compaction and condensin-mediated chromosome looping. Together, these processes generate a deeply compact, insulated, and transcriptionally restrained nuclear architecture tailored to the low-energy state of Q cells.

**Local chromatin remodeling through chromatin fiber compaction.** In budding yeast Q cells, neighboring nucleosomes engage in interactions dependent on the basic patch of the H4 histone tail, a process facilitated by global hypoacetylation of H4 [160]. This mode of fiber compaction is thought to broadly dampen transcription, largely independent of higher-order domain architecture. While mechanistically distinct, similar principles apply in human quiescent fibroblasts, where chromatin compaction is associated with increased H4K20me2/3 levels [92]. In *S. cerevisiae*, an additional contributor might be the histone H1 homolog Hho1. Hho1 chromatin occupancy is inversely correlated with gene expression, and it binds more strongly to chromatin in stationary-phase cells, with its loss leading to chromatin decompaction [161]. Notably, this decompaction does not result in widespread transcriptional derepression, indicating that chromatin compaction alone is not sufficient to enforce gene silencing [2,161].

**Global chromosome reorganization by condensin.** At a global scale, the quiescent genome is reorganized into a highly ordered structure defined by long-range interactions and condensin-mediated looping. Hi-C analyses of *S. cerevisiae* Q cells reveal a shift toward a more compact nuclear organization characterized by increased intrachromosomal contacts, reduced centromere clustering, and enhanced telomere–telomere interactions [156,162–164]. At the mesoscale, chromatin is organized into chromosomal interaction domains (CIDs), small genomic units encompassing 1–4 genes (0.5–8 kb) that engage in frequent local interactions and resemble metazoan topologically associating domains [165]. Superimposed on this architecture are larger chromatin interaction domains (L-CIDs) spanning 10–60 kb, which are present in both log and Q cell and whose boundaries remain

 https://doi.org/10.3390/biom16020203

largely unchanged. However, during quiescence, L-CIDs become more compact and topologically isolated, accompanied by enhanced loop formation [164]. Together, these global rearrangements indicate a transition to a condensed yet orderly chromatin state that is distinct from both log and stationary-phase cells.

A critical determinant in this process is the condensin complex, a ring-shaped protein complex that organizes higher-order chromosome structure and mediates sister chromatid cohesion [162,164]. During quiescence entry, condensin redistributes from its canonical locations at rDNA and centromeres to the 5′ ends of hundreds of stress-induced genes, aligning with L-CID boundaries. This relocalization is partially dependent on the transcription factor Msn2, which is involved in the activation of many stress-induced genes during the DS (see above). How Msn2 contributes to condensin binding is not well understood, but a role of transcription itself at Msn2-responsive genes has been discussed [59]. Loss of condensin in G0 leads to decompaction of L-CIDs, a reduction in loop formation, increased inter-domain contacts, and transcriptional derepression of nearby genes [164], underscoring its role as an architectural organizer that enforces domain insulation during quiescence. How L-CID boundaries are established, which occurs cohesin-independently in log cells, remains unresolved.

### 5.3. Telomere Reorganization

**Telomere hypercluster formation in *S. cerevisiae*.** In budding yeast, telomere organization undergoes extensive remodeling during metabolic transitions. Following the diauxic shift, telomere foci become fewer and brighter while remaining associated with the nuclear periphery. Upon complete carbon source exhaustion and entry into stationary phase, telomeres undergo a more pronounced reorganization in long-lived Q cells, coalescing into a large hypercluster positioned at the nuclear center [156]. This configuration is reversible: telomere hyperclusters are rapidly dismantled within 15 to 30 min upon return to growth, in a transcription-independent manner [57]. Importantly, telomere hypercluster formation correlates with survival upon quiescence exit and is specific to Q cells, as telomere organization remains unchanged in NQ cells [156].

The extent of telomere hyperclustering varies markedly between strain backgrounds. W303 strains exhibit robust hyperclustering, with telomere foci showing a moderate increase in intensity and remaining at the nuclear periphery in most cells of a stationary-phase culture [163,166]. Although early work proposed that elevated Sir3 levels might account for this difference [163], this model has been excluded: moderate Sir3 overexpression (2–3.5-fold) in BY fails to rescue the telomere hyperclustering defect, and Sir3 protein levels remain stable across metabolic transitions in both backgrounds [156,166]. Instead, reduced hyperclustering in BY strains likely reflects respiration defects [167], consistent with respiration being a key requirement for telomere hypercluster formation [156]. Notably, Sir3 post-translational modifications, including SUMOylation and phosphorylation, are dispensable for telomere hyperclustering in Q cells [166].

Recent physical modeling and genetic analyses identified telomere anchoring as a key negative regulator of telomere hyperclustering [166]. The inner nuclear membrane protein Esc1 acts as a major telomere anchor upon glucose exhaustion; in *esc1Δ* cells, telomere hyperclusters form prematurely, indicating that Esc1-mediated anchoring counteracts telomere clustering. Under glucose-replete conditions, however, hyperclusters do not form in *esc1Δ* cells, implying the existence of an Esc1-independent anchoring mechanism. This secondary anchoring pathway is regulated by glucose signaling, as glucose starvation or inactivation of the Ras/PKA pathway triggers hyperclustering in *esc1Δ* cells [166]. In wild-type Q cells, exhaustion of all carbon sources coincides with hypercluster formation,

 https://doi.org/10.3390/biom16020203

consistent with inactivation of Esc1 anchoring, whereas in NQ cells Esc1 remains active and prevents hypercluster formation [166].

Mechanistically, the Esc1-dependent anchoring pathway relies on phosphorylation of Esc1 at serine 1450, which enables interaction with the HBRCT domain of Sir4 [168,169]. Mutation of this residue phenocopies Esc1 loss, suggesting that reversible Esc1 phosphorylation controls telomere anchoring during metabolic transitions, although the upstream signaling cascade remains unknown [166].

Together, these findings indicate that budding yeast has evolved a fine-tuned mechanism to regulate telomere positioning in response to nutrient availability. While telomere hyperclustering promotes cell survival following exit from quiescence independently of transcriptional silencing, its precise protective function remains unknown [156]. Hyperclustering may enhance genome stability by protecting chromosome ends from degradation, end-to-end fusions, and aberrant recombination events. Alternatively, telomere cluster dynamics may influence the transcriptional regulation of subtelomeric genes, which are enriched in stress-response and metabolic genes, thereby coordinating cellular physiology with environmental conditions [170].

**Telomere hypercluster formation in *S. pombe*.** As in budding yeast, entry into quiescence in *S. pombe* is accompanied by a pronounced spatial reorganization of telomeres, most notably their collapse into a single hypercluster. In *S. pombe* log cells, telomeres form two to three nuclear envelope (NE)-associated clusters during interphase, an organization that persists through DNA replication [171–173]. The shelterin component Rap1 directly contributes to telomere tethering at the NE through interaction with the NE-associated Bqt3-Bqt4 complex [172,174]. NE localization is reinforced by subtelomeric anchoring mediated by the inner nuclear membrane LEM-domain proteins Lem2 and Man1, which associate with distinct subtelomeric regions [85,175,176]. Simultaneous loss of Bqt4 and the chromatin remodeler Fft3, which interacts with Man1, results in relocalization of telomeres toward the nuclear interior [175].

Upon quiescence entry, telomeres collapse into a single cluster. In contrast to *S. cerevisiae*, where telomere hyperclusters form in the center of the nucleus, fission yeast telomeres remain tightly anchored to the NE during nitrogen starvation, in a manner dependent on Rap1 and Bqt4. Nuclear polarity is maintained, with telomeres localizing opposite the spindle-pole body (SPB), where centromeres are anchored, thereby preserving the Rabl configuration [177]. Importantly, loss of Bqt4 does not alter the number of telomeric foci, indicating that NE tethering and telomere hyperclustering represent mechanistically distinct steps [177]. Disruption of NE attachment, either by loss of Bqt4 or telomere erosion, leads to increased telomeric transcription and predisposes cells to subtelomeric rearrangements via a quiescence-specific repair mechanism termed STEEx (STE1-expansion) [177,178]. Cells harboring rearranged telomeres fail to efficiently exit quiescence [178]. Telomere clustering has also been reported for quiescent mouse and human lymphocytes [179], supporting the notion that telomere repositioning during quiescence may represent a conserved strategy to preserve genome stability in non-proliferating cells [180].

### 5.4. Dynamic Relocalization of Stress-Induced Gene Clusters

In *S. pombe*, stress-response genes are frequently organized into physical clusters along chromosomes, consistent with coordinated regulation shaped by chromatin architecture and nuclear organization [181]. Several clusters that are rapidly induced upon nitrogen starvation are located near subtelomeres and associate with the nuclear periphery in proliferating cells, but relocalize to the nuclear interior upon starvation, coincident with their activation [182]. Because the nuclear periphery is generally considered a repressive environment enriched in silencing factors [183,184], these observations raised the question

 https://doi.org/10.3390/biom16020203

of whether gene repositioning is a cause or a consequence of transcriptional activation. Notably, similar behavior is observed for the *urg1–3* gene cluster, which is located in a euchromatic region rather than at subtelomeres. Despite this genomic context, the cluster associates with the nuclear periphery during growth and relocates to the nuclear interior upon nitrogen starvation, providing a useful model to study regulated gene repositioning independently of constitutive subtelomeric silencing.

Early experiments using the broad transcriptional inhibitor 1,10-phenanthroline suggested that cluster relocalization depends on transcription [182] However, given the pleiotropic effects of this compound, transcription-independent effects could not be excluded. More targeted perturbations support an alternative model, in which relocalization precedes transcription. The transcription factor Toe1 is required for the induction of genes within the *urg* cluster, yet deletion of *toe1* abolishes transcriptional activation without preventing relocalization of the cluster to the interior in starved cells, arguing that repositioning occurs upstream, or independent, from transcription (S.B. and Vishnu N. Suma Sreechakram, unpublished data).

Consistent with its euchromatic context, perinuclear tethering of the *urg* cluster does not require the inner nuclear membrane protein Lem2, a key factor in heterochromatin silencing and subtelomeric anchoring ([85] S.B. and Vishnu N. Suma Sreechakram, unpublished data). Instead, a putative NE-associated repressor complex, Ntu1-Ntu2, binds a promoter within this cluster, and loss of this complex phenocopies starvation-induced relocalization, implicating Ntu1-Ntu2 in anchoring the cluster at the nuclear periphery [185]. How nitrogen starvation signals trigger release of this tether, and whether additional NE proteins contribute to positioning of other stress-induced clusters, remain open questions.

### 5.5. Impact of Nuclear Reorganization on Genome Functions

As described above, nuclear reorganization is associated with genome stability, gene regulation, and the long-term viability of Q cells. However, establishing causal relationships between nuclear structures and functions remains challenging. Some structural changes arise as a consequence of functional transitions, while simultaneously conferring adaptive advantages. For example, chromatin and nucleolus compaction result, at least in part, from genome-wide deacetylation and transcriptional shutdown; however, these changes may also contribute to long-term genome stability. Likewise, nuclear microtubules become extensively reorganized upon carbon exhaustion in budding yeast, forming a single array of approximately 30 microtubules that emanate from the spindle pole body and traverse the nucleus, resulting in the displacement of the nucleolus [186]. This microtubule bundle is surprisingly stable and disappears upon quiescence exit. Mutants perturbing microtubule organization display defects in long-term survival and upon quiescence exit. While a role as a structural gatekeeper controlling quiescent exit has been proposed, the precise physiological function of this rearrangement remains unresolved [187]. Microtubule remodeling during quiescence has also been reported for *S. pombe*, plants and mammalian cells; however, these alterations predominantly involve cytoplasmic microtubules and the primary cilium [187,188]. Although genetic perturbations affecting nuclear reorganization have clear functional consequences, indirect effects are difficult to exclude and defects in quiescence entry, maintenance, or exit are often hard to disentangle. Inducible protein degradation strategies applied specifically in Q cells will therefore be essential to minimize pleiotropic effects and dissect the molecular requirements of these distinct phases.

## 6. Quiescence: Quo Vadis?

Work across yeast and metazoan models has firmly established quiescence as an actively regulated program rather than a passive cessation of proliferation. Nevertheless,


https://doi.org/10.3390/biom16020203

major challenges remain in resolving its underlying mechanisms. A central open question concerns the physiological function of the diverse nuclear rearrangements observed in Q cells: Do these changes act as causal drivers of quiescence or instead arise as downstream consequences of altered transcription, metabolism, and chromatin state? This question has not yet been resolved for the repositioning of stress-responsive gene clusters, telomere hyperclustering, global chromosome compaction, remodeling of nuclear bodies, and extensive reprogramming of RNA metabolism. For many of these phenomena, their functional contribution to quiescence entry, transcriptional adaptation, and long-term survival remain poorly defined.

A second major frontier is cellular heterogeneity, which complicates both mechanistic interpretation and conceptual definitions of quiescence. Quiescent populations typically comprise mixtures of Q and NQ cells, and even highly enriched Q populations likely span a continuum of states. This raises foundational questions including whether cells commit to quiescence long before proliferation ceases, whether previous experience of quiescence alters the speed of re-entry (is there a memory of quiescence?), when quiescence is firmly established, and whether there are differences in the "depth" of the quiescent state (see article in this special issue by Marek et al., Ref. [20]). Addressing these issues necessitates a shift from viewing quiescence as a single fixed endpoint toward a framework of layered or graded quiescence states, characterized by time-dependent changes in signaling output, chromatin dynamics, and gene expression, potentially relying on molecular timers that encode a physical memory of quiescence duration (see review in this special issue by Laporte and Sagot, Ref. [7]). Progress in this area will rely on single-cell and longitudinal approaches capable of resolving transition trajectories during entry, maintenance, and exit.

A third emerging theme centers on RNA-mediated nuclear regulation, including non-coding RNAs and post-transcriptional control. Q cells typically globally reduce transcriptional output while selectively accumulating subsets of long non-coding RNAs, raising the possibility that components of the "dark transcriptome" exert regulatory functions that become apparent only in non-proliferative states. Beyond intron retention, future studies should systematically examine isoform usage, RNA stability control, nuclear-cytoplasmatic partitioning, and RNA-protein condensate dynamics as mechanisms that preserve readiness for renewed proliferation while suppressing energetically costly biogenesis.

Finally, quiescence-specific nuclear dynamics connect to development and disease. Quiescence underpins adult stem cell maintenance and immune homeostasis, while quiescence-like programs contribute to both therapy resistance of cancer persister cells and to drug tolerance of chronic or recurrent fungal infections. Quiescence research offers potential new therapies for these important medical challenges. A comparative perspective remains essential: how conserved are nuclear quiescence programs across species and triggers, and which features represent cell type-specific adaptations rather than universal molecular pathways. Can this heterogeneity be selectively targeted for therapeutic purposes? Addressing these questions will require systematic cross-system analyses to identify condition- and disease-specific nuclear implementations of quiescence.

## 7. Conclusions

Quiescence is accompanied by extensive nuclear reorganization, encompassing the chromatin architecture, chromosome positioning, nuclear body dynamics, and RNA metabolism. While these changes are now well documented across eukaryotes, we are only beginning to understand their mechanistic basis and functional consequences. TOR signaling emerges as a central coordinator linking nutrient availability to nuclear architecture, though many mechanistic details remain unresolved. Understanding these mechanisms has direct clinical relevance: therapy-resistant persister cells in both cancers and fungal


https://doi.org/10.3390/biom16020203

infections exploit quiescence programs to evade treatment, and their distinctive nuclear features may represent targetable therapeutic vulnerabilities. Moving forward, resolving the broader interplay between nuclear structure and function during quiescence and understanding how nuclear dynamics contribute to quiescence entry, maintenance, and exit represents a major frontier for the field.

**Author Contributions:** Conceptualization, S.B. and C.K.; writing—original draft preparation, review and editing, S.B., C.K., A.B.-K., M.R., A.T., F.R. and D.W.-S.; visualization, S.B. and C.K. All authors have read and agreed to the published version of the manuscript.

**Funding:** This work was supported by the Agence Nationale pour la Recherche DeSynLE (ANR-22-CE120013-01) to A.T., the European Research Council (ERC; ERC-Starting Grant agreement no. 101117421, 'DOR CODE') and Max Planck Society to A.B.-K., the German Research Foundation (DFG) through the Heisenberg Program (project ID 464293512) to S.B., and the National Science Center, Poland, (OPUS grant no. 2024/53/B/NZ8/02259) to D.W.S.

**Institutional Review Board Statement:** Not applicable.

**Informed Consent Statement:** Not applicable.

**Data Availability Statement:** No new data were created or analyzed in this study.

**Acknowledgments:** We thank members of the 'NuQuest' consortium for fruitful discussions and valuable feedback. Schemes in figures were created with BioRender.com.

**Conflicts of Interest:** The authors declare no conflicts of interest.

## References

1. O'Farrell, P.H. Quiescence: Early Evolutionary Origins and Universality Do Not Imply Uniformity. *Philos. Trans. R. Soc. Lond. B Biol. Sci.* **2011**, *366*, 3498–3507. [CrossRef]
2. Breeden, L.L.; Tsukiyama, T. Quiescence in Saccharomyces Cerevisiae. *Annu. Rev. Genet.* **2022**, *56*, 253–278. [CrossRef]
3. Virgilio, C.D. The Essence of Yeast Quiescence. *FEMS Microbiol. Rev.* **2012**, *36*, 306–339. [CrossRef]
4. Gray, J.V.; Petsko, G.; Johnston, G.C.; Ringe, D.; Singer, R.A.; Werner-Washburne, M. "Sleeping Beauty": Quiescence in Saccharomyces Cerevisiae. *Microbiol. Mol. Biol. Rev.* **2004**, *68*, 187–206. [CrossRef]
5. Laribee, R.N.; Weisman, R. Nuclear Functions of TOR: Impact on Transcription and the Epigenome. *Genes* **2020**, *11*, 641. [CrossRef]
6. Bulut-Karslioglu, A.; Biechele, S.; Jin, H.; Macrae, T.A.; Hejna, M.; Gertsenstein, M.; Song, J.S.; Ramalho-Santos, M. Inhibition of mTOR Induces a Paused Pluripotent State. *Nature* **2016**, *540*, 119–123. [CrossRef] [PubMed]
7. Laporte, D.; Sagot, I. Quiescence Multiverse. *Biomolecules* **2025**, *15*, 960. [CrossRef] [PubMed]
8. Sagot, I.; Laporte, D. The Cell Biology of Quiescent Yeast—A Diversity of Individual Scenarios. *J. Cell Sci.* **2019**, *132*, jcs213025. [CrossRef]
9. Jacquel, B.; Aspert, T.; Laporte, D.; Sagot, I.; Charvin, G. Monitoring Single-Cell Dynamics of Entry into Quiescence during an Unperturbed Life Cycle. *Elife* **2021**, *10*, e73186. [CrossRef]
10. Medina-Menéndez, C.; Tirado-Melendro, P.; Li, L.; Rodríguez-Martín, P.; Melgarejo-de la Peña, E.; Díaz-García, M.; Valdés-Bescós, M.; López-Sansegundo, R.; Morales, A.V. Sox5 Controls the Establishment of Quiescence in Neural Stem Cells during Postnatal Development. *PLoS Biol.* **2025**, *23*, e3002654. [CrossRef]
11. Liu, B.; Wang, X.; Jiang, L.; Xu, J.; Zohar, Y.; Yao, G. Extracellular Fluid Flow Induces Shallow Quiescence through Physical and Biochemical Cues. *Front. Cell Dev. Biol.* **2022**, *10*, 792719. [CrossRef]
12. Kwon, J.S.; Everetts, N.J.; Wang, X.; Wang, W.; Della Croce, K.; Xing, J.; Yao, G. Controlling Depth of Cellular Quiescence by an Rb-E2F Network Switch. *Cell Rep.* **2017**, *20*, 3223–3235. [CrossRef] [PubMed]
13. Yang, N.; Occean, J.R.; Melters, D.P.; Shi, C.; Wang, L.; Stransky, S.; Doyle, M.E.; Cui, C.-Y.; Delannoy, M.; Fan, J.; et al. A Hyper-Quiescent Chromatin State Formed during Aging Is Reversed by Regeneration. *Mol. Cell* **2023**, *83*, 1659–1676.e11. [CrossRef]
14. Sajiki, K.; Hatanaka, M.; Nakamura, T.; Takeda, K.; Shimanuki, M.; Yoshida, T.; Hanyu, Y.; Hayashi, T.; Nakaseko, Y.; Yanagida, M. Genetic Control of Cellular Quiescence in S. Pombe. *J. Cell Sci.* **2009**, *122*, 1418–1429. [CrossRef]
15. Sideri, T.; Rallis, C.; Bitton, D.A.; Lages, B.M.; Suo, F.; Rodríguez-López, M.; Du, L.-L.; Bähler, J. Parallel Profiling of Fission Yeast Deletion Mutants for Proliferation and for Lifespan during Long-Term Quiescence. *G3 Genes Genomes Genet.* **2014**, *5*, 145–155. [CrossRef]

 https://doi.org/10.3390/biom16020203

16. Yanagida, M. Cellular Quiescence: Are Controlling Genes Conserved? *Trends Cell Biol.* **2009**, *19*, 705–715. [CrossRef] [PubMed]
17. Zahedi, Y.; Durand-Dubief, M.; Ekwall, K. High-Throughput Flow Cytometry Combined with Genetic Analysis Brings New Insights into the Understanding of Chromatin Regulation of Cellular Quiescence. *Int. J. Mol. Sci.* **2020**, *21*, 9022. [CrossRef]
18. Rutherford, K.M.; Lera-Ramírez, M.; Wood, V. PomBase: A Global Core Biodata Resource-Growth, Collaboration, and Sustainability. *Genetics* **2024**, *227*, iyae007. [CrossRef]
19. Sun, S.; Gresham, D. Cellular Quiescence in Budding Yeast. *Yeast* **2021**, *38*, 12–29. [CrossRef]
20. Marek, A.; Tomala, K.; Wloch-Salamon, D. Ecological History Shapes Transcriptome Variation in Quiescent Saccharomyces Cerevisiae. *Biomolecules* **2025**, *15*, 1588. [CrossRef]
21. de Morree, A.; Rando, T.A. Regulation of Adult Stem Cell Quiescence and Its Functions in the Maintenance of Tissue Integrity. *Nat. Rev. Mol. Cell Biol.* **2023**, *24*, 334–354. [CrossRef]
22. Gangloff, S.; Arcangioli, B. DNA Repair and Mutations during Quiescence in Yeast. *FEMS Yeast Res.* **2017**, *17*, fox002. [CrossRef]
23. Johnston, G.C.; Pringle, J.R.; Hartwell, L.H. Coordination of Growth with Cell Division in the Yeast Saccharomyces Cerevisiae. *Exp. Cell Res.* **1977**, *105*, 79–98. [CrossRef] [PubMed]
24. Lillie, S.H.; Pringle, J.R. Reserve Carbohydrate Metabolism in Saccharomyces Cerevisiae: Responses to Nutrient Limitation. *J. Bacteriol.* **1980**, *143*, 1384–1394. [CrossRef]
25. Werner-Washburne, M.; Braun, E.; Johnston, G.C.; Singer, R.A. Stationary Phase in the Yeast Saccharomyces Cerevisiae. *Microbiol. Rev.* **1993**, *57*, 383–401. [CrossRef]
26. Allen, C.P.; Büttner, S.; Aragon, A.; Thomas, J.A.; Meirelles, O.; Jaetao, J.E.; Benn, D.; Ruby, S.; Veenhuis, M.; Madeo, F.; et al. Isolation of Quiescent and Nonquiescent Cells from Yeast Stationary-Phase Cultures. *J. Cell Biol.* **2006**, *174*, 89–100. [CrossRef]
27. Yang, H.; Ren, Q.; Zhang, Z. Chromosome or Chromatin Condensation Leads to Meiosis or Apoptosis in Stationary Yeast (Saccharomyces Cerevisiae) Cells. *FEMS Yeast Res.* **2006**, *6*, 1254–1263. [CrossRef]
28. Opalek, M.; Tutaj, H.; Pirog, A.; Smug, B.J.; Rutkowska, J.; Wloch-Salamon, D. A Systematic Review on Quiescent State Research Approaches in S. Cerevisiae. *Cells* **2023**, *12*, 1608. [CrossRef]
29. Li, L.; Miles, S.; Melville, Z.; Prasad, A.; Bradley, G.T.; Breeden, L. Key Events during the Transition from Rapid Growth to Quiescence in Budding Yeast Require Posttranscriptional Regulators. *Mol. Biol. Cell* **2013**, *24*, 3697–3709. [CrossRef] [PubMed]
30. Davidson, G.S.; Joe, R.M.; Roy, S.; Meirelles, O.; Allen, C.P.; Wilson, M.R.; Tapia, P.H.; Manzanilla, E.E.; Dodson, A.E.; Chakraborty, S.; et al. The Proteomics of Quiescent and Nonquiescent Cell Differentiation in Yeast Stationary-Phase Cultures. *Mol. Biol. Cell* **2011**, *22*, 988–998. [CrossRef] [PubMed]
31. Miles, S.; Li, L.; Davison, J.; Breeden, L.L. Xbp1 Directs Global Repression of Budding Yeast Transcription during the Transition to Quiescence and Is Important for the Longevity and Reversibility of the Quiescent State. *PLoS Genet.* **2013**, *9*, e1003854. [CrossRef]
32. Miles, S.; Lee, C.; Breeden, L. BY4741 Cannot Enter Quiescence from Rich Medium. *MicroPubl. Biol.* **2023**, *2023*, 000742. [CrossRef]
33. Su, S.S.; Tanaka, Y.; Samejima, I.; Tanaka, K.; Yanagida, M. A Nitrogen Starvation-Induced Dormant G0 State in Fission Yeast: The Establishment from Uncommitted G1 State and Its Delay for Return to Proliferation. *J. Cell Sci.* **1996**, *109*, 1347–1357. [CrossRef]
34. Roche, B.; Arcangioli, B.; Martienssen, R. Transcriptional Reprogramming in Cellular Quiescence. *RNA Biol.* **2017**, *14*, 843–853. [CrossRef]
35. Hubbard, E.J.A.; Schedl, T. Biology of the Caenorhabditis Elegans Germline Stem Cell System. *Genetics* **2019**, *213*, 1145–1188. [CrossRef]
36. Rashid, S.; Wong, C.; Roy, R. Developmental Plasticity and the Response to Nutrient Stress in Caenorhabditis Elegans. *Dev. Biol.* **2021**, *475*, 265–276. [CrossRef] [PubMed]
37. Gujar, M.R.; Wang, H. Signaling Mechanisms in the Reactivation of Quiescent Neural Stem Cells in Drosophila. *Curr. Opin. Cell Biol.* **2025**, *96*, 102566. [CrossRef] [PubMed]
38. Cheung, T.H.; Rando, T. Molecular Regulation of Stem Cell Quiescence. *Nat. Rev. Mol. Cell Biol.* **2013**, *14*, 329–340. [CrossRef]
39. Trumpp, A.; Essers, M.; Wilson, A. Awakening Dormant Haematopoietic Stem Cells. *Nat. Rev. Immunol.* **2010**, *10*, 201–209. [CrossRef] [PubMed]
40. Höfer, T.; Busch, K.; Klapproth, K.; Rodewald, H.-R. Fate Mapping and Quantitation of Hematopoiesis in Vivo. *Annu. Rev. Immunol.* **2016**, *34*, 449–478. [CrossRef]
41. Wilson, A.; Laurenti, E.; Oser, G.; van der Wath, R.C.; Blanco-Bose, W.; Jaworski, M.; Offner, S.; Dunant, C.F.; Eshkind, L.; Bockamp, E.; et al. Hematopoietic Stem Cells Reversibly Switch from Dormancy to Self-Renewal during Homeostasis and Repair. *Cell* **2008**, *135*, 1118–1129. [CrossRef]
42. Busch, K.; Klapproth, K.; Barile, M.; Flossdorf, M.; Holland-Letz, T.; Schlenner, S.M.; Reth, M.; Höfer, T.; Rodewald, H.-R. Fundamental Properties of Unperturbed Haematopoiesis from Stem Cells in Vivo. *Nature* **2015**, *518*, 542–546. [CrossRef]
43. Kucinski, I.; Campos, J.; Barile, M.; Severi, F.; Bohin, N.; Moreira, P.N.; Allen, L.; Lawson, H.; Haltalli, M.L.R.; Kinston, S.J.; et al. A Time- and Single-Cell-Resolved Model of Murine Bone Marrow Hematopoiesis. *Cell Stem Cell* **2024**, *31*, 244–259.e10. [CrossRef] [PubMed]


 https://doi.org/10.3390/biom16020203

44. van der Weijden, V.A.; Bulut-Karslioglu, A. Molecular Regulation of Paused Pluripotency in Early Mammalian Embryos and Stem Cells. *Front. Cell Dev. Biol.* **2021**, *9*, 708318. [CrossRef]
45. Renfree, M.B.; Fenelon, J.C. The Enigma of Embryonic Diapause. *Development* **2017**, *144*, 3199–3210. [CrossRef]
46. Hunter, S.M.; Evans, M. Non-Surgical Method for the Induction of Delayed Implantation and Recovery of Viable Blastocysts in Rats and Mice by the Use of Tamoxifen and Depo-Provera. *Mol. Reprod. Dev.* **1999**, *52*, 29–32. [CrossRef]
47. Paria, B.C.; Huet-Hudson, Y.M.; Dey, S.K. Blastocyst's State of Activity Determines the "Window" of Implantation in the Receptive Mouse Uterus. *Proc. Natl. Acad. Sci. USA* **1993**, *90*, 10159–10162. [CrossRef] [PubMed]
48. Iyer, D.P.; Khoei, H.H.; van der Weijden, V.A.; Kagawa, H.; Pradhan, S.J.; Novatchkova, M.; McCarthy, A.; Rayon, T.; Simon, C.S.; Dunkel, I.; et al. mTOR Activity Paces Human Blastocyst Stage Developmental Progression. *Cell* **2024**, *187*, 6566–6583.e22. [CrossRef]
49. Boroviak, T.; Loos, R.; Lombard, P.; Okahara, J.; Behr, R.; Sasaki, E.; Nichols, J.; Smith, A.; Bertone, P. Lineage-Specific Profiling Delineates the Emergence and Progression of Naive Pluripotency in Mammalian Embryogenesis. *Dev. Cell* **2015**, *35*, 366–382. [CrossRef]
50. Scognamiglio, R.; Cabezas-Wallscheid, N.; Thier, M.C.; Altamura, S.; Reyes, A.; Prendergast, Á.M.; Baumgärtner, D.; Carnevalli, L.S.; Atzberger, A.; Haas, S.; et al. Myc Depletion Induces a Pluripotent Dormant State Mimicking Diapause. *Cell* **2016**, *164*, 668–680. [CrossRef]
51. Özgüldez, H.Ö.; Bulut-Karslioğlu, A. Dormancy, Quiescence, and Diapause: Savings Accounts for Life. *Annu. Rev. Cell Dev. Biol.* **2024**, *40*, 25–49. [CrossRef]
52. McKnight, J.N.; Boerma, J.W.; Breeden, L.L.; Tsukiyama, T. Global Promoter Targeting of a Conserved Lysine Deacetylase for Transcriptional Shutoff during Quiescence Entry. *Mol. Cell* **2015**, *59*, 732–743. [CrossRef]
53. Young, C.P.; Hillyer, C.; Hokamp, K.; Fitzpatrick, D.J.; Konstantinov, N.K.; Welty, J.S.; Ness, S.A.; Werner-Washburne, M.; Fleming, A.B.; Osley, M.A. Distinct Histone Methylation and Transcription Profiles Are Established during the Development of Cellular Quiescence in Yeast. *BMC Genom.* **2017**, *18*, 107. [CrossRef]
54. Mews, P.; Zee, B.M.; Liu, S.; Donahue, G.; Garcia, B.A.; Berger, S.L. Histone Methylation Has Dynamics Distinct from Those of Histone Acetylation in Cell Cycle Reentry from Quiescence. *Mol. Cell. Biol.* **2014**, *34*, 3968–3980. [CrossRef]
55. Marguerat, S.; Schmidt, M.A.; Codlin, S.; Chen, W.-W.; Aebersold, R.; Bähler, J. Quantitative Analysis of Fission Yeast Transcriptomes and Proteomes in Proliferating and Quiescent Cells. *Cell* **2012**, *151*, 671–683. [CrossRef]
56. Greenlaw, A.; Dell, R.; Tsukiyama, T. Initial Acidic Media Promotes Quiescence Entry in Saccharomyces Cerevisiae. *MicroPubl. Biol.* **2024**, *2024*, 001071. [CrossRef]
57. Baquero Pérez, M.; Laenen, G.; Loïodice, I.; Garnier, M.; Szachnowski, U.; Morillon, A.; Ruault, M.; Taddei, A. Intergenic Accumulation of RNA Polymerase II Maintains the Potential for Swift Transcriptional Restart upon Release from Quiescence. *Genome Res.* **2025**, *35*, 2226–2239. [CrossRef] [PubMed]
58. Radonjic, M.; Andrau, J.-C.; Lijnzaad, P.; Kemmeren, P.; Kockelkorn, T.T.J.P.; van Leenen, D.; van Berkum, N.L.; Holstege, F.C.P. Genome-Wide Analyses Reveal RNA Polymerase II Located Upstream of Genes Poised for Rapid Response upon S. Cerevisiae Stationary Phase Exit. *Mol. Cells* **2005**, *18*, 171–183. [CrossRef] [PubMed]
59. Swygert, S.G.; Tsukiyama, T. Unraveling Quiescence-Specific Repressive Chromatin Domains. *Curr. Genet.* **2019**, *65*, 1145–1151. [CrossRef]
60. Aragon, A.D.; Rodriguez, A.L.; Meirelles, O.; Roy, S.; Davidson, G.S.; Tapia, P.H.; Allen, C.; Joe, R.; Benn, D.; Werner-Washburne, M. Characterization of Differentiated Quiescent and Nonquiescent Cells in Yeast Stationary-Phase Cultures. *Mol. Biol. Cell* **2008**, *19*, 1271–1280. [CrossRef]
61. Shah, K.H.; Zhang, B.; Ramachandran, V.; Herman, P.K. Processing Body and Stress Granule Assembly Occur by Independent and Differentially Regulated Pathways in Saccharomyces Cerevisiae. *Genetics* **2013**, *193*, 109–123. [CrossRef]
62. Zeng, S.; Ekwall, K. Epigenome Mapping in Quiescent Cells Reveals a Key Role for H3K4me3 in Regulation of RNA Polymerase II Activity. *Epigenomes* **2024**, *8*, 39. [CrossRef]
63. Joh, R.I.; Khanduja, J.S.; Calvo, I.A.; Mistry, M.; Palmieri, C.M.; Savol, A.J.; Sui, S.J.H.; Sadreyev, R.I.; Aryee, M.J.; Motamedi, M. Survival in Quiescence Requires the Euchromatic Deployment of Clr4/SUV39H by Argonaute-Associated Small RNAs. *Mol. Cell* **2016**, *64*, 1088–1101. [CrossRef] [PubMed]
64. Zahedi, Y.; Zeng, S.; Ekwall, K. An Essential Role for the Ino80 Chromatin Remodeling Complex in Regulation of Gene Expression during Cellular Quiescence. *Chromosome Res.* **2023**, *31*, 14. [CrossRef]
65. Westholm, J.O.; Nordberg, N.; Murén, E.; Ameur, A.; Komorowski, J.; Ronne, H. Combinatorial Control of Gene Expression by the Three Yeast Repressors Mig1, Mig2 and Mig3. *BMC Genom.* **2008**, *9*, 601. [CrossRef] [PubMed]
66. Treitel, M.A.; Kuchin, S.; Carlson, M. Snf1 Protein Kinase Regulates Phosphorylation of the Mig1 Repressor in Saccharomyces Cerevisiae. *Mol. Cell. Biol.* **1998**, *18*, 6273–6280. [CrossRef] [PubMed]

67. Martínez-Pastor, M.T.; Marchler, G.; Schüller, C.; Marchler-Bauer, A.; Ruis, H.; Estruch, F. The Saccharomyces Cerevisiae Zinc Finger Proteins Msn2p and Msn4p Are Required for Transcriptional Induction through the Stress Response Element (STRE). *EMBO J.* **1996**, *15*, 2227–2235. [CrossRef]
68. Santhanam, A.; Hartley, A.; Düvel, K.; Broach, J.R.; Garrett, S. PP2A Phosphatase Activity Is Required for Stress and Tor Kinase Regulation of Yeast Stress Response Factor Msn2p. *Eukaryot. Cell* **2004**, *3*, 1261–1271. [CrossRef]
69. Mai, B.; Breeden, L. Xbp1, a Stress-Induced Transcriptional Repressor of the Saccharomyces Cerevisiae Swi4/Mbp1 Family. *Mol. Cell. Biol.* **1997**, *17*, 6491–6501. [CrossRef]
70. Huber, A.; French, S.L.; Tekotte, H.; Yerlikaya, S.; Stahl, M.; Perepelkina, M.P.; Tyers, M.; Rougemont, J.; Beyer, A.L.; Loewith, R. Sch9 Regulates Ribosome Biogenesis via Stb3, Dot6 and Tod6 and the Histone Deacetylase Complex RPD3L. *EMBO J.* **2011**, *30*, 3052–3064. [CrossRef]
71. Hirai, H.; Ohta, K. Comparative Research: Regulatory Mechanisms of Ribosomal Gene Transcription in Saccharomyces Cerevisiae and Schizosaccharomyces Pombe. *Biomolecules* **2023**, *13*, 288. [CrossRef]
72. Bonitto, K.; Sarathy, K.; Atai, K.; Mitra, M.; Coller, H. Is There a Histone Code for Cellular Quiescence? *Front. Cell Dev. Biol.* **2021**, *9*, 739780. [CrossRef]
73. Tao, R.; Chen, H.; Gao, C.; Xue, P.; Yang, F.; Han, J.-D.J.; Zhou, B.; Chen, Y.-G. Xbp1-Mediated Histone H4 Deacetylation Contributes to DNA Double-Strand Break Repair in Yeast. *Cell Res.* **2011**, *21*, 1619–1633. [CrossRef]
74. Nirello, V.D.; Rodrigues de Paula, D.; Araújo, N.V.P.; Varga-Weisz, P.D. Does Chromatin Function as a Metabolite Reservoir? *Trends Biochem. Sci.* **2022**, *47*, 732–735. [CrossRef]
75. Friis, R.M.N.; Wu, B.P.; Reinke, S.N.; Hockman, D.J.; Sykes, B.D.; Schultz, M.C. A Glycolytic Burst Drives Glucose Induction of Global Histone Acetylation by picNuA4 and SAGA. *Nucleic Acids Res.* **2009**, *37*, 3969–3980. [CrossRef] [PubMed]
76. Cucinotta, C.E.; Dell, R.H.; Braceros, K.C.; Tsukiyama, T. RSC Primes the Quiescent Genome for Hypertranscription upon Cell-Cycle Re-Entry. *eLife* **2021**, *10*, e67033. [CrossRef] [PubMed]
77. Mitra, M.; Batista, S.L.; Coller, H.A. Transcription Factor Networks in Cellular Quiescence. *Nat. Cell Biol.* **2025**, *27*, 14–27. [CrossRef] [PubMed]
78. Montes, M.; Moreira-Ramos, S.; Rojas, D.A.; Urbina, F.; Käufer, N.F.; Maldonado, E. RNA Polymerase II Components and Rrn7 Form a Preinitiation Complex on the HomolD Box to Promote Ribosomal Protein Gene Expression in Schizosaccharomyces Pombe. *FEBS J.* **2017**, *284*, 615–633. [CrossRef]
79. Allshire, R.C.; Madhani, H.D. Ten Principles of Heterochromatin Formation and Function. *Nat. Rev. Mol. Cell Biol.* **2018**, *19*, 229–244. [CrossRef]
80. Zofall, M.; Yamanaka, S.; Reyes-Turcu, F.E.; Zhang, K.; Rubin, C.; Grewal, S.I.S. RNA Elimination Machinery Targeting Meiotic mRNAs Promotes Facultative Heterochromatin Formation. *Science* **2012**, *335*, 96–100. [CrossRef]
81. Roche, B.; Arcangioli, B.; Martienssen, R. RNA Interference Is Essential for Cellular Quiescence. *Science* **2016**, *354*, aah5651. [CrossRef]
82. Oya, E.; Durand-Dubief, M.; Cohen, A.; Maksimov, V.; Schurra, C.; Nakayama, J.-I.; Weisman, R.; Arcangioli, B.; Ekwall, K. Leo1 Is Essential for the Dynamic Regulation of Heterochromatin and Gene Expression during Cellular Quiescence. *Epigenet. Chromatin* **2019**, *12*, 45. [CrossRef]
83. Yuan, X.; Feng, W.; Imhof, A.; Grummt, I.; Zhou, Y. Activation of RNA Polymerase I Transcription by Cockayne Syndrome Group B Protein and Histone Methyltransferase G9a. *Mol. Cell* **2007**, *27*, 585–595. [CrossRef]
84. Tadeo, X.; Wang, J.; Kallgren, S.P.; Liu, J.; Reddy, B.D.; Qiao, F.; Jia, S. Elimination of Shelterin Components Bypasses RNAi for Pericentric Heterochromatin Assembly. *Genes Dev.* **2013**, *27*, 2489–2499. [CrossRef]
85. Barrales, R.R.; Forn, M.; Georgescu, P.R.; Sarkadi, Z.; Braun, S. Control of Heterochromatin Localization and Silencing by the Nuclear Membrane Protein Lem2. *Genes Dev.* **2016**, *30*, 133–148. [CrossRef] [PubMed]
86. Muhammad, A.; Sarkadi, Z.; Mazumder, A.; Ait Saada, A.; van Emden, T.; Capella, M.; Fekete, G.; Suma Sreechakram, V.N.; Al-Sady, B.; Lambert, S.A.E.; et al. A Systematic Quantitative Approach Comprehensively Defines Domain-Specific Functional Pathways Linked to Schizosaccharomyces Pombe Heterochromatin Regulation. *Nucleic Acids Res.* **2024**, *52*, 13665–13689. [CrossRef] [PubMed]
87. Deng, X.; Yao, Q.; Horvath, A.; Jiang, Z.; Zhao, J.; Fischer, T.; Sugiyama, T. The Fission Yeast Ortholog of Coilin, Mug174, Forms Cajal Body-like Nuclear Condensates and Is Essential for Cellular Quiescence. *Nucleic Acids Res.* **2024**, *52*, 9174–9192. [CrossRef]
88. Liu, L.; Cheung, T.H.; Charville, G.W.; Hurgo, B.M.C.; Leavitt, T.; Shih, J.; Brunet, A.; Rando, T.A. Chromatin Modifications as Determinants of Muscle Stem Cell Quiescence and Chronological Aging. *Cell Rep.* **2013**, *4*, 189–204. [CrossRef]
89. Margueron, R.; Li, G.; Sarma, K.; Blais, A.; Zavadil, J.; Woodcock, C.L.; Dynlacht, B.D.; Reinberg, D. Ezh1 and Ezh2 Maintain Repressive Chromatin through Different Mechanisms. *Mol. Cell* **2008**, *32*, 503–518. [CrossRef]
90. Flora, P.; Dalal, G.; Cohen, I.; Ezhkova, E. Polycomb Repressive Complex(es) and Their Role in Adult Stem Cells. *Genes* **2021**, *12*, 1485. [CrossRef] [PubMed]



https://doi.org/10.3390/biom16020203

91. Boonsanay, V.; Zhang, T.; Georgieva, A.; Kostin, S.; Qi, H.; Yuan, X.; Zhou, Y.; Braun, T. Regulation of Skeletal Muscle Stem Cell Quiescence by Suv4-20h1-Dependent Facultative Heterochromatin Formation. *Cell Stem Cell* **2016**, *18*, 229–242. [CrossRef] [PubMed]
92. Evertts, A.G.; Manning, A.L.; Wang, X.; Dyson, N.; Garcia, B.; Coller, H. H4K20 Methylation Regulates Quiescence and Chromatin Compaction. *Mol. Biol. Cell* **2013**, *24*, 3025–3037. [CrossRef] [PubMed]
93. Bierhoff, H.; Dammert, M.A.; Brocks, D.; Dambacher, S.; Schotta, G.; Grummt, I. Quiescence-Induced LncRNAs Trigger H4K20 Trimethylation and Transcriptional Silencing. *Mol. Cell* **2014**, *54*, 675–682. [CrossRef]
94. Ramponi, V.; Richart, L.; Kovatcheva, M.; Stephan-Otto Attolini, C.; Capellades, J.; Lord, A.E.; Yanes, O.; Ficz, G.; Serrano, M. H4K20me3-Mediated Repression of Inflammatory Genes Is a Characteristic and Targetable Vulnerability of Persister Cancer Cells. *Cancer Res.* **2025**, *85*, 32–51. [CrossRef] [PubMed]
95. Chen, R.; Fan, R.; Chen, F.; Govindasamy, N.; Brinkmann, H.; Stehling, M.; Adams, R.H.; Jeong, H.-W.; Bedzhov, I. Analyzing Embryo Dormancy at Single-Cell Resolution Reveals Dynamic Transcriptional Responses and Activation of Integrin-Yap/Taz Prosurvival Signaling. *Cell Stem Cell* **2024**, *31*, 1262–1279.e8. [CrossRef]
96. Fu, Z.; Wang, B.; Wang, S.; Wu, W.; Wang, Q.; Chen, Y.; Kong, S.; Lu, J.; Tang, Z.; Ran, H.; et al. Integral Proteomic Analysis of Blastocysts Reveals Key Molecular Machinery Governing Embryonic Diapause and Reactivation for Implantation in Mice. *Biol. Reprod.* **2014**, *90*, 52. [CrossRef]
97. Hussein, A.M.; Wang, Y.; Mathieu, J.; Margaretha, L.; Song, C.; Jones, D.C.; Cavanaugh, C.; Miklas, J.W.; Mahen, E.; Showalter, M.R.; et al. Metabolic Control over mTOR-Dependent Diapause-like State. *Dev. Cell* **2020**, *52*, 236–250.e7. [CrossRef]
98. van der Weijden, V.A.; Bulut-Karslioğlu, A. Embryos Burn Fat in Standby. *Trends Cell Biol.* **2024**, *34*, 700–702. [CrossRef]
99. Stötzel, M.; Cheng, C.-Y.; Ilik, I.A.; Kumar, A.S.; Omgba, P.A.; van der Weijden, V.A.; Zhang, Y.; Vingron, M.; Meissner, A.; Aktaş, T.; et al. TET Activity Safeguards Pluripotency throughout Embryonic Dormancy. *Nat. Struct. Mol. Biol.* **2024**, *31*, 1625–1639. [CrossRef]
100. Fan, R.; Kim, Y.S.; Wu, J.; Chen, R.; Zeuschner, D.; Mildner, K.; Adachi, K.; Wu, G.; Galatidou, S.; Li, J.; et al. Wnt/Beta-catenin/Esrrb Signaling Controls the Tissue-Scale Reorganization and Maintenance of the Pluripotent Lineage during Murine Embryonic Diapause. *Nat. Commun.* **2020**, *11*, 5499. [CrossRef]
101. Furlan, G.; Martin, S.B.; Cho, B.; McClymont, S.A.; Robertson, E.J.; Collignon, E.; Ramalho-Santos, M. Nodal/Smad2 Signaling Sustains Developmental Pausing by Repressing Pparg-Mediated Lipid Metabolism. *bioRxiv* **2025**. [CrossRef]
102. Farooq, Z.; Banday, S.; Pandita, T.K.; Altaf, M. The Many Faces of Histone H3K79 Methylation. *Mutat. Res. Rev. Mutat. Res.* **2016**, *768*, 46–52. [CrossRef]
103. Sajiki, K.; Tahara, Y.; Uehara, L.; Sasaki, T.; Pluskal, T.; Yanagida, M. Genetic Regulation of Mitotic Competence in G0 Quiescent Cells. *Sci. Adv.* **2018**, *4*, eaat5685. [CrossRef] [PubMed]
104. Sadeghi, L.; Prasad, P.; Ekwall, K.; Cohen, A.; Svensson, J.P. The Paf1 Complex Factors Leo1 and Paf1 Promote Local Histone Turnover to Modulate Chromatin States in Fission Yeast. *EMBO Rep.* **2015**, *16*, 1673–1687. [CrossRef] [PubMed]
105. Cohen, A.; Habib, A.; Laor, D.; Yadav, S.; Kupiec, M.; Weisman, R. TOR Complex 2 in Fission Yeast Is Required for Chromatin-Mediated Gene Silencing and Assembly of Heterochromatic Domains at Subtelomeres. *J. Biol. Chem.* **2018**, *293*, 8138–8150. [CrossRef]
106. Oya, T.; Tanaka, M.; Hayashi, A.; Yoshimura, Y.; Nakamura, R.; Arita, K.; Murakami, Y.; Nakayama, J.-I. Characterization of the Swi6/HP1 Binding Motif in Its Partner Protein Reveals the Basis for the Functional Divergence of the HP1 Family Proteins in Fission Yeast. *FASEB J.* **2025**, *39*, e70387. [CrossRef]
107. Jacob, A.G.; Smith, C.W.J. Intron Retention as a Component of Regulated Gene Expression Programs. *Hum. Genet.* **2017**, *136*, 1043–1057. [CrossRef]
108. Pleiss, J.A.; Whitworth, G.B.; Bergkessel, M.; Guthrie, C. Rapid, Transcript-Specific Changes in Splicing in Response to Environmental Stress. *Mol. Cell* **2007**, *27*, 928–937. [CrossRef]
109. Parenteau, J.; Tsang, J.; Downs, S.R.; Zhou, D.; Scott, M.S.; Pleiss, J.A.; Abou Elela, S. Cells Resist Starvation through a Nutrient Stress Splice Switch. *Nucleic Acids Res.* **2025**, *53*, gkaf525. [CrossRef]
110. Parenteau, J.; Maignon, L.; Berthoumieux, M.; Catala, M.; Gagnon, V.; Abou Elela, S. Introns Are Mediators of Cell Response to Starvation. *Nature* **2019**, *565*, 612–617. [CrossRef]
111. Morgan, J.T.; Fink, G.R.; Bartel, D.P. Excised Linear Introns Regulate Growth in Yeast. *Nature* **2019**, *565*, 606–611. [CrossRef]
112. Yue, L.; Wan, R.; Luan, S.; Zeng, W.; Cheung, T.H. Dek Modulates Global Intron Retention during Muscle Stem Cells Quiescence Exit. *Dev. Cell* **2020**, *53*, 661–676.e6. [CrossRef] [PubMed]
113. Mitra, M.; Johnson, E.L.; Swamy, V.S.; Nersesian, L.E.; Corney, D.C.; Robinson, D.G.; Taylor, D.G.; Ambrus, A.M.; Jelinek, D.; Wang, W.; et al. Alternative Polyadenylation Factors Link Cell Cycle to Migration. *Genome Biol.* **2018**, *19*, 176. [CrossRef] [PubMed]
114. Lukačišin, M.; Espinosa-Cantú, A.; Bollenbach, T. Intron-Mediated Induction of Phenotypic Heterogeneity. *Nature* **2022**, *605*, 113–118. [CrossRef]


 https://doi.org/10.3390/biom16020203

115. Kilchert, C.; Wittmann, S.; Passoni, M.; Shah, S.; Granneman, S.; Vasiljeva, L. Regulation of mRNA Levels by Decay-Promoting Introns That Recruit the Exosome Specificity Factor Mmi1. *Cell Rep.* **2015**, *13*, 2504–2515. [CrossRef] [PubMed]
116. Harigaya, Y.; Tanaka, H.; Yamanaka, S.; Tanaka, K.; Watanabe, Y.; Tsutsumi, C.; Chikashige, Y.; Hiraoka, Y.; Yamashita, A.; Yamamoto, M. Selective Elimination of Messenger RNA Prevents an Incidence of Untimely Meiosis. *Nature* **2006**, *442*, 45–50. [CrossRef]
117. Lee, G.; Zheng, Y.; Cho, S.; Jang, C.; England, C.; Dempsey, J.M.; Yu, Y.; Liu, X.; He, L.; Cavaliere, P.M.; et al. Post-Transcriptional Regulation of DE Novo Lipogenesis by mTORC1-S6K1-SRPK2 Signaling. *Cell* **2017**, *171*, 1545–1558.e18. [CrossRef]
118. Sandberg, R.; Neilson, J.R.; Sarma, A.; Sharp, P.A.; Burge, C.B. Proliferating Cells Express mRNAs with Shortened 3′ Untranslated Regions and Fewer microRNA Target Sites. *Science* **2008**, *320*, 1643–1647. [CrossRef]
119. Ji, Z.; Lee, J.Y.; Pan, Z.; Jiang, B.; Tian, B. Progressive Lengthening of 3′ Untranslated Regions of mRNAs by Alternative Polyadenylation during Mouse Embryonic Development. *Proc. Natl. Acad. Sci. USA* **2009**, *106*, 7028–7033. [CrossRef]
120. Liu, X.; Hoque, M.; Larochelle, M.; Lemay, J.-F.; Yurko, N.; Manley, J.L.; Bachand, F.; Tian, B. Comparative Analysis of Alternative Polyadenylation in S. Cerevisiae and S. Pombe. *Genome Res.* **2017**, *27*, 1685–1695. [CrossRef]
121. Schlackow, M.; Marguerat, S.; Proudfoot, N.J.; Bähler, J.; Erban, R.; Gullerova, M. Genome-Wide Analysis of poly(A) Site Selection in Schizosaccharomyces Pombe. *RNA* **2013**, *19*, 1617–1631. [CrossRef]
122. Kaczmarek Michaels, K.; Mohd Mostafa, S.; Ruiz Capella, J.; Moore, C.L. Regulation of Alternative Polyadenylation in the Yeast Saccharomyces Cerevisiae by Histone H3K4 and H3K36 Methyltransferases. *Nucleic Acids Res.* **2020**, *48*, 5407–5425. [CrossRef]
123. Tian, B.; Manley, J.L. Alternative Polyadenylation of mRNA Precursors. *Nat. Rev. Mol. Cell Biol.* **2017**, *18*, 18–30. [CrossRef]
124. Geisberg, J.V.; Moqtaderi, Z.; Struhl, K. Chromatin Regulates Alternative Polyadenylation via the RNA Polymerase II Elongation Rate. *Proc. Natl. Acad. Sci. USA* **2024**, *121*, e2405827121. [CrossRef]
125. Chang, J.-W.; Zhang, W.; Yeh, H.-S.; de Jong, E.P.; Jun, S.; Kim, K.-H.; Bae, S.S.; Beckman, K.; Hwang, T.H.; Kim, K.-S.; et al. mRNA 3′-UTR Shortening Is a Molecular Signature of mTORC1 Activation. *Nat. Commun.* **2015**, *6*, 7218. [CrossRef] [PubMed]
126. Geisberg, J.V.; Moqtaderi, Z.; Struhl, K. Condition-Specific 3′ mRNA Isoform Half-Lives and Stability Elements in Yeast. *Proc. Natl. Acad. Sci. USA* **2023**, *120*, e2301117120. [CrossRef] [PubMed]
127. Iwakawa, H.-O.; Tomari, Y. Life of RISC: Formation, Action, and Degradation of RNA-Induced Silencing Complex. *Mol. Cell* **2022**, *82*, 30–43. [CrossRef]
128. Zhang, Q.; Han, Z.; Zhu, Y.; Chen, J.; Li, W. Role of Hypoxia Inducible Factor-1 in Cancer Stem Cells (Review). *Mol. Med. Rep.* **2021**, *23*, 17. [CrossRef]
129. Macias, S.; Cordiner, R.A.; Gautier, P.; Plass, M.; Cáceres, J.F. DGCR8 Acts as an Adaptor for the Exosome Complex to Degrade Double-Stranded Structured RNAs. *Mol. Cell* **2015**, *60*, 873–885. [CrossRef]
130. Cheung, T.H.; Quach, N.L.; Charville, G.W.; Liu, L.; Park, L.; Edalati, A.; Yoo, B.; Hoang, P.; Rando, T.A. Maintenance of Muscle Stem-Cell Quiescence by microRNA-489. *Nature* **2012**, *482*, 524–528. [CrossRef] [PubMed]
131. Liu, W.-M.; Pang, R.T.K.; Cheong, A.W.Y.; Ng, E.H.Y.; Lao, K.; Lee, K.-F.; Yeung, W.S.B. Involvement of microRNA Lethal-7a in the Regulation of Embryo Implantation in Mice. *PLoS ONE* **2012**, *7*, e37039. [CrossRef] [PubMed]
132. Wang, Y.; Hussein, A.M.; Somasundaram, L.; Sankar, R.; Detraux, D.; Mathieu, J.; Ruohola-Baker, H. MicroRNAs Regulating Human and Mouse Naïve Pluripotency. *Int. J. Mol. Sci.* **2019**, *20*, 5864. [CrossRef]
133. Wang, Y.; Medvid, R.; Melton, C.; Jaenisch, R.; Blelloch, R. DGCR8 Is Essential for microRNA Biogenesis and Silencing of Embryonic Stem Cell Self-Renewal. *Nat. Genet.* **2007**, *39*, 380–385. [CrossRef]
134. Liu, W.M.; Cheng, R.R.; Niu, Z.R.; Chen, A.C.; Ma, M.Y.; Li, T.; Chiu, P.C.; Pang, R.T.; Lee, Y.L.; Ou, J.P.; et al. Let-7 Derived from Endometrial Extracellular Vesicles Is an Important Inducer of Embryonic Diapause in Mice. *Sci. Adv.* **2020**, *6*, eaaz7070. [CrossRef]
135. Kilchert, C. RNA Exosomes and Their Cofactors. *Methods Mol. Biol.* **2020**, *2062*, 215–235.
136. Li, J.-N.; Wang, M.-Y.; Ruan, J.-W.; Lyu, Y.-J.; Weng, Y.-H.; Brindangnanam, P.; Coumar, M.S.; Chen, P.-S. A Transcription-Independent Role for HIF-1$\alpha$ in Modulating Microprocessor Assembly. *Nucleic Acids Res.* **2024**, *52*, 11806–11821. [CrossRef] [PubMed]
137. Greenlaw, A.C.; Alavattam, K.G.; Tsukiyama, T. Post-Transcriptional Regulation Shapes the Transcriptome of Quiescent Budding Yeast. *Nucleic Acids Res.* **2023**, *52*, 1043–1063. [CrossRef]
138. Atkinson, S.R.; Marguerat, S.; Bitton, D.A.; Rodríguez-López, M.; Rallis, C.; Lemay, J.-F.; Cotobal, C.; Malecki, M.; Smialowski, P.; Mata, J.; et al. Long Noncoding RNA Repertoire and Targeting by Nuclear Exosome, Cytoplasmic Exonuclease, and RNAi in Fission Yeast. *RNA* **2018**, *24*, 1195–1213. [CrossRef] [PubMed]
139. Han, X.; Xing, L.; Hong, Y.; Zhang, X.; Hao, B.; Lu, J.Y.; Huang, M.; Wang, Z.; Ma, S.; Zhan, G.; et al. Nuclear RNA Homeostasis Promotes Systems-Level Coordination of Cell Fate and Senescence. *Cell Stem Cell* **2024**, *31*, 694–716.e11. [CrossRef]
140. Birot, A.; Kus, K.; Priest, E.; Al Alwash, A.; Castello, A.; Mohammed, S.; Vasiljeva, L.; Kilchert, C. RNA-Binding Protein Mub1 and the Nuclear RNA Exosome Act to Fine-Tune Environmental Stress Response. *Life Sci. Alliance* **2022**, *5*, e202101111. [CrossRef]


 https://doi.org/10.3390/biom16020203

141. Shah, S.; Wittmann, S.; Kilchert, C.; Vasiljeva, L. lncRNA Recruits RNAi and the Exosome to Dynamically Regulate pho1 Expression in Response to Phosphate Levels in Fission Yeast. *Genes Dev.* **2014**, *28*, 231–244. [CrossRef] [PubMed]
142. Liu, Y.; Zhang, P.; Wu, Q.; Fang, H.; Wang, Y.; Xiao, Y.; Cong, M.; Wang, T.; He, Y.; Ma, C.; et al. Long Non-Coding RNA NR2F1-AS1 Induces Breast Cancer Lung Metastatic Dormancy by Regulating NR2F1 and ΔNp63. *Nat. Commun.* **2021**, *12*, 5232. [CrossRef]
143. Anver, S.; Sumit, A.F.; Sun, X.-M.; Hatimy, A.; Thalassinos, K.; Marguerat, S.; Alic, N.; Bähler, J. Ageing-Associated Long Non-Coding RNA Extends Lifespan and Reduces Translation in Non-Dividing Cells. *EMBO Rep.* **2024**, *25*, 4921–4949. [CrossRef]
144. Hirose, T.; Fujiwara, N.; Ninomiya, K.; Yamamoto, T.; Nakagawa, S.; Yamazaki, T. Architectural RNAs: Blueprints for Functional Membraneless Organelle Assembly. *Trends Genet.* **2025**, *41*, 919–933. [CrossRef] [PubMed]
145. Hirose, T.; Ninomiya, K.; Nakagawa, S.; Yamazaki, T. A Guide to Membraneless Organelles and Their Various Roles in Gene Regulation. *Nat. Rev. Mol. Cell Biol.* **2023**, *24*, 288–304. [CrossRef]
146. Ding, D.-Q.; Okamasa, K.; Katou, Y.; Oya, E.; Nakayama, J.-I.; Chikashige, Y.; Shirahige, K.; Haraguchi, T.; Hiraoka, Y. Chromosome-Associated RNA-Protein Complexes Promote Pairing of Homologous Chromosomes during Meiosis in Schizosaccharomyces Pombe. *Nat. Commun.* **2019**, *10*, 5598. [CrossRef]
147. Aydin, E.; Schreiner, S.; Böhme, J.; Keil, B.; Weber, J.; Žunar, B.; Glatter, T.; Kilchert, C. DEAD-Box ATPase Dbp2 Is the Key Enzyme in an mRNP Assembly Checkpoint at the 3′-End of Genes and Involved in the Recycling of Cleavage Factors. *Nat. Commun.* **2024**, *15*, 6829. [CrossRef]
148. Shuman, S. Transcriptional Interference at Tandem lncRNA and Protein-Coding Genes: An Emerging Theme in Regulation of Cellular Nutrient Homeostasis. *Nucleic Acids Res.* **2020**, *48*, 8243–8254. [CrossRef]
149. Nguyen, L.A.C.; Mori, M.; Yasuda, Y.; Galipon, J. Functional Consequences of Shifting Transcript Boundaries in Glucose Starvation. *Mol. Cell. Biol.* **2023**, *43*, 611–628. [CrossRef]
150. Nishizawa, M.; Komai, T.; Katou, Y.; Shirahige, K.; Ito, T.; Toh-E, A. Nutrient-Regulated Antisense and Intragenic RNAs Modulate a Signal Transduction Pathway in Yeast. *PLoS Biol.* **2008**, *6*, 2817–2830. [CrossRef]
151. Miki, A.; Galipon, J.; Sawai, S.; Inada, T.; Ohta, K. RNA Decay Systems Enhance Reciprocal Switching of Sense and Antisense Transcripts in Response to Glucose Starvation. *Genes Cells* **2016**, *21*, 1276–1289. [CrossRef]
152. Lazorthes, S.; Vallot, C.; Briois, S.; Aguirrebengoa, M.; Thuret, J.-Y.; St Laurent, G.; Rougeulle, C.; Kapranov, P.; Mann, C.; Trouche, D.; et al. A vlincRNA Participates in Senescence Maintenance by Relieving H2AZ-Mediated Repression at the INK4 Locus. *Nat. Commun.* **2015**, *6*, 5971. [CrossRef]
153. Mutzel, V.; Schwämmle, T.; Oeverdieck, S.; Librenjak, L.; Boesen, B.; Bothe, M.; Gjaltema, R.A.F.; Dunkel, I.; Noviello, G.; Schulz, E.G. Antisense Transcription Can Induce Expression Memory via Stable Promoter Repression. *Genome Biol.* **2025**, *26*, 430. [CrossRef]
154. Hirota, K.; Miyoshi, T.; Kugou, K.; Hoffman, C.S.; Shibata, T.; Ohta, K. Stepwise Chromatin Remodeling by a Cascade of Transcription Initiation of Non-Coding RNAs. *Nature* **2008**, *456*, 130–134. [CrossRef]
155. Oda, A.; Takemata, N.; Hirata, Y.; Miyoshi, T.; Suzuki, Y.; Sugano, S.; Ohta, K. Dynamic Transition of Transcription and Chromatin Landscape during Fission Yeast Adaptation to Glucose Starvation. *Genes Cells* **2015**, *20*, 392–407. [CrossRef] [PubMed]
156. Guidi, M.; Ruault, M.; Marbouty, M.; Loïodice, I.; Cournac, A.; Billaudeau, C.; Hocher, A.; Mozziconacci, J.; Koszul, R.; Taddei, A. Spatial Reorganization of Telomeres in Long-Lived Quiescent Cells. *Genome Biol.* **2015**, *16*, 206. [CrossRef]
157. Tan, Z.Y.; Cai, S.J.; Paithankar, S.A.; Liu, T.; Nie, X.; Shi, J.; Gan, L. Macromolecular and Cytological Changes in Fission Yeast G0 Nuclei. *J. Cell Sci.* **2025**, *138*, jcs263654. [CrossRef] [PubMed]
158. Tiku, V.; Jain, C.; Raz, Y.; Nakamura, S.; Heestand, B.; Liu, W.; Späth, M.; Suchiman, H.E.D.; Müller, R.-U.; Slagboom, P.E.; et al. Small Nucleoli Are a Cellular Hallmark of Longevity. *Nat. Commun.* **2017**, *8*, 16083. [CrossRef] [PubMed]
159. Schilling, M.; Prusty, A.B.; Boysen, B.; Oppermann, F.S.; Riedel, Y.L.; Husedzinovic, A.; Rasouli, H.; König, A.; Ramanathan, P.; Reymann, J.; et al. TOR Signaling Regulates Liquid Phase Separation of the SMN Complex Governing snRNP Biogenesis. *Cell Rep.* **2021**, *35*, 109277. [CrossRef]
160. Swygert, S.G.; Lin, D.; Portillo-Ledesma, S.; Lin, P.-Y.; Hunt, D.R.; Kao, C.; Schlick, T.; Noble, W.S.; Tsukiyama, T. Local Chromatin Fiber Folding Represses Transcription and Loop Extrusion in Quiescent Cells. *eLife* **2021**, *10*, e72062. [CrossRef]
161. Schäfer, G.; McEvoy, C.; Patterton, H. The Saccharomyces Cerevisiae Linker Histone Hho1p Is Essential for Chromatin Compaction in Stationary Phase and Is Displaced by Transcription. *Proc. Natl. Acad. Sci. USA* **2008**, *105*, 14838–14843. [CrossRef]
162. Rutledge, M.T.; Russo, M.; Belton, J.M.; Dekker, J.; Broach, J.R. The Yeast Genome Undergoes Significant Topological Reorganization in Quiescence. *Nucleic Acids Res.* **2015**, *43*, 8299–8313. [CrossRef]
163. Laporte, D.; Courtout, F.; Tollis, S.; Sagot, I. Quiescent Saccharomyces Cerevisiae Forms Telomere Hyperclusters at the Nuclear Membrane Vicinity through a Multifaceted Mechanism Involving Esc1, the Sir Complex, and Chromatin Condensation. *Mol. Biol. Cell* **2016**, *27*, 1875–1884. [CrossRef]

 https://doi.org/10.3390/biom16020203

164. Swygert, S.G.; Kim, S.; Wu, X.; Fu, T.; Hsieh, T.-H.S.; Rando, O.J.; Eisenman, R.N.; Shendure, J.; McKnight, J.N.; Tsukiyama, T. Condensin-Dependent Chromatin Compaction Represses Transcription Globally during Quiescence. *Mol. Cells* **2019**, *73*, 533–546. [CrossRef]
165. Hsieh, T.-H.S.; Weiner, A.; Lajoie, B.; Dekker, J.; Friedman, N.; Rando, O.J. Mapping Nucleosome Resolution Chromosome Folding in Yeast by Micro-C. *Cell* **2015**, *162*, 108–119. [CrossRef] [PubMed]
166. Ruault, M.; Loïodice, I.; Keister, B.D.; Even, A.; Garnier, M.; Baquero-Pérez, M.; Waterman, D.; Haber, J.E.; Blagoev, K.B.; Scolari, V.F.; et al. Esc1-Mediated Anchoring Regulates Telomere Clustering in Response to Metabolic Changes. *bioRxiv* **2025**. [CrossRef]
167. Ocampo, A.; Liu, J.; Schroeder, E.A.; Shadel, G.S.; Barrientos, A. Mitochondrial Respiratory Thresholds Regulate Yeast Chronological Life Span and Its Extension by Caloric Restriction. *Cell Metab.* **2012**, *16*, 55–67. [CrossRef]
168. Faure, G.; Jézéquel, K.; Roisné-Hamelin, F.; Bitard-Feildel, T.; Lamiable, A.; Marcand, S.; Callebaut, I. Discovery and Evolution of New Domains in Yeast Heterochromatin Factor Sir4 and Its Partner Esc1. *Genome Biol. Evol.* **2019**, *11*, 572–585. [CrossRef] [PubMed]
169. Deshpande, I.; Keusch, J.J.; Challa, K.; Iesmantavicius, V.; Gasser, S.M.; Gut, H. The Sir4 H-BRCT Domain Interacts with Phospho-Proteins to Sequester and Repress Yeast Heterochromatin. *EMBO J.* **2020**, *39*, e103976. [CrossRef]
170. Hocher, A.; Taddei, A. Subtelomeres as Specialized Chromatin Domains. *Bioessays* **2020**, *42*, e1900205. [CrossRef]
171. Funabiki, H.; Hagan, I.; Uzawa, S.; Yanagida, M. Cell Cycle-Dependent Specific Positioning and Clustering of Centromeres and Telomeres in Fission Yeast. *J. Cell Biol.* **1993**, *121*, 961–976. [CrossRef]
172. Chikashige, Y.; Yamane, M.; Okamasa, K.; Tsutsumi, C.; Kojidani, T.; Sato, M.; Haraguchi, T.; Hiraoka, Y. Membrane Proteins Bqt3 and -4 Anchor Telomeres to the Nuclear Envelope to Ensure Chromosomal Bouquet Formation. *J. Cell Biol.* **2009**, *187*, 413–427. [CrossRef]
173. Ebrahimi, H.; Masuda, H.; Jain, D.; Cooper, J.P. Distinct "Safe Zones" at the Nuclear Envelope Ensure Robust Replication of Heterochromatic Chromosome Regions. *Elife* **2018**, *7*, e32911. [CrossRef]
174. Hu, C.; Inoue, H.; Sun, W.; Takeshita, Y.; Huang, Y.; Xu, Y.; Kanoh, J.; Chen, Y. The Inner Nuclear Membrane Protein Bqt4 in Fission Yeast Contains a DNA-Binding Domain Essential for Telomere Association with the Nuclear Envelope. *Structure* **2019**, *27*, 335–343.e3. [CrossRef] [PubMed]
175. Steglich, B.; Strålfors, A.; Khorosjutina, O.; Persson, J.; Smialowska, A.; Javerzat, J.-P.; Ekwall, K. The Fun30 Chromatin Remodeler Fft3 Controls Nuclear Organization and Chromatin Structure of Insulators and Subtelomeres in Fission Yeast. *PLoS Genet.* **2015**, *11*, e1005101. [CrossRef]
176. Tange, Y.; Chikashige, Y.; Takahata, S.; Kawakami, K.; Higashi, M.; Mori, C.; Kojidani, T.; Hirano, Y.; Asakawa, H.; Murakami, Y.; et al. Inner Nuclear Membrane Protein Lem2 Augments Heterochromatin Formation in Response to Nutritional Conditions. *Genes Cells* **2016**, *21*, 812–832. [CrossRef] [PubMed]
177. Maestroni, L.; Reyes, C.; Vaurs, M.; Gachet, Y.; Tournier, S.; Géli, V.; Coulon, S. Nuclear Envelope Attachment of Telomeres Limits TERRA and Telomeric Rearrangements in Quiescent Fission Yeast Cells. *Nucleic Acids Res.* **2020**, *48*, 3029–3041. [CrossRef] [PubMed]
178. Maestroni, L.; Audry, J.; Matmati, S.; Arcangioli, B.; Géli, V.; Coulon, S. Eroded Telomeres Are Rearranged in Quiescent Fission Yeast Cells through Duplications of Subtelomeric Sequences. *Nat. Commun.* **2017**, *8*, 1684. [CrossRef]
179. Weierich, C.; Brero, A.; Stein, S.; von Hase, J.; Cremer, C.; Cremer, T.; Solovei, I. Three-Dimensional Arrangements of Centromeres and Telomeres in Nuclei of Human and Murine Lymphocytes. *Chromosome Res.* **2003**, *11*, 485–502. [CrossRef]
180. Pennarun, G.; Picotto, J.; Bertrand, P. Close Ties between the Nuclear Envelope and Mammalian Telomeres: Give Me Shelter. *Genes* **2023**, *14*, 775. [CrossRef]
181. Mata, J.; Lyne, R.; Burns, G.; Bähler, J. The Transcriptional Program of Meiosis and Sporulation in Fission Yeast. *Nat. Genet.* **2002**, *32*, 143–147. [CrossRef] [PubMed]
182. Alfredsson-Timmins, J.; Kristell, C.; Henningson, F.; Lyckman, S.; Bjerling, P. Reorganization of Chromatin Is an Early Response to Nitrogen Starvation in Schizosaccharomyces Pombe. *Chromosoma* **2009**, *118*, 99–112. [CrossRef]
183. Taddei, A.; Gasser, S.M. Structure and Function in the Budding Yeast Nucleus. *Genetics* **2012**, *192*, 107–129. [CrossRef]
184. Braun, S.; Barrales, R.R. Beyond Tethering and the LEM Domain: MSCellaneous Functions of the Inner Nuclear Membrane Lem2. *Nucleus* **2016**, *7*, 523–531. [CrossRef]
185. Skribbe, M.; Soneson, C.; Stadler, M.B.; Schwaiger, M.; Suma Sreechakram, V.N.; Iesmantavicius, V.; Hess, D.; Moreno, E.P.F.; Braun, S.; Seebacher, J.; et al. A Comprehensive Schizosaccharomyces Pombe Atlas of Physical Transcription Factor Interactions with Proteins and Chromatin. *Mol. Cell* **2025**, *85*, 1426–1444.e8. [CrossRef]
186. Laporte, D.; Courtout, F.; Salin, B.; Ceschin, J.; Sagot, I. An Array of Nuclear Microtubules Reorganizes the Budding Yeast Nucleus during Quiescence. *J. Cell Biol.* **2013**, *203*, 585–594. [CrossRef]


 https://doi.org/10.3390/biom16020203

187. Laporte, D.; Sagot, I. Microtubule Reorganization and Quiescence: An Intertwined Relationship. *Physiology* **2025**, *40*, 145–157. [CrossRef]
188. Laporte, D.; Courtout, F.; Pinson, B.; Dompierre, J.; Salin, B.; Brocard, L.; Sagot, I. A Stable Microtubule Array Drives Fission Yeast Polarity Reestablishment upon Quiescence Exit. *J. Cell Biol.* **2015**, *210*, 99–113. [CrossRef] [PubMed]




https://doi.org/10.3390/biom16020203

biomolecules

MDPI

*Review*

# Quiescence Multiverse

**Damien Laporte and Isabelle Sagot** *

CNRS, Institut de Biochimie et Génétique Cellulaires, UMR5095, Université de Bordeaux, 33077 Bordeaux, France
* Correspondence: isabelle.sagot@ibgc.cnrs.fr

**Abstract:** Cellular quiescence is operationally defined as a temporary and reversible cessation of proliferation. This state encompasses a wide range of physiological situations since most cells, from microbes to cells composing complex tissues, spend most of their lives non-dividing, waiting for signals to reproliferate. As such, individual quiescent cells must withstand the effects of time not only to survive but also to maintain their ability to divide. These capacities are shaped by a combination of deterministic factors relying on cell history and cumulative stochastic events linked to the environment but also to time. In addition, with time, quiescence deepens, the quiescence exit process being extended. Yet, this deepening is not necessarily sensed evenly by each individual quiescent cell, and some cells exit quiescence faster than others. Hence, time generates heterogeneity within quiescent cell populations, heterogeneity that, in turn, increases cell population resilience and robustness to time. In this review, we discuss some of the loops that link quiescence and time.

**Keywords:** ageing; quiescence; stem cells

## 1. Introduction

In the Cambridge Dictionary, quiescence is defined as "the state of being temporary quiet or not active". In biology, quiescence is a term that embraces many physiological situations. At the cellular level, quiescence is operationally defined as temporary and reversible arrest of proliferation. This cellular state has the greatest prevalence on the planet since most cells—from microbes to cells within complex tissues—spend the majority of their life non-dividing, waiting for signals to reproliferate. In fact, in mammals, quiescent cells are much more numerous than their proliferating counterparts [1]. While research on cellular quiescence was rather confidential until the end of the 20th century, the development of new techniques allowing the study of this amazingly complex cellular state and the recognition of its implication in several human diseases, including cancer [2–4], have tremendously stimulated the field. It is now widely acknowledged that quiescence is not simply a passive resting condition but rather an actively initiated and maintained cellular state [5–7].

In unicellular organisms, quiescence is most of the time triggered by adverse environmental conditions, such as nutrient exhaustion [8–10]. In contrast, in multicellular organisms, cellular quiescence is not associated with scarcity but is rather controlled by complex physicochemical signals emanating from both the niche and the macroenvironment [4,11]. In fact, depending on the environmental input that triggers quiescence entry, quiescent cell properties drastically vary. For example, in budding yeast, the nature of the limiting nutrient significantly impacts cell viability in quiescence [12]. Furthermore, the transcriptome, proteome, and metabolome of quiescent yeast cells differ depending on whether they have ceased to proliferate following nitrogen, carbon, or phosphate starvation [13]. Similarly, human fibroblasts that have entered quiescence through contact inhibition, mitogen withdrawal or loss of adhesion display gene expression profiles that are different from one another [14]. Hence, both the micro- and macro- environment shape quiescence. Consequently, in starved unicellular organisms, quiescence is most of the time associated with low metabolic activity. In comparison, in multicellular organisms, quiescent cells are not necessarily "inactive", some of them being even more "active" metabolically than their proliferating counterparts [15]. In this case, the metabolic "activity" of quiescent cells depends on their nutritional status and the biological functions that cells have to perform. Thus, both the input (the reason of entering quiescence) and the output (the cellular consequences of entering quiescence) are highly diverse depending on the cell type.

The signalling pathways, including the transcription factors, involved in the quiescence establishment and maintenance, and the quiescent cell's metabolism properties have been recently described elsewhere and will not be discussed here [3,4,6,15]. In this review, our aim is to question the complex relationships between cellular quiescence and time. We will refer to time as a chronological concept, although time can be envisioned differently [16], and refer to ageing as the biological processes impacted by the effect of time [17–19]. We will try to untangle how time influences individual quiescent cell properties and quiescent population robustness, and reciprocally, how quiescence flexibility and heterogeneity play a pivotal role in facing ageing.

## 2. The Multiple Scales of Ageing

The effects of time on biological life operate at multiple scales. The extent to which individual cell ageing is connected to and responsible for ageing at the whole organism level is a complex question. As soon as in the mid-1950s, D. Harman proposed that ageing is caused by the harmful effects of free radicals that build up during life. He thus introduced the notion of a direct and linear relationship between time and ageing [20] The following 60 years of studies have shown that in multicellular organisms, many processes are at work during ageing. The dynamic field of geroscience is now focusing on understanding how the molecular mechanisms of ageing contribute to age-related diseases [21,22]. In 2013, ten hallmarks of ageing have been posited: genomic instability, telomere attrition, epigenetic alteration, loss of proteostasis, deregulation of nutrient sensing, mitochondrial dysfunction, cellular senescence, autophagy, stem cell exhaustion and alteration of intercellular communications [23]. Some of these hallmarks concern intracellular molecular processes, others rather concern tissue homeostasis, but all support the idea of ageing as a continuous and unidirectional process of decline.

In humans, it is now widely recognized that the occurrence of ageing-related diseases does not follow the linear trend of a proportional increase with age but rather accelerates at specific points throughout the lifespan. At the molecular level, many studies have reported the nonlinear alterations of macromolecules with the age of the individual. This has been particularly well documented for DNA methylation. Indeed, it was shown that this epigenetic mark exhibited nonlinear changes in intensity with age at DNA specific sites, some accelerating for the young and others slowing down for the elderly [24], with sudden hypermethylation events being observed at specific stages of life [25]. Recent deep multi-omics profiling on a longitudinal human cohort has shown that many molecular markers exhibited a nonlinear pattern throughout the ageing process [26], suggesting that ageing occurs in discrete steps rather than as a continuous process. More and more comprehensive multi-omics approaches have explored molecular changes that occur during ageing, and, helped by artificial intelligence, highly complex combinatory methods have allowed the building of "ageing machines" that are powerful enough to consider many molecular parameters, even those that vary nonlinearly with time [27].

The development of single cell multi-omics approaches with an extraordinary precision has opened up further new avenues for exploring the ageing process at the cellular level. Indeed, a combination of mass spectrometry, DNA sequencing, transcriptomic, epigenetic marks, and 3D genome organization analyses are now performed at the single cell level. Combined with high resolution imaging-based methods, these multi-omics approaches can take into account both space and time [28]. All these incredibly complex multimodal analyses converge to demonstrate that within a single body, organs do not age at the same rate [29,30] and that within the same organ, cell type composition and proportion vary with time [31]. These works revealed significant cell type-specific variations that extend far beyond differences in turnover rates and thus demonstrate that not all cell types age at the same speed. This is exemplified in the brain, where different types of neurons exhibit substantial variation in age-related changes in gene expression [32]. In an amazing study, Lu and colleagues have followed all these parameters and have launched an ageing atlas of *Drosophila*. By studying a single-nucleus transcriptomic, they tracked more than 150 distinct cell types and analyzed changes in tissues of male and female flies across their life span and as such portrayed ageing at the single cell level in a full organism [33]. Together, these studies illuminate the nonlinearity of ageing at the organism, organ, tissue, and cell type scales.

## 3. Time Generates Heterogeneity

What could be the cause of this nonlinearity? Could this be the result of heterogeneity in the properties of individual cells in facing age? To tackle these questions, a valuable model is quiescent cell populations. Yet, even when we narrow the focus to clonal cells, these populations are not as homogeneous as they appear, and time itself is generating heterogeneity.

In clonal quiescent cell populations, all cells may not have the same history. In 1961, Hayflick and Moorhead showed that human fetal fibroblasts in culture do not divide indefinitely but undergo a limited number of divisions, then enter senescence [34]. This phenomenon called "replicative ageing" has now been documented from yeast to human. It is acknowledged that a major causal feature of replicative senescence is telomere erosion, a process in which the telomeres lose 50–200 base pairs at each division, due to the inability of DNA polymerase to completely replicate the 3′ end of the DNA strand. In addition, dividing cells accumulate epigenetic marks, DNA mutations, and other damaged macromolecules. Consequently, at the individual cell level, replicative ageing was theorized as linear. Multi-omics studies have indeed recently demonstrated that senescence entry is a gradual process [35]. Whether replicative ageing is relevant for ageing at the whole-body level remains controversial [36], but the cumulative nature of the replicative age does not call into question as the two phenomena are conceptually linked in cellular models in which cell division is symmetric and conservative. However, when cells divide asymmetrically, the linearity may be broken as new daughter cells can be born "young", the "mother" cell keeping all the damages. This process called rejuvenation, somehow resets time and confers to the population a non-uniform ageing [37,38]. Therefore, replicative age, while cumulative, can possibly generate quiescent cells with different history and as such, different individual properties.

Another layer of heterogeneity comes from the fact that cells do not enter quiescence from the same stage in the cell cycle. The concept that posits that quiescent cells are all arrested at the same point in G1 (the restriction point), proposed by Pardee's was widely spread by the textbooks [39]. However, it is now demonstrated that cells can be quiescent not only in G1, but also in other cell cycle phases depending on the species, the cell type, or the quiescence input signal. For example, *Saccharomyces cerevisiae* cells enter quiescence mainly from G1, but they can accidentally arrest in other cell cycle phases [40]. Furthermore, *Schizosaccharomyces pombe*, *Cryptococcus neoformans*, *Tetrahymena pyriformis*, *Zea mays* shoot and root meristem cells, *Xenopus* neural progenitors, and many mammalian cell types, such as neural stem cells, can enter quiescence in G1 or G2 depending on external stimuli [41–46]. Furthermore, in mammals, oocytes are arrested in the diplotene stage of prophase I of meiosis. This arrest occurs during fetal development and persists until puberty [47]. In addition, for mammalian cell types entering quiescence from G1, it was proposed that cells do not halt synchronously at the restriction point but rather form a cohort of cells arrested as a continuum throughout G1 [41,48–50]. These asynchronous cell cycle arrests, thus, participate in the non-uniformity of quiescent cell populations.

Regardless of their history, quiescent cells must face directly the harmful effects of time, a process referred to as "chronological ageing". Indeed, with time, quiescent cells stochastically build up various metabolic by-products and wastes. As they cannot be diluted via cell division, these compounds have the potential to damage macromolecules and, thus, cellular machineries [51]. For example, quiescent cells may accumulate DNA mutations and/or epigenetic marks that could modify gene expression and hence protein quantity and/or quality. Proteins can also be directly damaged with a panel of so-called "degenerative protein modifications", such as oxidation, carbonylation, deamidation, etc., that may alter protein structure and functions [52]. Lipids are also age-targets, and their alteration may not only have direct impacts on membrane composition but can also produce by-products that are harmful for other cellular components [53–55]. Stochastic macromolecule alteration is obviously greatly influenced by the environment, since harsh conditions such as UV or toxic compound exposure somehow accelerate the deleterious effect of time. The niche also plays a crucial role, as extracellular matrix biophysical parameters and signals secreted by neighbouring lineages can influence the ageing process [11,56,57]. Thus, with time, stochastic macromolecule alterations shape cell individual experience and are another source of heterogeneity within clonal quiescent cell populations.

Overall, the diversity of individual cell properties is shaped by a combination of both deterministic factors relying on cell history and cumulative stochastic events linked to time and the environment. In fact, studying the appearance of heterogeneity within clonal cell populations through in silico approaches predicts that extrinsic factors (macro- and micro-environments) generate distinct subgroups in which cells share common features, while intrinsic factors (stochastic mutations, epigenetic cues, and macromolecule damages) create a broad spectrum of cells with different individual properties [58]. Obviously, quiescent population heterogeneity evolves with time, as described in muscle stem cells [59], illustrating that population heterogeneity itself is dynamic (Figure 1).

## 4. Heterogeneity and Robustness in Facing Time

As exemplified above, time influences the individual quiescent cell's characteristics, thus participating in the generation and the complexification of heterogeneous cell populations. But does this heterogeneity translate into different individual behaviour in facing time?

In *Drosophila*, cell cycle arrest heterogeneity directly influences the cells' ability to exit quiescence, as G2-arrested neural stem cells tend to reactivate more rapidly than those arrested in G1 [41]. In fact, quiescence entry in G2 has been associated with efficient stem cell regeneration in a variety of organisms, such as zebrafish, hydra, and axolotl [59]. Thus, the cell cycle stage at which cells enter quiescence does influence quiescent cell properties.

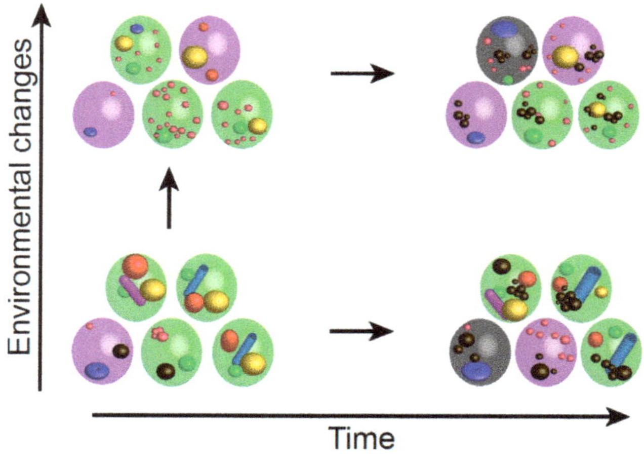


**Figure 1.** Each cell displays unique features, such as the organization of specific cellular machineries or cellular damages—as schematized by the different objects drawn inside the cells—thereby conferring to each cellspecific individual properties. These individual cell features evolve with time and/or with the environmental fluctuations, hence impacting cell fate (quiescent cells are green, senescent cells are violet, and dead cells are grey).

In *S. cerevisiae*, quiescent cell populations are tremendously heterogenous, this heterogeneity evolving with time. In this species, carbon exhaustion causes proliferation cessation that goes with a multitude of cellular reorganizations. For example, cells entirely reshape their nucleus, assembling a stable microtubule bundle form their centrosome, compacting their DNA and clustering their telomeres [60–64]. Cells may also assemble various cytoplasmic foci in which specific enzymes are embedded in inactive forms, such as RNA processing enzyme containing-P-bodies [65], or proteasome storage granules [66]. Some of these cellular reorganizations do form in the majority of the cells regardless of their history, while other are rather rare. This complex combination of reorganizations leads to a population made up of individuals with highly different phenotypes [67]. Importantly, some reorganizations, such as microtubule stabilization, are required for quiescent cell survival upon chronological ageing [68]. The underlying molecular reasons are still unknown, but this phenomenon has been documented in several species [69]. Interestingly, some reorganization are predictors of cell fate. For example, cells that will be capable of surviving in quiescence for weeks reorganize their mitochondria into small cortical vesicles that will reform a dynamic tubular network within few minutes upon quiescence exit. By contrast, quiescent cells that will enter senescence within a couple of days display globular mitochondria that progressively lose their oxphos abilities [70]. Although the interdependence between each cellular remodeling and how they are modulated by time and environmental factors remain mostly unclear, the ability of yeast cells to maintain quiescence and withstand ageing in a given environment is underpinned by their individual cell phenotype (Figure 1).

At the population level, this diversity of phenotypes could be crucial for the survival of the species. Indeed, some quiescent cells with given properties may be more capable of surviving in a cognate environment, while others may be more resilient in other kinds of harsh conditions. The most famous example is persisters. Within a genetically homogeneous population of quiescent bacterial cells, persister cells can survive to environmental stressors such as massive antibiotic exposure. When the stress is removed, persisters generate offspring that are susceptible to the same stress, demonstrating that "persistence" is not genetically encoded [71]. Intriguingly, in several species like *Escherichia coli*, *Salmonella typhimurium*, or *Shigella flexneri*, persister cells assembled a specific subcellular structure called regrowth-delay body, made of multiple proteins such as FtsZ (the tubulin homologue) and FtsA (the actin homologue) that favour re-proliferation [72]. As such, bacterial populations' heterogeneity can provide selective advantages in changing environments [73]. This is reminiscent of what is known for T-cell populations. Indeed, after activation by an infectious agent, few activated T-cells remain quiescent, creating an "immune memory" that provides a rapid response in case of an eventual re-infection by the same pathogenic agent. This collection of quiescent T-cells with different properties laid the foundation of an effective immune response in changing environments [74]. Interestingly, Muscle Stem Cells (MuSC) in mice display a functional heterogeneity in their responses to environmental stress, such as pollutant exposure, that depends on the level of expression of the transcription factor Pax3 [75]. Thus, heterogeneity in quiescent cell phenotype confers robustness to the population when external conditions are non-optimal or fluctuating.

## 5. Quiescence Deepening

Quiescent cells have not only to survive the deleterious effect of age, but with time, they also have to keep their capacity to reproliferate. In the 1960s, Nancy Bucher was focusing on liver regeneration. She observed in both rats and mice, that "young animals restore tissue mass and cell populations more rapidly than adults" and that "DNA synthesis gets underway more slowly as age increases" [76]. In 1974, Augenlicht and Baserga reported that the longer WI-38 fibroblasts spent in quiescence, the longer was the pre-replicative phase after re-stimulation. They proposed that "G0 is apparently composed of several different states, since the cells go 'deeper' into G0 with time", thereby introducing the notion of quiescence deepening [77]. As early as in 1976, Baserga proposed that "prolonged quiescence may bear some resemblance to the process of aging." Then, studies from the Baserga, the Soprano, and the O'Farrell laboratories repeatedly reported that the longer the time in quiescence, the longer it takes for quiescent cells to return into the proliferation cycle [78–81]. More recently, similar observations were made in *E. coli* [82], in budding [83] and fission yeast [84], in *C. elegans* [85] and in many mammalian models, including rat embryonic fibroblasts [86], hematopoietic stem cells, neural stem cells, and muscle stem cells [87–90].

Another characteristic of quiescence deepening is the "desensitization" to the re-proliferation stimulus. For example, with increasingly longer serum starvation, rat embryonic fibroblasts require a greater serum stimulation strength (concentration) to re-enter the cell cycle than cells in shallower quiescence [86,89]. This is reminiscent of the primary observation reporting that short-term quiescent WI-38 cells bind substantially higher amounts of EGF [91] and internalize EGF receptor complexes more rapidly [92] than long-term quiescent cells. Therefore, in several mammalian models, quiescence deepening is characterized by both a reduced sensitivity to stimulation and slower re-entry into the cell cycle.

The molecular basis of the quiescence deepening process is only beginning to be understood. It has long been observed that both the transcription and translation rates decrease and that the chromatin condensation increases with time in quiescence [9,80,93]. It was also shown that gene expression and metabolic profiles evolved with time in quiescence, as exemplified by studies in fibroblasts [14,89]. Hence, a gene signature, derived from transcriptomic approaches and called the quiescence depth score, has been proposed to predict the deepness of quiescence in both human fibroblast and mouse neural stem cells [94]. While we can speculate that the molecular processes underlying quiescence deepening likely involve a complex interplay between various molecular pathways, converging themes start to emerge. In fibroblasts, neural and muscle stem cells, decreased lysosomal autophagy drives progressively cells into deeper quiescence, likely by controlling the level of protein aggregates [89,95–98]. Furthermore, quiescence depth is influenced by the Rb-E2F network that acts as a bi-stable switch. Through experiments in fibroblasts and computational modeling, it was shown that different components of the Rb-E2F network can be manipulated to fine-tune quiescence depth. In fact, deep quiescent cells have a higher activation threshold of the Rb-E2F switch, requiring stronger stimulation to re-enter the cell cycle [86,93,97]. In addition, circadian proteins, such as cryptochrome (Cry) and Rev-erb, also regulate quiescence depth in rat embryonic fibroblasts, by downregulating cyclin D/Cdk4,6 activity, leading to an increased need for Rb-E2F activation threshold [99]. As with circadian proteins, the Rb–E2F switch is connected to many other regulatory pathways, including Notch, Hes1, p21, etc., all being involved in quiescence establishment. Hence, it was proposed that the Rb–E2F switch serves as a hub that integrates signals that quantitatively tune quiescence depth [93]. The bone morphogenetic protein (BMP) also plays a crucial role in regulating the depth of quiescence, notably in adult neural stem cells, where it was shown that BMP4 in combination with FGF2 induces a shallow quiescent state, whereas BMP4 alone induces a deeper quiescent state [100,101]. The implication of BMP in quiescence has recently been discussed in [102]. Finally, recent deep learning approaches leveraging both metabolic and epigenetic data have identified small molecules and genetic modulators that accurately predict quiescence depth in diverse biological contexts [103].

Taken together, all these studies converge on the idea that quiescence is not a uniform state but exists as a continuum that deepens with time. In some models, this trajectory may end in senescence, as evidenced by the significant similarity between the transcriptomic profiles of deep quiescent cells and senescent cells [104]. This idea has been elegantly discussed in [93]. Yet, our understanding of the molecular mechanisms that regulate the transitions between proliferation, quiescence, senescence, and cell death still remains rather limited and may be very different depending on the cell type and the organism.

## 6. Quiescence Exit Efficiency

In many stem cell lineages, it has been observed that quiescence exit is not homogenous and, as such, defines subpopulations displaying various swiftness in their ability to re-enter the proliferation cycle [4]. In adult muscles, quiescent MuSCs must respond to environmental signals in order to ensure the regeneration of the tissue. MuSCs quiescence is actively maintained through a combination of intrinsic factors and by the niche, including myofibers and other muscle-resident cells together with the basal lamina and adhesion molecules [105]. MuSCs are highly heterogeneous, and subpopulations can be distinguished by different gene expression profiles or surface markers [106]. MuSCs that display features of "deeper" quiescence display dynamic cellular projections. These structures retract following muscle injury in order to allow MuSC reproliferation, a reorganization occurring through a rapid cytoskeletal remodelling mediated by Rac/Rho GTPase signalling [107]. In fact, MuSCs exist in two different quiescent states: a dormant state that is thought to preserve the long-term regenerative capacity of muscle tissue, and a more responsive, primed state called G(Alert) [108]. MuSCs in the G(Alert) state exhibit an increased cell size, elevated transcriptional and mitochondrial activity, and higher ATP levels compared to their deeply quiescent counterparts [2]. High levels of PAX7 and low levels of MYF5 delineate the deeply quiescent MuSC subpopulation [109]. The transition from deep quiescence to G(Alert) is orchestrated by systemic signals released upon tissue injury that involve the hepatocyte growth factor activator (HGFA) and HMGB1, which activate the mTORC1 and the CXCL12/CXCR4 pathways [108].

Similarly, among hematopoietic stem cell populations (HSCs), long-term HSCs (LT-HSCs) are deeply quiescent and can be "activated" to become short-term HSCs (ST-HSCs), which, while still quiescent, are primed for cell cycle entry [88,110]. This transition is tightly controlled by extrinsic cues from the bone marrow niche and intrinsic regulators, such as the expression of cell cycle regulators like CDK6 or CD38 [111,112]. Since LT-HSCs are characterized by their potent self-renewal capacity over the entire lifespan of an organism, it was proposed that LT-HSCs ensure a reservoir for long-term hematopoietic integrity, while ST-HSC allow the rapid expansion of HSCs in response to the physiological demands or stress [113,114].

Likewise, primed quiescent neural stem cells (p-qNSC) characterized by a high level of LRIG1 and CD9 have been described in adult mice brains [115]. These cells display a reduced glycolytic metabolism, a downregulated Notch pathway, and upregulation of protein synthesis [116,117]. Primed qNSC are poised for rapid cell cycle re-entry and can efficiently engraft into the adult subventricular zone niche, thereby contributing to tissue regeneration and repair. Interestingly, qNSC from young and old mice have no major differences, neither in their levels of DNA damage nor in their transcriptional profile. Yet, young qNSCs transit more efficiently to a primed-quiescent state than older qNSCs, but once activated, both young and old qNSCs display a similar capacity to undergo several rounds of division [118].

In the above examples, quiescence is described as a binary state with "deep/slow" or "primed/fast" states, the latter being induced via an extrinsic pre-stimulation from the niche. Yet, could quiescence "responsiveness" be more gradual and exit as a continuum within a population? Could intrinsic individual cell features be sufficient to support a variation in quiescence exit efficiency, independently of a pre-stimulation by the niche? In quiescent *S. cerevisiae* population, a continuum in cell size distribution correlates with a continuum in quiescence exit efficacy, the tendency being that the bigger the cell, the more prone it is to exit quiescence efficiently. Yet this is not an absolute rule, and while daughter cell quiescence exit efficiency is mostly controlled by a sizer, in mother cells, it is neither influenced by the cell volume nor by the cell replicative age, but rather depends on a "timer" [83]. In quiescent rat embryonic fibroblast populations, some quiescent cells are capable of responding to a short serum stimulation, while others need to be in contact with the stimulus for a longer period of time to be able to exit quiescence [119]. Thus, a continuum of quiescent cell responsiveness does exist in a clonal cell population in which cells have experienced the same environmental history [119]. This heterogeneity in quiescence exit efficiency seems to be influenced by a "memory" combining deterministic factors, such as the cell size at the time of quiescence induction, with stochastic elements, such as noise in the Rb-E2F bi-stable switch [119].

Altogether, these observations underscore that heterogeneity in quiescence exit efficiency is a common feature of quiescent populations of many cell types. Depending on the context, this heterogeneity may be the consequence of stochastic individual features or can be induced by deterministic events that confer cells a predictable behaviour. This diversity may enhance the robustness and the resilience of population facing age and/or environmental fluctuations. In tissues, this may safeguard against acute depletion of the quiescent cell reservoir in response to a single activating stimulus. In single cell species, such as bacteria, such intrapopulation heterogeneity has been proposed to provide a bet-hedging advantage, ensuring that not all cells respond identically to reactivation signals and resume growth simultaneously. This strategy may protect the population from sudden environmental threats and support long-term survival [120]. More generally, heterogeneity in quiescence exit efficiency could enhance the long-term survival and adaptability of the population in fluctuating environments.

## 7. Conclusions and Model

While it is experimentally challenging to distinguish quiescence deepening and priming—as deepening is synonymous with delayed quiescence exit—these two processes are not conceptually equivalent. By definition, quiescence is reversible and, as such, is an "aller-retour". The "aller" part of the ticket corresponds to the length of time a cell has spent in quiescence. This time is incompressible. During that trip, cells accumulate damages, not necessarily with a linear trend, but that impact quiescent cell individual properties and, as such, may contribute to the rising of population heterogeneity. The "retour" is the time a cell needs to exit quiescence id est to re-enter the proliferation cycle. The length of the "retour" trip depends on the time spent in quiescence, the environment and the cell individual features (Figure 2A). The "retour" is related to the "aller" in that the longer a cell has been in quiescence, the more time it needs to return into the proliferation cycle. But for a similar "aller" time, the time needed for the retour differs depending on the quiescence depth (Figure 2B(a)), the cell capacity to sense and react to the quiescence exit signal (Figure 2B(b)), and the speed at which the cell will exit quiescence (Figure 2B(c)). Together, these models highlight the multifaceted nature of quiescence and its relationships to time, demonstrating that a quiescent cell population can encompass a collection of cellular universes that co-exist at the same time.

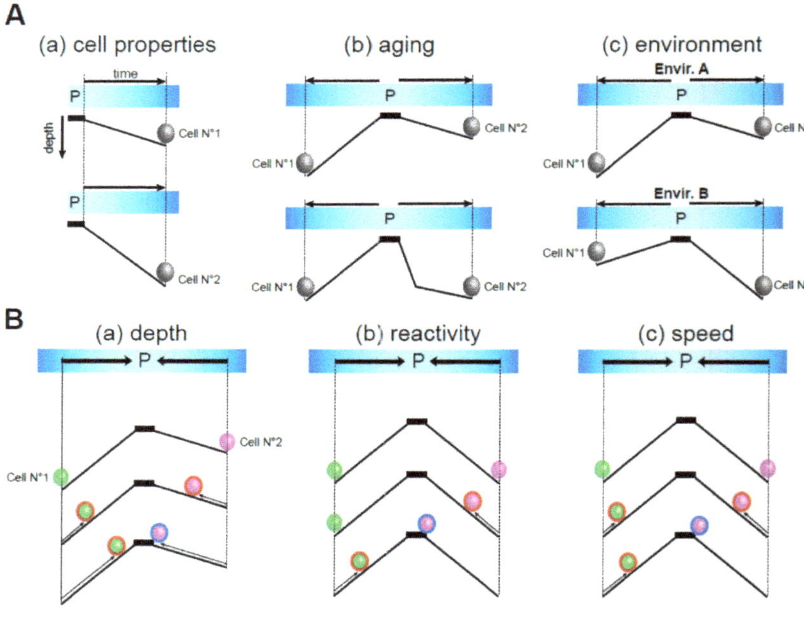


**Figure 2.** (**A**) Depending on their individual properties, time, and environment, cells deepen into quiescence with different slopes. (P stands for proliferation). (**B**) The quiescence exit process is unique to each individual cell and depends on the depth of quiescence, the quiescence exit speed, and the reactivity of quiescent cells to the quiescence exit signal. Schemes represent two individual cells with (a) the same reactivity, the same quiescence exit speed but different quiescence depth; or (b) same reactivity, same quiescence depth but different quiescence exit speed; or (c) same quiescence depth, the same quiescence exit speed but different reactivity to the quiescence exit signal. Quiescent cells that are "activated for exit" are circled in red and cells entering into the proliferation cycle are circled in blue.

**Funding:** This work was funded by the Agence Nationale pour la Recherche (ANR-21-CE13-0023-01), the University of Bordeaux, the Ligue Contre le Cancer de la Dordogne and the Centre National de la Recherche Scientifique.

**Institutional Review Board Statement:** Not applicable.

**Informed Consent Statement:** Not applicable.

**Data Availability Statement:** No new data were created or analyzed in this study.

**Acknowledgments:** We express our gratitude to Bertrand Daignan-Fornier for suggestions on the manuscript. We are really thankful to Y. Charbonnel for helps with Figure 2.

**Conflicts of Interest:** The authors declare no conflicts of interest.

## Abbreviations

| | |
|---|---|
| EGF | Epidermal Growth Factor |
| MuSC | Muscle Stem Cell |
| HSC | Hematopoietic Stem Cell |
| NSC | Neural Stem Cell |

## References

1. Sender, R.; Milo, R. The Distribution of Cellular Turnover in the Human Body. *Nat. Med.* **2021**, *27*, 45–48. [CrossRef] [PubMed]
2. van Velthoven, C.T.J.; Rando, T.A. Stem Cell Quiescence: Dynamism, Restraint, and Cellular Idling. *Cell Stem Cell* **2019**, *24*, 213–225. [CrossRef]
3. Marescal, O.; Cheeseman, I.M. Cellular Mechanisms and Regulation of Quiescence. *Dev. Cell* **2020**, *55*, 259–271. [CrossRef] [PubMed]
4. de Morree, A.; Rando, T.A. Regulation of Adult Stem Cell Quiescence and Its Functions in the Maintenance of Tissue Integrity. *Nat. Rev. Mol. Cell Biol.* **2023**, *24*, 334–354. [CrossRef] [PubMed]
5. Cheung, T.H.; Rando, T.A. Molecular Regulation of Stem Cell Quiescence. *Nat. Rev. Mol. Cell Biol.* **2013**, *14*, 329–340. [CrossRef]
6. Mitra, M.; Batista, S.L.; Coller, H.A. Transcription Factor Networks in Cellular Quiescence. *Nat. Cell Biol.* **2025**, *27*, 14–27. [CrossRef]
7. Sang, L.; Coller, H.A.; Roberts, J.M. Control of the Reversibility of Cellular Quiescence by the Transcriptional Repressor HES1. *Science* **2008**, *321*, 1095–1100. [CrossRef]
8. De Virgilio, C. The Essence of Yeast Quiescence. *FEMS Microbiol. Rev.* **2012**, *36*, 306–339. [CrossRef]
9. Gray, J.V.; Petsko, G.A.; Johnston, G.C.; Ringe, D.; Singer, R.A.; Werner-Washburne, M. "Sleeping Beauty": Quiescence in Saccharomyces cerevisiae. *Microbiol. Mol. Biol. Rev. MMBR* **2004**, *68*, 187–206. [CrossRef]
10. Daignan-Fornier, B.; Laporte, D.; Sagot, I. Quiescence Through the Prism of Evolution. *Front. Cell Dev. Biol.* **2021**, *9*, 745069. [CrossRef]
11. Fiore, A.P.Z.P.; de Ribeiro, P.F.; Bruni-Cardoso, A. Sleeping Beauty and the Microenvironment Enchantment: Microenvironmental Regulation of the Proliferation-Quiescence Decision in Normal Tissues and in Cancer Development. *Front. Cell Dev. Biol.* **2018**, *6*, 59. [CrossRef] [PubMed]
12. Boer, V.M.; Crutchfield, C.A.; Bradley, P.H.; Botstein, D.; Rabinowitz, J.D. Growth-Limiting Intracellular Metabolites in Yeast Growing under Diverse Nutrient Limitations. *Mol. Biol. Cell* **2010**, *21*, 198–211. [CrossRef] [PubMed]
13. Klosinska, M.M.; Crutchfield, C.A.; Bradley, P.H.; Rabinowitz, J.D.; Broach, J.R. Yeast Cells Can Access Distinct Quiescent States. *Genes Dev.* **2011**, *25*, 336–349. [CrossRef]
14. Coller, H.A.; Sang, L.; Roberts, J.M. A New Description of Cellular Quiescence. *PLoS Biol.* **2006**, *4*, e83. [CrossRef] [PubMed]
15. Coller, H.A. The Paradox of Metabolism in Quiescent Stem Cells. *FEBS Lett.* **2019**, *593*, 2817–2839. [CrossRef]
16. Toynbee, A.J.; Markowitz, W.; Smart, J.J.C. Time and its role in the history of thought and action. *Encyclopedia Britannica*, 1 January 2025.
17. Lemoine, M. Defining Aging. *Biol. Philos.* **2020**, *35*, 46. [CrossRef]
18. Blasimme, A.; Boniolo, G.; Nathan, M.J. Rethinking Ageing: Introduction. *Hist. Philos. Life Sci.* **2021**, *43*, 95. [CrossRef]
19. Saborido, C.; García-Barranquero, P. Is Aging a Disease? The Theoretical Definition of Aging in the Light of the Philosophy of Medicine. *J. Med. Philos. Forum Bioeth. Philos. Med.* **2022**, *47*, 770–783. [CrossRef]
20. Harman, D. Aging: A Theory Based on Free Radical and Radiation Chemistry. *J. Gerontol.* **1956**, *11*, 298–300. [CrossRef]

21. Kennedy, B.K.; Berger, S.L.; Brunet, A.; Campisi, J.; Cuervo, A.M.; Epel, E.S.; Franceschi, C.; Lithgow, G.J.; Morimoto, R.I.; Pessin, J.E.; et al. Geroscience: Linking Aging to Chronic Disease. *Cell* **2014**, *159*, 709–713. [CrossRef]
22. Kroemer, G.; Maier, A.B.; Cuervo, A.M.; Gladyshev, V.N.; Ferrucci, L.; Gorbunova, V.; Kennedy, B.K.; Rando, T.A.; Seluanov, A.; Sierra, F.; et al. From Geroscience to Precision Geromedicine: Understanding and Managing Aging. *Cell* **2025**, *188*, 2043–2062. [CrossRef]
23. López-Otín, C.; Blasco, M.A.; Partridge, L.; Serrano, M.; Kroemer, G. The Hallmarks of Aging. *Cell* **2013**, *153*, 1194–1217. [CrossRef] [PubMed]
24. Vershinina, O.; Bacalini, M.G.; Zaikin, A.; Franceschi, C.; Ivanchenko, M. Disentangling Age-Dependent DNA Methylation: Deterministic, Stochastic, and Nonlinear. *Sci. Rep.* **2021**, *11*, 9201. [CrossRef] [PubMed]
25. Olecka, M.; van Bömmel, A.; Best, L.; Haase, M.; Foerste, S.; Riege, K.; Dost, T.; Flor, S.; Witte, O.W.; Franzenburg, S.; et al. Nonlinear DNA Methylation Trajectories in Aging Male Mice. *Nat. Commun.* **2024**, *15*, 3074. [CrossRef]
26. Shen, X.; Wang, C.; Zhou, X.; Zhou, W.; Hornburg, D.; Wu, S.; Snyder, M.P. Nonlinear Dynamics of Multi-Omics Profiles during Human Aging. *Nat. Aging* **2024**, *4*, 1619–1634. [CrossRef] [PubMed]
27. Rutledge, J.; Oh, H.; Wyss-Coray, T. Measuring Biological Age Using Omics Data. *Nat. Rev. Genet.* **2022**, *23*, 715–727. [CrossRef]
28. Vandereyken, K.; Sifrim, A.; Thienpont, B.; Voet, T. Methods and Applications for Single-Cell and Spatial Multi-Omics. *Nat. Rev. Genet.* **2023**, *24*, 494–515. [CrossRef]
29. Schaum, N.; Lehallier, B.; Hahn, O.; Pálovics, R.; Hosseinzadeh, S.; Lee, S.E.; Sit, R.; Lee, D.P.; Losada, P.M.; Zardeneta, M.E.; et al. Ageing Hallmarks Exhibit Organ-Specific Temporal Signatures. *Nature* **2020**, *583*, 596–602. [CrossRef]
30. Nie, C.; Li, Y.; Li, R.; Yan, Y.; Zhang, D.; Li, T.; Li, Z.; Sun, Y.; Zhen, H.; Ding, J.; et al. Distinct Biological Ages of Organs and Systems Identified from a Multi-Omics Study. *Cell Rep.* **2022**, *38*, 110459. [CrossRef]
31. Izgi, H.; Han, D.; Isildak, U.; Huang, S.; Kocabiyik, E.; Khaitovich, P.; Somel, M.; Dönertaş, H.M. Inter-Tissue Convergence of Gene Expression during Ageing Suggests Age-Related Loss of Tissue and Cellular Identity. *eLife* **2022**, *11*, e68048. [CrossRef]
32. Chien, J.-F.; Liu, H.; Wang, B.-A.; Luo, C.; Bartlett, A.; Castanon, R.; Johnson, N.D.; Nery, J.R.; Osteen, J.; Li, J.; et al. Cell-Type-Specific Effects of Age and Sex on Human Cortical Neurons. *Neuron* **2024**, *112*, 2524–2539.e5. [CrossRef] [PubMed]
33. Lu, T.-C.; Brbić, M.; Park, Y.-J.; Jackson, T.; Chen, J.; Kolluru, S.S.; Qi, Y.; Katheder, N.S.; Cai, X.T.; Lee, S.; et al. Aging Fly Cell Atlas Identifies Exhaustive Aging Features at Cellular Resolution. *Science* **2023**, *380*, eadg0934. [CrossRef] [PubMed]
34. Hayflick, L.; Moorhead, P.S. The Serial Cultivation of Human Diploid Cell Strains. *Exp. Cell Res.* **1961**, *25*, 585–621. [CrossRef] [PubMed]
35. Chan, M.; Yuan, H.; Soifer, I.; Maile, T.M.; Wang, R.Y.; Ireland, A.; O'Brien, J.J.; Goudeau, J.; Chan, L.J.G.; Vijay, T.; et al. Novel Insights from a Multiomics Dissection of the Hayflick Limit. *eLife* **2022**, *11*, e70283. [CrossRef]
36. Wright, W.E.; Shay, J.W. Historical Claims and Current Interpretations of Replicative Aging. *Nat. Biotechnol.* **2002**, *20*, 682–688. [CrossRef]
37. Hormoz, S. Stem Cell Population Asymmetry Can Reduce Rate of Replicative Aging. *J. Theor. Biol.* **2013**, *331*, 19–27. [CrossRef]
38. Sheldrake, A.R. Cellular Senescence, Rejuvenation and Potential Immortality. *Proc. Biol. Sci.* **2022**, *289*, 20212434. [CrossRef]
39. Pardee, A.B. A Restriction Point for Control of Normal Animal Cell Proliferation. *Proc. Natl. Acad. Sci. USA* **1974**, *71*, 1286–1290. [CrossRef]
40. Laporte, D.; Lebaudy, A.; Sahin, A.; Pinson, B.; Ceschin, J.; Daignan-Fornier, B.; Sagot, I. Metabolic Status Rather than Cell Cycle Signals Control Quiescence Entry and Exit. *J. Cell Biol.* **2011**, *192*, 949–957. [CrossRef]
41. Otsuki, L.; Brand, A.H. Cell Cycle Heterogeneity Directs the Timing of Neural Stem Cell Activation from Quiescence. *Science* **2018**, *360*, 99–102. [CrossRef]
42. Costello, G.; Rodgers, L.; Beach, D. Fission Yeast Enters the Stationary Phase G0 State from Either Mitotic G1 or G2. *Curr. Genet.* **1986**, *11*, 119–125. [CrossRef]
43. Takeo, K.; Tanaka, R.; Miyaji, M.; Nishimura, K. Unbudded G2 as Well as G1 Arrest in the Stationary Phase of the Basidiomycetous Yeast Cryptococcus neoformans. *FEMS Microbiol. Lett.* **1995**, *129*, 231–235. [PubMed]
44. Cameron, I.L.; Bols, N.C. Effect of Cell Population Density on G2 Arrest in Tetrahymena. *J. Cell Biol.* **1975**, *67*, 518–522. [CrossRef] [PubMed]
45. Conger, B.V.; Carabia, J.V. Microspectrophotometric Determination of the 2C and 4C Nuclear Complement in the Root and Shoot of the Dormant Maize Embryo. *Environ. Exp. Bot.* **1976**, *16*, 171–175. [CrossRef]
46. McKeown, C.R.; Cline, H.T. Nutrient Restriction Causes Reversible G2 Arrest in Xenopus Neural Progenitors. *Dev. Camb. Engl.* **2019**, *146*, dev178871. [CrossRef]
47. Tripathi, A.; Kumar, K.V.P.; Chaube, S.K. Meiotic Cell Cycle Arrest in Mammalian Oocytes. *J. Cell. Physiol.* **2010**, *223*, 592–600. [CrossRef]
48. Cooper, S. On the Proposal of a G0 Phase and the Restriction Point. *FASEB J. Off. Publ. Fed. Am. Soc. Exp. Biol.* **1998**, *12*, 367–373. [CrossRef]

49. Cooper, S. Reappraisal of Serum Starvation, the Restriction Point, G0, and G1 Phase Arrest Points. *FASEB J. Off. Publ. Fed. Am. Soc. Exp. Biol.* **2003**, *17*, 333–340. [CrossRef]
50. Matson, J.P.; Cook, J.G. Cell Cycle Proliferation Decisions: The Impact of Single Cell Analyses. *FEBS J.* **2017**, *284*, 362–375. [CrossRef]
51. Ogrodnik, M.; Salmonowicz, H.; Gladyshev, V.N. Integrating Cellular Senescence with the Concept of Damage Accumulation in Aging: Relevance for Clearance of Senescent Cells. *Aging Cell* **2019**, *18*, e12841. [CrossRef]
52. Gallart-Palau, X.; Tan, L.M.; Serra, A.; Gao, Y.; Ho, H.H.; Richards, A.M.; Kandiah, N.; Chen, C.P.; Kalaria, R.N.; Sze, S.K. Degenerative Protein Modifications in the Aging Vasculature and Central Nervous System: A Problem Shared Is Not Always Halved. *Ageing Res. Rev.* **2019**, *53*, 100909. [CrossRef] [PubMed]
53. Ayala, A.; Muñoz, M.F.; Argüelles, S. Lipid Peroxidation: Production, Metabolism, and Signaling Mechanisms of Malondialdehyde and 4-Hydroxy-2-Nonenal. *Oxid. Med. Cell. Longev.* **2014**, *2014*, 360438. [CrossRef]
54. Gille, B.; Galuska, C.E.; Fuchs, B.; Peleg, S. Recent Advances in Studying Age-Associated Lipids Alterations and Dietary Interventions in Mammals. *Front. Aging* **2021**, *2*, 773795. [CrossRef]
55. Gęgotek, A.; Skrzydlewska, E. Biological Effect of Protein Modifications by Lipid Peroxidation Products. *Chem. Phys. Lipids* **2019**, *221*, 46–52. [CrossRef] [PubMed]
56. Brunet, A.; Goodell, M.A.; Rando, T.A. Ageing and Rejuvenation of Tissue Stem Cells and Their Niches. *Nat. Rev. Mol. Cell Biol.* **2023**, *24*, 45–62. [CrossRef]
57. Malecki, K.M.C.; Andersen, J.K.; Geller, A.M.; Harry, G.J.; Jackson, C.L.; James, K.A.; Miller, G.W.; Ottinger, M.A. Integrating Environment and Aging Research: Opportunities for Synergy and Acceleration. *Front. Aging Neurosci.* **2022**, *14*, 824921. [CrossRef] [PubMed]
58. Stockholm, D.; Benchaouir, R.; Picot, J.; Rameau, P.; Neildez, T.M.A.; Landini, G.; Laplace-Builhé, C.; Paldi, A. The Origin of Phenotypic Heterogeneity in a Clonal Cell Population in Vitro. *PLoS ONE* **2007**, *2*, e394. [CrossRef]
59. Sutcu, H.H.; Ricchetti, M. Loss of Heterogeneity, Quiescence, and Differentiation in Muscle Stem Cells. *Stem Cell Investig.* **2018**, *5*, 9. [CrossRef]
60. Laporte, D.; Courtout, F.; Salin, B.; Ceschin, J.; Sagot, I. An Array of Nuclear Microtubules Reorganizes the Budding Yeast Nucleus during Quiescence. *J. Cell Biol.* **2013**, *203*, 585–594. [CrossRef]
61. Laporte, D.; Courtout, F.; Tollis, S.; Sagot, I. Quiescent Saccharomyces cerevisiae Forms Telomere Hyperclusters at the Nuclear Membrane Vicinity through a Multifaceted Mechanism Involving Esc1, the Sir Complex, and Chromatin Condensation. *Mol. Biol. Cell* **2016**, *27*, 1875–1884. [CrossRef]
62. Laporte, D.; Sagot, I. Microtubules Move the Nucleus to Quiescence. *Nucleus* **2014**, *5*, 113–118. [CrossRef] [PubMed]
63. Guidi, M.; Ruault, M.; Marbouty, M.; Loïodice, I.; Cournac, A.; Billaudeau, C.; Hocher, A.; Mozziconacci, J.; Koszul, R.; Taddei, A. Spatial Reorganization of Telomeres in Long-Lived Quiescent Cells. *Genome Biol.* **2015**, *16*, 206. [CrossRef]
64. Rutledge, M.T.; Russo, M.; Belton, J.-M.; Dekker, J.; Broach, J.R. The Yeast Genome Undergoes Significant Topological Reorganization in Quiescence. *Nucleic Acids Res.* **2015**, *43*, 8299–8313. [CrossRef]
65. Sheth, U.; Parker, R. Decapping and Decay of Messenger RNA Occur in Cytoplasmic Processing Bodies. *Science* **2003**, *300*, 805–808. [CrossRef] [PubMed]
66. Laporte, D.; Salin, B.; Daignan-Fornier, B.; Sagot, I. Reversible Cytoplasmic Localization of the Proteasome in Quiescent Yeast Cells. *J. Cell Biol.* **2008**, *181*, 737–745. [CrossRef]
67. Sagot, I.; Laporte, D. The Cell Biology of Quiescent Yeast—A Diversity of Individual Scenarios. *J. Cell Sci.* **2019**, *132*, jcs213025. [CrossRef] [PubMed]
68. Laporte, D.; Massoni-Laporte, A.; Lefranc, C.; Dompierre, J.; Mauboules, D.; Nsamba, E.T.; Royou, A.; Gal, L.; Schuldiner, M.; Gupta, M.L.; et al. A Stable Microtubule Bundle Formed through an Orchestrated Multistep Process Controls Quiescence Exit. *eLife* **2024**, *12*, RP89958. [CrossRef]
69. Laporte, D.; Sagot, I. Microtubule Reorganization and Quiescence: An Intertwined Relationship. *Physiol. Bethesda Md.* **2025**, *40*, 145–157. [CrossRef]
70. Laporte, D.; Gouleme, L.; Jimenez, L.; Khemiri, I.; Sagot, I. Mitochondria Reorganization upon Proliferation Arrest Predicts Individual Yeast Cell Fate. *eLife* **2018**, *7*, e35685. [CrossRef]
71. Niu, H.; Gu, J.; Zhang, Y. Bacterial Persisters: Molecular Mechanisms and Therapeutic Development. *Signal Transduct. Target. Ther.* **2024**, *9*, 174. [CrossRef]
72. Yu, J.; Liu, Y.; Yin, H.; Chang, Z. Regrowth-Delay Body as a Bacterial Subcellular Structure Marking Multidrug-Tolerant Persisters. *Cell Discov.* **2019**, *5*, 8. [CrossRef] [PubMed]
73. Davis, K.M.; Isberg, R.R. Defining Heterogeneity within Bacterial Populations via Single Cell Approaches. *BioEssays News Rev. Mol. Cell. Dev. Biol.* **2016**, *38*, 782–790. [CrossRef] [PubMed]
74. Arias, C.F.; Herrero, M.A.; Acosta, F.J.; Fernandez-Arias, C. Population Mechanics: A Mathematical Framework to Study T Cell Homeostasis. *Sci. Rep.* **2017**, *7*, 9511. [CrossRef] [PubMed]

75. Vartanian, A.D.; Quétin, M.; Michineau, S.; Auradé, F.; Hayashi, S.; Dubois, C.; Rocancourt, D.; Drayton-Libotte, B.; Szegedi, A.; Buckingham, M.; et al. PAX3 Confers Functional Heterogeneity in Skeletal Muscle Stem Cell Responses to Environmental Stress. *Cell Stem Cell* **2019**, *24*, 958–973.e9. [CrossRef]
76. Bucher, N.L.R. Regeneration of mammalian liver. *Int. Rev. Cytol.* **1963**, *15*, 245–300. [CrossRef]
77. Augenlicht, L.H.; Baserga, R. Changes in the G0 State of WI-38 Fibroblasts at Different Times after Confluence. *Exp. Cell Res.* **1974**, *89*, 255–262. [CrossRef]
78. Owen, T.A.; Soprano, D.R.; Soprano, K.J. Analysis of the Growth Factor Requirements for Stimulation of WI-38 Cells after Extended Periods of Density-Dependent Growth Arrest. *J. Cell. Physiol.* **1989**, *139*, 424–431. [CrossRef]
79. Owen, T.A.; Cosenza, S.C.; Soprano, D.R.; Soprano, K.J. Time of C-Fos and c-Myc Expression in Human Diploid Fibroblasts Stimulated to Proliferate after Prolonged Periods in Quiescence. *J. Biol. Chem.* **1987**, *262*, 15111–15117. [CrossRef]
80. Rossini, M.; Lin, J.-C.; Baserga, R. Effects of Prolonged Quiescence on Nuclei and Chromatin of WI-38 Fibroblasts. *J. Cell. Physiol.* **1976**, *88*, 1–11. [CrossRef]
81. Yanez, I.; O'Farrell, M. Variation in the Length of the Lag Phase Following Serum Restimulation of Mouse 3T3 Cells. *Cell Biol. Int. Rep.* **1989**, *13*, 453–462. [CrossRef]
82. Levin-Reisman, I.; Gefen, O.; Fridman, O.; Ronin, I.; Shwa, D.; Sheftel, H.; Balaban, N.Q. Automated Imaging with ScanLag Reveals Previously Undetectable Bacterial Growth Phenotypes. *Nat. Methods* **2010**, *7*, 737–739. [CrossRef] [PubMed]
83. Laporte, D.; Jimenez, L.; Gouleme, L.; Sagot, I. Yeast Quiescence Exit Swiftness Is Influenced by Cell Volume and Chronological Age. *Microb. Cell* **2017**, *5*, 104–111. [CrossRef] [PubMed]
84. Liu, Q.; Sheng, N.; Zhang, Z.; He, C.; Zhao, Y.; Sun, H.; Chen, J.; Yang, X.; Tang, C. Initial Nutrient Condition Determines the Recovery Speed of Quiescent Cells in Fission Yeast. *Heliyon* **2024**, *10*, e26558. [CrossRef] [PubMed]
85. Olmedo, M.; Mata-Cabana, A.; Rodríguez-Palero, M.J.; García-Sánchez, S.; Fernández-Yañez, A.; Merrow, M.; Artal-Sanz, M. Prolonged Quiescence Delays Somatic Stem Cell-like Divisions in Caenorhabditis Elegans and Is Controlled by Insulin Signaling. *Aging Cell* **2020**, *19*, e13085. [CrossRef]
86. Kwon, J.S.; Everetts, N.J.; Wang, X.; Wang, W.; Della Croce, K.; Xing, J.; Yao, G. Controlling Depth of Cellular Quiescence by an Rb-E2F Network Switch. *Cell Rep.* **2017**, *20*, 3223–3235. [CrossRef]
87. Urbán, N.; Cheung, T.H. Stem Cell Quiescence: The Challenging Path to Activation. *Development* **2021**, *148*, dev165084. [CrossRef]
88. Cabezas-Wallscheid, N.; Buettner, F.; Sommerkamp, P.; Klimmeck, D.; Ladel, L.; Thalheimer, F.B.; Pastor-Flores, D.; Roma, L.P.; Renders, S.; Zeisberger, P.; et al. Vitamin A-Retinoic Acid Signaling Regulates Hematopoietic Stem Cell Dormancy. *Cell* **2017**, *169*, 807–823.e19. [CrossRef]
89. Fujimaki, K.; Li, R.; Chen, H.; Della Croce, K.; Zhang, H.H.; Xing, J.; Bai, F.; Yao, G. Graded Regulation of Cellular Quiescence Depth between Proliferation and Senescence by a Lysosomal Dimmer Switch. *Proc. Natl. Acad. Sci. USA* **2019**, *116*, 22624–22634. [CrossRef]
90. Audesse, A.J.; Webb, A.E. Mechanisms of Enhanced Quiescence in Neural Stem Cell Aging. *Mech. Ageing Dev.* **2020**, *191*, 111323. [CrossRef]
91. Donigan, A.M.; Cavalli, R.C.; Pena, A.A.; Savage, C.R.; Soprano, D.R.; Soprano, K.J. Epidermal Growth Factor Receptors Lose Ligand Binding Ability as WI-38 Cells Progress from Short-Term to Long-Term Quiescence. *J. Cell. Physiol.* **1993**, *155*, 164–170. [CrossRef]
92. Soprano, K.J. WI-38 Cell Long-Term Quiescence Model System: A Valuable Tool to Study Molecular Events That Regulate Growth. *J. Cell. Biochem.* **1994**, *54*, 405–414. [CrossRef] [PubMed]
93. Yao, G. Quiescence-Origin Senescence: A New Paradigm in Cellular Aging. *Biomedicines* **2024**, *12*, 1837. [CrossRef]
94. Wei, M.Y.; Yao, G. Modeling the Depth of Cellular Dormancy from RNA-Sequencing Data. *Methods Mol. Biol.* **2024**, *2811*, 123–135. [CrossRef] [PubMed]
95. Murley, A.; Popovici, A.C.; Hu, X.S.; Lund, A.; Wickham, K.; Durieux, J.; Joe, L.; Koronyo, E.; Zhang, H.; Genuth, N.R.; et al. Quiescent Cell Re-Entry Is Limited by Macroautophagy-Induced Lysosomal Damage. *Cell* **2025**, *188*, 2670–2686.e14. [CrossRef]
96. Murley, A.; Dillin, A. Macroautophagy in Quiescent and Senescent Cells: A Pathway to Longevity? *Trends Cell Biol.* **2023**, *33*, 495–504. [CrossRef]
97. Fujimaki, K.; Yao, G. Cell Dormancy Plasticity: Quiescence Deepens into Senescence through a Dimmer Switch. *Physiol. Genom.* **2020**, *52*, 558–562. [CrossRef]
98. Leeman, D.S.; Hebestreit, K.; Ruetz, T.; Webb, A.E.; McKay, A.; Pollina, E.A.; Dulken, B.W.; Zhao, X.; Yeo, R.W.; Ho, T.T.; et al. Lysosome Activation Clears Aggregates and Enhances Quiescent Neural Stem Cell Activation during Aging. *Science* **2018**, *359*, 1277–1283. [CrossRef] [PubMed]
99. Wang, X.; Liu, B.; Pan, Q.; Kwon, J.S.; Miller, M.A.; Croce, K.D.; Yao, G. Circadian Proteins Cry and Rev-Erb Converge to Deepen Cellular Quiescence by Downregulating Cyclin D and Cdk4,6. *bioRxiv* **2021**. [CrossRef]

100. Mira, H.; Andreu, Z.; Suh, H.; Lie, D.C.; Jessberger, S.; Consiglio, A.; San Emeterio, J.; Hortigüela, R.; Marqués-Torrejón, M.A.; Nakashima, K.; et al. Signaling through BMPR-IA Regulates Quiescence and Long-Term Activity of Neural Stem Cells in the Adult Hippocampus. *Cell Stem Cell* **2010**, *7*, 78–89. [CrossRef]
101. Mira, H.; Morante, J. Neurogenesis From Embryo to Adult—Lessons From Flies and Mice. *Front. Cell Dev. Biol.* **2020**, *8*, 533. [CrossRef]
102. Elkin, A.M.; Robbins, S.; Barros, C.S.; Bossing, T. The Critical Balance Between Quiescence and Reactivation of Neural Stem Cells. *Biomolecules* **2025**, *15*, 672. [CrossRef] [PubMed]
103. Eames, A.; Chandrasekaran, S. Leveraging Metabolic Modeling and Machine Learning to Uncover Modulators of Quiescence Depth. *PNAS Nexus* **2024**, *3*, pgae013. [CrossRef] [PubMed]
104. Hernandez-Segura, A.; de Jong, T.V.; Melov, S.; Guryev, V.; Campisi, J.; Demaria, M. Unmasking Transcriptional Heterogeneity in Senescent Cells. *Curr. Biol. CB* **2017**, *27*, 2652–2660.e4. [CrossRef]
105. Goel, A.J.; Rieder, M.-K.; Arnold, H.-H.; Radice, G.L.; Krauss, R.S. Niche Cadherins Control the Quiescence-to-Activation Transition in Muscle Stem Cells. *Cell Rep.* **2017**, *21*, 2236–2250. [CrossRef] [PubMed]
106. Guilhot, C.; Catenacci, M.; Lofaro, S.; Rudnicki, M.A. The Satellite Cell in Skeletal Muscle: A Story of Heterogeneity. *Curr. Top. Dev. Biol.* **2024**, *158*, 15–51. [CrossRef]
107. Kann, A.P.; Hung, M.; Wang, W.; Nguyen, J.; Gilbert, P.M.; Wu, Z.; Krauss, R.S. An Injury-Responsive Rac-to-Rho GTPase Switch Drives Activation of Muscle Stem Cells through Rapid Cytoskeletal Remodeling. *Cell Stem Cell* **2022**, *29*, 933–947.e6. [CrossRef]
108. Rodgers, J.T.; King, K.Y.; Brett, J.O.; Cromie, M.J.; Charville, G.W.; Maguire, K.K.; Brunson, C.; Mastey, N.; Liu, L.; Tsai, C.-R.; et al. mTORC1 Controls the Adaptive Transition of Quiescent Stem Cells from G0 to G(Alert). *Nature* **2014**, *510*, 393–396. [CrossRef]
109. Ancel, S.; Michaud, J.; Sizzano, F.; Tauzin, L.; Oliveira, M.; Migliavacca, E.; Schüler, S.C.; Raja, S.; Dammone, G.; Karaz, S.; et al. A Dual-Color PAX7 and MYF5 in Vivo Reporter to Investigate Muscle Stem Cell Heterogeneity in Regeneration and Aging. *Stem Cell Rep.* **2024**, *19*, 1024–1040. [CrossRef]
110. Wilson, A.; Laurenti, E.; Oser, G.; van der Wath, R.C.; Blanco-Bose, W.; Jaworski, M.; Offner, S.; Dunant, C.F.; Eshkind, L.; Bockamp, E.; et al. Hematopoietic Stem Cells Reversibly Switch from Dormancy to Self-Renewal during Homeostasis and Repair. *Cell* **2008**, *135*, 1118–1129. [CrossRef]
111. Laurenti, E.; Frelin, C.; Xie, S.; Ferrari, R.; Dunant, C.F.; Zandi, S.; Neumann, A.; Plumb, I.; Doulatov, S.; Chen, J.; et al. CDK6 Levels Regulate Quiescence Exit in Human Hematopoietic Stem Cells. *Cell Stem Cell* **2015**, *16*, 302–313. [CrossRef]
112. Ibneeva, L.; Singh, S.P.; Sinha, A.; Eski, S.E.; Wehner, R.; Rupp, L.; Kovtun, I.; Pérez-Valencia, J.A.; Gerbaulet, A.; Reinhardt, S.; et al. CD38 Promotes Hematopoietic Stem Cell Dormancy. *PLoS Biol.* **2024**, *22*, e3002517. [CrossRef] [PubMed]
113. Mann, Z.; Sengar, M.; Verma, Y.K.; Rajalingam, R.; Raghav, P.K. Hematopoietic Stem Cell Factors: Their Functional Role in Self-Renewal and Clinical Aspects. *Front. Cell Dev. Biol.* **2022**, *10*, 664261. [CrossRef] [PubMed]
114. Ruffinatto, L.; Groult, Y.; Iacono, J.; Sarrazin, S.; de Laval, B. Hematopoietic Stem Cell a Reservoir of Innate Immune Memory. *Front. Immunol.* **2024**, *15*, 1491729. [CrossRef] [PubMed]
115. Marqués-Torrejón, M.Á.; Williams, C.A.C.; Southgate, B.; Alfazema, N.; Clements, M.P.; Garcia-Diaz, C.; Blin, C.; Arranz-Emparan, N.; Fraser, J.; Gammoh, N.; et al. LRIG1 Is a Gatekeeper to Exit from Quiescence in Adult Neural Stem Cells. *Nat. Commun.* **2021**, *12*, 2594. [CrossRef]
116. Meng, H.; Huan, Y.; Zhang, K.; Yi, X.; Meng, X.; Kang, E.; Wu, S.; Deng, W.; Wang, Y. Quiescent Adult Neural Stem Cells: Developmental Origin and Regulatory Mechanisms. *Neurosci. Bull.* **2024**, *40*, 1353–1363. [CrossRef]
117. Llorens-Bobadilla, E.; Zhao, S.; Baser, A.; Saiz-Castro, G.; Zwadlo, K.; Martin-Villalba, A. Single-Cell Transcriptomics Reveals a Population of Dormant Neural Stem Cells That Become Activated upon Brain Injury. *Cell Stem Cell* **2015**, *17*, 329–340. [CrossRef]
118. Kalamakis, G.; Brüne, D.; Ravichandran, S.; Bolz, J.; Fan, W.; Ziebell, F.; Stiehl, T.; Catalá-Martinez, F.; Kupke, J.; Zhao, S.; et al. Quiescence Modulates Stem Cell Maintenance and Regenerative Capacity in the Aging Brain. *Cell* **2019**, *176*, 1407–1419.e14. [CrossRef]
119. Wang, X.; Fujimaki, K.; Mitchell, G.C.; Kwon, J.S.; Della Croce, K.; Langsdorf, C.; Zhang, H.H.; Yao, G. Exit from Quiescence Displays a Memory of Cell Growth and Division. *Nat. Commun.* **2017**, *8*, 321. [CrossRef]
120. de Jong, I.G.; Haccou, P.; Kuipers, O.P. Bet Hedging or Not? A Guide to Proper Classification of Microbial Survival Strategies. *BioEssays* **2011**, *33*, 215–223. [CrossRef]

biomolecules

MDPI

*Review*

# Molecular and Biophysical Perspectives on Dormancy Breaking: Lessons from Yeast Spore

**Keiichiro Sakai [1], Yohei Kondo [2], Kazuhiro Aoki [3,4,5,6] and Yuhei Goto [3,4,*]**

1 Department of Biology, Brooklyn College, The City University of New York, 2900 Bedford Avenue, Brooklyn, NY 11210, USA; sakai.keiichiro@brooklyn.cuny.edu
2 Center for One Medicine Innovative Translational Research (COMIT), Nagoya University, 65 Tsurumai-cho, Showa-ku, Nagoya 466-8550, Japan; kondo.yohei.e8@f.mail.nagoya-u.ac.jp
3 Laboratory of Cell Cycle Regulation, Graduate School of Biostudies, Kyoto University, Yoshidakonoe-cho, Sakyo-ku, Kyoto 606-8501, Japan; aoki.kazuhiro.6v@kyoto-u.ac.jp
4 Center for Living Systems Information Science, Graduate School of Biostudies, Kyoto University, Yoshidakonoe-cho, Sakyo-ku, Kyoto 606-8501, Japan
5 Division of Quantitative Biology, National Institute for Basic Biology, National Institutes of Natural Sciences, 5-1 Higashiyama, Myodaiji-cho, Okazaki 444-8787, Aichi, Japan
6 Quantitative Biology Research Group, Exploratory Research Center on Life and Living Systems (ExCELLS), National Institutes of Natural Sciences, 5-1 Higashiyama, Myodaiji-cho, Okazaki 444-8787, Aichi, Japan
* Correspondence: goto.yuhei.4c@kyoto-u.ac.jp; Tel.: +81+75-753-4195

**Abstract:** Dormancy is a physiological state that enables cells to survive under adverse conditions by halting their proliferation while retaining the capacity to resume growth when conditions become favorable. This remarkable transition between dormant and proliferative states occurs across a wide range of species, including bacteria, fungi, plants, and tardigrades. Among these organisms, yeast cells have emerged as powerful model systems for elucidating the molecular and biophysical principles governing dormancy and dormancy breaking. In this review, we provide a comprehensive summary of current knowledge on the molecular mechanisms underlying cellular dormancy, with particular focus on the two major model yeasts: *Saccharomyces cerevisiae* and *Schizosaccharomyces pombe*. Recent advances in multifaceted approaches—such as single-cell RNA-seq, proteomic analysis, and live-cell imaging—have revealed dynamic changes in gene expression, proteome composition, and viability. Furthermore, insights into the biophysical properties of the cytoplasm have offered new understanding of dormant cell regulation through changes in cytoplasmic fluidity. These properties contribute to both the remarkable stability of dormant cells and their capacity to exit dormancy upon environmental cues, deepening our understanding of fundamental cellular survival strategies across diverse species.

**Keywords:** dormancy; dormancy breaking; spore; germination; fission yeast; budding yeast; cytoplasmic fluidity

## 1. Introduction

Living organisms have evolved diverse strategies to survive under unfavorable environmental conditions. Among these strategies, dormancy represents a remarkable adaptation that enables organisms to temporarily suspend their growth and metabolism while maintaining the capacity to resume proliferation when conditions become favorable [1,2]. This survival strategy is widely conserved across living organisms, from prokaryotes to eukaryotes. In bacteria, endospore formation, most notably studied in *Bacillus subtilis*, allows long-term survival under extreme conditions, including resistance to heat, radiation, and chemical exposure [3]. Among fungi, various species produce specialized dormant structures: *Aspergillus* and *Penicillium* form highly resistant conidia [4], while pathogenic fungi like *Cryptococcus* produce spores crucial for both survival and host infection [5]. In the plant kingdom, seeds represent a complex form of dormancy, allowing embryos to survive for years or even centuries while awaiting suitable germination conditions [6]. Even more remarkably, several animals have evolved sophisticated dormancy strategies. Tardigrades enter cryptobiosis (meaning 'hidden life'), a term coined by David Keilin [7] and defined as 'the state of an organism when it shows no visible signs of life and when its metabolic activity becomes hardly measurable, or comes reversibly to a standstill' [8]. The cryptobiosis enables their survival in extreme environments, including space exposure [9]. Similarly, the larvae of the African midge *Polypedilum vanderplanki*, known as "sleeping chironomids," can survive complete desiccation through anhydrobiosis (meaning 'life without water'), a state of cryptobiosis that is induced by desiccation, maintaining their viable state for over one year through the accumulation of trehalose and heat shock proteins [10,11]. The ubiquity of dormancy across such diverse taxa underscores its fundamental importance as a survival strategy. Dormancy is characterized by several distinct features: (1) dramatic reduction in metabolic activity, (2) enhanced resistance to various environmental stresses, (3) reversible transition between dormant and proliferative states (dormancy breaking), and (4) long-term survival capability. These characteristics provide significant evolutionary advantages, allowing organisms to persist through periods of nutrient limitation, temperature extremes, or other environmental challenges. The ability to enter and exit dormancy in response to environmental cues is crucial for the survival and propagation of living organisms in fluctuating environments.

*Saccharomyces cerevisiae* and *Schizosaccharomyces pombe* have emerged as powerful models for studying cellular dormancy. These two yeast species diverged approximately 330–420 million years ago, with an evolutionary distance comparable to that between yeasts and mammals [12,13]. Accordingly, comparative studies of these phylogenetically distant yeasts are expected to provide insights into both the evolutionary conservation and diversification of cellular dormancy. Nitrogen starvation induces the formation of spores, a dormant state, in these yeasts when they exist as diploid cells [14]. During this sporulation process, diploid yeast cells undergo meiosis to form haploid spores, which are dormant cells equipped with specialized cell walls and distinct metabolic properties [15,16]. These spores remain viable for extended periods and break dormancy, i.e., germination, to resume vegetative growth when favorable conditions return. Beyond sporulation, yeasts are also able to enter quiescence during the stationary phase of vegetative growth [17,18]. Both dormancy and quiescence refer to reversible growth-arrested states, but dormancy is generally thought to be a "deeper" form of quiescence [19]. Quiescent cells typically retain higher levels of intracellular activity compared to dormant cells. Furthermore, they differ in their formation mechanisms, stress resistance properties, and metabolic characteristics [20]. This review primarily focuses on spore dormancy and germination.

Here, we discuss the molecular and biophysical mechanisms underlying yeast spore dormancy and germination, with particular emphasis on recent findings from *S. cerevisiae* and *S. pombe*. Table 1 provides a comparative overview of the key features of dormancy and germination in these two yeast species, highlighting their similarities and differences. We further explore emerging questions in this rapidly evolving field, with particular focus on molecular, biophysical, and ecological perspectives.

**Table 1.** Comparison of Dormancy and Germination between *S. cerevisiae* and *S. pombe.*

| Feature | *S. cerevisiae* (Budding Yeast) | Refs | *S. pombe* (Fission Yeast) | Refs |
|---|---|---|---|---|
| Life Cycle Characteristics | | | | |
| Natural state | Predominantly diploid | [21] | Predominantly haploid | |
| Mating type | MATa and MATα | | h- and h+ | |
| Sporulation Conditions | | | | |
| Primary trigger | Nitrogen depletion with non-fermentable carbon source | [22] | Nitrogen depletion | |
| Spore Structure | | | | |
| Ascus wall after sporulation | Remains intact | | Digested by glucanases (Agn2, Eng2) | [23–25] |
| Spore wall composition | Four layers: mannan, β-1,3-glucan, chitosan, dityrosine | [15,26,27] | Primarily glucan and chitosan (exact composition not fully characterized) | [28–30] |
| Spore connection | Interspore bridges connect sibling spores | [31] | No interspore bridges | |
| Spore surface | Ridged proteinaceous layer | [32] | Characteristic outward projection | [33,34] |
| Metabolic Features | | | | |
| Trehalose accumulation | Significant increase | [35,36] | ~1000-fold increase compared to vegetative cells | [37–39] |
| Glycogen accumulation | Present | [35,36] | ~40-fold increase compared to vegetative cells | [37,39] |
| Adenosine triphosphate (ATP) levels | Substantial (~3–4 mM) | [35] | Relatively high compared to residual ascus | [40] |
| Transcription activity | ~5% of vegetative cells | [41] | Not precisely quantified | |
| Protein filament formation | Acetyl-CoA synthetase Acs1 forms filaments | [42] | Not known | |
| Germination Process | | | | |
| Typical duration | 4–6 h | [43–46] | 10–12 h | [37,47–49] |
| Primary trigger | Glucose (Cyclic adenosine monophosphate-protein kinase A (cAMP-PKA) pathway) | [50] | Glucose (cAMP-PKA pathway) | [37,47,51] |
| Initial stages | Spore uncoating, polarized growth | [43–46] | Bright-to-dark transition, isotropic swelling | [37,47,48] |
| Polarization mechanism | Prepolarized to grow away from interspore bridges | [52] | Random polarization with dynamic polar cap movement | [48] |
| Polarization proteins | Cdc10, Bud8, Bud5 | [52,53] | GTP-bound Cdc42, Bud6, Bgs4, Cdc42 GAP (Rga6) | [48,54,55] |

**Table 1.** *Cont.*

| Feature | *S. cerevisiae* (Budding Yeast) | Refs | *S. pombe* (Fission Yeast) | Refs |
|---|---|---|---|---|
| Cell growth pattern | Polarized growth -> Non-polarized growth -> Budding | [43–46] | Isotropic swelling -> Germ tube formation (outgrowth) | [37,47–49] |
| Required Nutrients for Complete Germination | | | | |
| Glucose | Required for fast response | [44,50,56,57] | Essential | [58] |
| Additional nutrients | Required for later stages | [44,50,57] | Copper and iron ions required for outgrowth | [59,60] |
| Molecular Regulators | | | | |
| Key signaling pathway | cAMP-PKA pathway | [50] | cAMP-PKA pathway | [37,47,51] |
| Trehalase | Nth1, Nth2 | | Ntp1 | [37,38] |
| Cell cycle regulators | Not required for early germination stage (Cdc28, Cdc37, Cdc4, Cdc34, Cdc7, Cdc24) | [50] | Not known | |
| Actin role | Essential for polarized growth | [46] | Essential for germ tube formation | [47] |
| Histone dynamics | Not known | | Hht1 (H3) expression decreases during germination | [61] |
| Heat shock protein | Hsp42 | [62] | Not known | |
| Biophysical Properties | | | | |
| Particle mobility in dormant spores | Restricted (~50–150 nm particles) | [62] | Restricted (~40 nm and ~50–150 nm particles) | [37] |
| Small protein diffusion | Not fully documented | | Relatively free diffusion | [37] |
| pH changes during germination | Dormant spores: ~5.9 -> Vegetative cells: ~7.4 | [62,63] | Not known | |
| Ecological Context | | | | |
| Natural habitat | Fruits, insect vectors, forest niches | [64–66] | Not well characterized (honey?) | [67] |
| Spore survival advantage | High survival in the insect gut | [64] | High survival in the insect gut | [64] |
| Germination pattern | Commonly sibling spore mating | [52,68] | Single spores | |

## 2. Dormancy in Yeast Cells

### 2.1. *S. pombe*

The life cycle of the fission yeast *S. pombe* is shown in Figure 1A. *S. pombe* naturally exists as haploid cells with two mating types; minus (h−) and plus (h+). In response to nitrogen depletion, *S. pombe* cells arrest in the G1 phase, and heterothallic haploid cells of opposite mating types conjugate to form diploid zygotes. In contrast, when haploid cells fail to encounter the opposite mating type, they enter a quiescent state. These diploid zygotes undergo sporulation, following chromosome recombination and meiosis. Transcriptome, proteome, and phosphoproteome profiles drastically change throughout meiosis and sporulation [69–71]. A genome-wide screening using transposon insertion sequencing identified 532 genes contributing to meiosis and sporulation, and a screening using a non-essential gene deletion library identified 34 genes contributing to them [72,73].

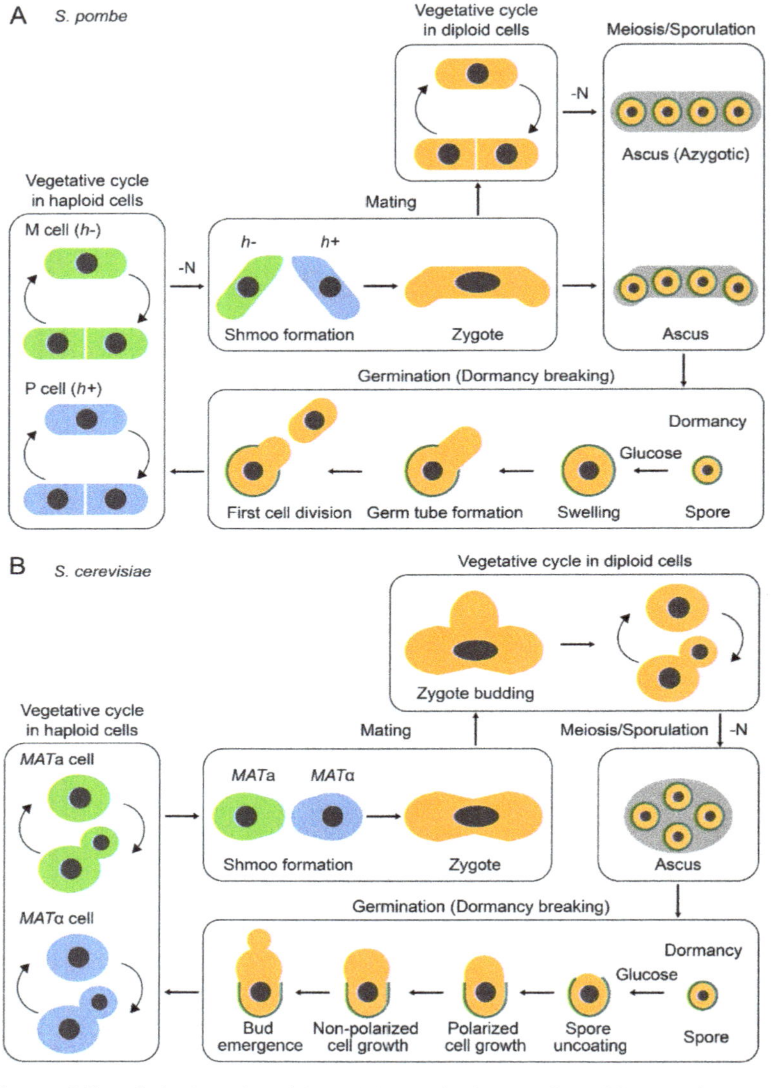


**Figure 1.** Life cycle for *S. pombe* and *S. cerevisiae*. Haploid yeast cells proliferate under nutrient-rich conditions. In *S. pombe* (**A**), nitrogen starvation (N) induces the mating of opposite mating types (h− and h+), followed by meiosis and sporulation, resulting in ascus formation. If zygotes receive nutrients before progressing through these processes, they can proliferate as diploid cells. These diploid cells undergo meiosis and sporulation in response to nitrogen starvation, forming an azygotic ascus, which has a distinct morphology from the normal ascus. These azygotic ascus and zygotic ascus naturally rupture, releasing dormant spores. When spores sense the presence of glucose, they initiate the germination process (dormancy breaking), followed by spore swelling and germ tube formation (outgrowth). After the first cell division, they re-enter the vegetative cycle. In *S. cerevisiae* (**B**), even under nutrient-rich conditions, cells of opposite mating types (*MATa* and *MATα*) can mate and

proliferate as diploid cells. In response to nitrogen starvation, diploid cells undergo meiosis and sporulation, forming a zygotic ascus. The ascus remains intact for an extended period, unlike *S. pombe*. During germination, spores undergo several morphological changes, such as spore uncoating, polarized cell growth, non-polarized cell growth, and bud emergence.

#### 2.1.1. Enhanced Stress Resistance of *S. pombe* Spores

Sporulation yields four identical spores with a characteristic outward projection as a surface structure [33,34]. These spores possess a spore wall, a structure thicker than the cell wall of vegetative cells. Although its composition remains unknown in *S. pombe* spores, it is thought to be primarily composed of glucan and chitosan [28–30]. *S. pombe* spores exhibit high resistance to various environmental stresses, such as heat, organic solvents (e.g., ethanol), digestive enzymes (e.g., glusulase), and nutrient starvation [16,53,74]. They can survive for several months even in an aqueous solution with almost no nutrients, demonstrating higher survival rates compared to cells in the stationary phase and quiescence [16]. The spore coat protein Isp3, localized to the outermost layer of the spore wall, appears to confer this remarkable stress tolerance [74]. This enhanced stress resistance provides a survival advantage in natural environments; for example, spores exhibit significantly higher viability than vegetative cells when present inside the body of flies [64].

#### 2.1.2. Metabolic Changes in *S. pombe* Spores

Although global metabolic changes during *S. pombe* sporulation remain uncharacterized, specific metabolites show dramatic alteration in spores; trehalose and glycogen levels increase 1000-fold and 40-fold, respectively, compared to vegetative cells [37–39]. These changes in carbohydrate levels are proposed to regulate cytoplasmic fluidity in dormant spores [37] (refer to Section 4.2. Dormancy and Dormancy Breaking). Spore cytoplasm maintains relatively high ATP levels compared to the residual ascus [40]. Furthermore, lipid droplets, primarily composed of triacylglycerols and sterol esters, are actively redistributed from mother cells to their nascent spores. This redistribution depends on actin polymerization and is essential for proper spore germination, as the absence of lipid droplets results in germination defects [75].

### 2.2. *S. cerevisiae*

The life cycle of the budding yeast *S. cerevisiae* is shown in Figure 1B. In the natural environment, the budding yeast *S. cerevisiae* predominantly exists as diploid cells [21] with two mating types; *MATa* and *MATα*. These a/α diploid cells undergo meiosis and sporulation when exposed to nitrogen-depleted conditions in the presence of a nonfermentable carbon source, such as acetate, ethanol, and pyruvate [22]. By forming dormant spores, yeast cells acquire resistance to a variety of environmental stresses, enhancing their survival in harsh conditions [15]. Notably, sporulation functions as a rejuvenation by eliminating age-associated cellular damage, including protein aggregation and nucleolar aberrations [76]. Consequently, spores derived from aged cells exhibit replicative potential equivalent to those from young cells [76].

#### 2.2.1. Enhanced Stress Resistance of *S. cerevisiae* Spores

The budding yeast spores exhibit remarkable resistance to various stresses, including heat, desiccation, organic solvents, and digestive enzymes [15,77,78]. This stress resistance is largely mediated through the spore wall [15,79–81]. The spore wall, which is thicker than the vegetative cell wall, consists of four distinct layers; mannan, β-1,3-glucan, chitosan, and dityrosine layers. The two inner layers are composed of mannan and β-1,3-glucan, and the third and fourth layers are composed of chitosan and dityrosine, respectively [15,26,27]. While mannan and β-1,3-glucan layers are present in both spore and vegetative cell walls, the chitosan and dityrosine layers are unique to the spore wall [15,79,81]. Mutation of DIT1, which encodes a dityrosine precursor enzyme, specifically disrupts the dityrosine layer without affecting the chitosan layer [79]. The mutant of the chitin synthase CHS3 leads to the loss of both chitosan and dityrosine layers while preserving the inner layers of the spore wall [81]. Spores lacking either the dityrosine layer alone or both the chitosan and dityrosine layers show increased stress sensitivity, indicating the critical role of these layers in stress resistance [79,81]. The dityrosine layer appears to function as a diffusion barrier, preventing the entry of protein-sized molecules into the spore wall, thereby potentially explaining its role in conferring resistance to digestive enzymes [15,80]. A recent study reported that *S. cerevisiae* spores possess an additional surface layer above the dityrosine layer [32]. This ridged proteinaceous layer is suggested to function as a protective structure.

#### 2.2.2. Reduced Metabolic Activities in *S. cerevisiae* Spores

Dormant spores are thought to be metabolically inactive. Transcription and translation activities in spores at 30 °C incubation are approximately 5% of those observed in vegetative cells [41]. In addition, mRNAs remain stable for several months in spores stored at 4 °C, suggesting that both synthesis and degradation of cellular components are markedly suppressed during dormancy. Using a doxycycline-inducible synthetic gene-circuit, it was demonstrated that GFP expression reaches a plateau after approximately 20 h in spores, compared to 8 h in vegetative cells [82]. This observation suggests a slowdown in gene expression in spores, although the delay may also reflect the unique physiological environment within spores that affects oxygen-dependent GFP chromophore maturation. This reduced metabolic activity likely promotes spore survival under harsh environmental conditions, particularly during nutrient starvation where energy generation and biosynthesis are limited. Notably, pharmacological inhibition of either ATP synthesis or translation for approximately 24 h does not compromise spore viability [82]. In contrast, transcriptional inhibition results in the death of almost all spores, suggesting that minimal transcriptional activity is necessary for spore survival [82]. Since transcription initiation requires ATP as an energy source [83], it is speculated that dormant spores retain sufficient ATP to sustain minimal transcription for at least 24 h in the absence of ATP synthesis. Another study identified the assembly of metabolic enzymes into filaments within spores by combining cryo-electron tomography (cryoET) and single-particle cryo-electron microscopy (cryoEM) [42]. Specifically, acetyl-coenzyme A (CoA) synthetase Acs1 forms filaments that suppress acetyl-CoA production in the spore [42]. This filament-mediated inactivation of Acs1, a major ATP-consuming enzyme, may help maintain ATP levels necessary for minimal metabolic activity [42]. Indeed, metabolomic analyses have revealed that spores maintain substantial amounts of ATP (~3–4 mM) [35], comparable to those found in vegetative cells [40,84].

#### 2.2.3. Carbohydrate Accumulation in *S. cerevisiae* Spores

Carbon storage represents a conserved feature of cells in dormant and quiescent states [2]. Similar to *S. pombe* spores, *S. cerevisiae* spores accumulate specific carbohydrates, particularly trehalose and glycogen. During meiosis and sporulation in budding yeast, trehalose synthesis is actively induced [35,36]. This process is catalyzed by Tps1, a subunit of the trehalose-6-phosphate synthase/phosphatase complex, which serves as the key enzyme for trehalose biosynthesis [85]. Diploid cells lacking Tps1 exhibit significantly reduced sporulation efficiency compared to wild-type cells [86], indicating that trehalose accumulation is essential for faithful sporulation. While glycogen also accumulates in mature spores [35,36], it undergoes transient degradation following key meiotic events, potentially providing precursors for spore wall assembly [35]. Similar patterns of trehalose and glycogen accumulation occur in quiescent and G1-arrested states in *S. cerevisiae* cells [87–89], where these carbohydrates likely provide energy to fuel cell cycle re-entry upon growth resumption. By analogy, stored carbohydrates in spores may serve as a crucial energy source during germination progression after release from dormancy.

#### 2.2.4. Ecological Significance of *S. cerevisiae* Spores

The ecological significance of sporulation in budding yeast has been revealed through various studies. *S. cerevisiae* has been isolated from a variety of natural habitats, especially from fruits, while also establishing associations with insect vectors, such as social wasps, flies, and other insects [64–66]. In insect vectors such as the fruitfly *Drosophila melanogaster*, spores exhibit remarkable resistance to digestive processes in their gut [64]. The paper demonstrates that spores pass through the fly gut with significantly higher survival rates compared to vegetative cells. Notably, when tetrads (spores enclosed within the ascus) pass through the fly gut, the ascus walls are digested, causing the dissemination of each spore [90]. This dissemination of spores enhances outbreeding rates more than 10-fold relative to intact tetrads, which increases genetic recombination between different yeast strains and promotes genetic diversity in the resulting progeny [90]. Beyond insects, *S. cerevisiae* has been isolated from forest niches, such as soil and tree bark. Although direct evidence for spores in these forest niches remains elusive, experimental studies have demonstrated that soil environments promote sporulation in *S. cerevisiae* [65]. These observations led to the development of the 'fruit forest-reservoir hypothesis', which proposes a tripartite ecological cycle. According to this model, *S. cerevisiae* primarily inhabits three distinct niches: fruits, insect vectors, and forest environments. During fruit season, yeast cells proliferate through mitotic division and fermentation within fruits. As fruits become scarce, some fraction of the cells migrate to forest environments via insect vectors or fallen fruits. In forest niches, cells adopt a sporulated state, enabling survival through nutrient-poor winter conditions.

## 3. Dormancy Breaking in Yeast Cells

### 3.1. *S. pombe*

Upon exposure to favorable conditions, *S. pombe* spores exit dormancy and initiate cell cycle progression and vegetative growth. Germination onset can be monitored using phase contrast microscopy [37,47,48]. Dormant spores appear as highly refractile, bright cells under phase contrast microscopic observation. When transferred to a nutrient-rich medium, spores begin germination, marked by a characteristic bright-to-dark transition of the cytoplasm 1–2 h after induction—this represents the earliest detectable hallmark of germination onset (germination sensu stricto) [37,47,48]. The germination process continues with isotropic swelling at 4 h, followed by germ tube elongation (outgrowth) at 6 h post-induction [37,47–49]. Subsequently, the first nuclear division occurs at 10 h with cytokinesis (septation) following at 12 h post-induction [37,47–49]. The entire sequence constitutes germination sensu lato [49]. The major events during spore germination are illustrated in Figure 2.

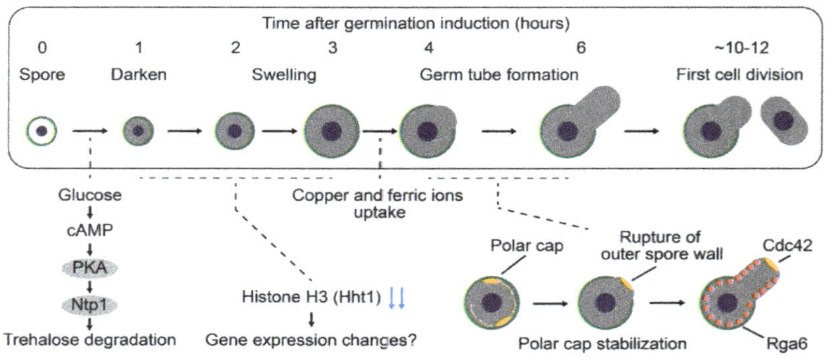


**Figure 2.** The events during spore germination in *S. pombe*. First, fission yeast spores sense and respond to glucose through the cAMP-PKA signaling pathway, which leads to Ntp1-mediated trehalose degradation. Subsequently, the expression level of histone H3 Hht1 is decreased, potentially facilitating chromatin relaxation and inducing global changes in gene expression. Additionally, the uptake of copper and ferric ions is indispensable for germ tube formation (outgrowth). In swollen spores, the polar cap, including a Rho-type GTPase Cdc42, moves dynamically along the cell surface before stabilizing at a specific site, where the outer spore wall ruptures to initiate germ tube formation (orange dots). This monopolar outgrowth requires Rga6, a Cdc42 GAP (red dots), which determines the tip localization of Cdc42.

#### 3.1.1. Glucose-Sensing and Trehalose Degradation Pathway During Germination in *S. pombe*

Spore germination is typically triggered by a glucose-containing medium [47,58,91]. In *S. pombe*, glucose serves dual functions; it acts as both a carbon source supporting cell cycle progression and an activator of the glucose-sensing cAMP-PKA pathway [37,47,92]. Glucose recognition occurs through the G-protein coupled receptor, Git3, at the cell membrane, which triggers activation of the adenylate cyclase, Cyr1, via heterotrimeric G proteins, Gpa2 (Gα), Git5 (Gβ), and Git11 (Gγ). Activated Cyr1 produces cAMP as a second messenger, which binds to the PKA regulatory subunit Cgs1 and leads to the activation of the PKA catalytic subunit Pka1 [51,93,94]. Genetic analysis of spores lacking *git3*, *gpa2*, *cyr1*, or *pka1* genes revealed defective germination, demonstrating the essential role of the cAMP-PKA pathway in germination initiation [37,47,51]. PKA activation also promotes trehalose degradation during spore germination; wild-type spores degrade trehalose within one hour of germination induction, whereas *pka1*Δ spores maintain constant trehalose levels [37–39]. Trehalose degradation requires natural trehalase Ntp1, which hydrolyzes trehalose into two molecules of glucose [95]. Spores lacking Ntp1 fail to degrade trehalose upon germination induction, phenocopying *pka1*Δ [37,38]. Furthermore, the *ntp1*Δ spores also exhibit impaired germination initiation [37,38]. The germination defects in *gpa2*Δ spores are partially rescued by Ntp1 overexpression, indicating that Ntp1 is one of the major downstream factors in the cAMP-PKA pathway during germination [37]. Thus, the glucose-activated cAMP-PKA-Ntp1 signaling axis plays a central role in spore germination progression in *S. pombe*. While trehalose degradation is crucial, trehalose synthesis also influences the spore germination process: deletion of the *tps1*, encoding trehalose-6-phosphate synthase, prevents germination initiation [96]. Given the importance of trehalose degradation for spore germination, trehalose synthesis likely functions primarily to ensure spore survival under adverse environments rather than directly promoting the germination process. In the future, it will be of particular interest to identify downstream factors of the cAMP-PKA pathway and to elucidate their specific roles during germination. In *S. pombe*, a comprehensive analysis of PKA substrates remains unavailable, even for vegetative growth conditions.

Consequently, our understanding of molecular pathways activated downstream of glucose-induced PKA signaling is limited, with only a few identified targets such as Ntp1. A promising approach would involve integrated transcriptome, proteome, and phosphoproteome analysis of the *pka1*Δ strain in vegetative cells, as synchronizing germination in *S. pombe* spores presents significant technical challenges (see Section 3.1.2. Gene Expression Landscape during Germination in *S. pombe*). Such multi-omics would likely reveal the comprehensive molecular networks that orchestrate the initiation of germination.

#### 3.1.2. Gene Expression Landscape During Germination in *S. pombe*

Recent single-cell transcriptomic analysis has revealed the gene expression landscape during spore germination in *S. pombe* [61]. The single-cell transcriptome was necessary because synchronizing germination progression in *S. pombe* spores remains technically challenging, in contrast to *S. cerevisiae* spores [43]. In this article, the authors identified 167 genes exhibiting dynamic expression changes during the early stage of germination. The enrichment analysis revealed associations with mRNA metabolic process, lipid metabolic process, nucleotide-containing small molecule metabolic process, ascospore formation, and detoxification. Among these, *hht1*, encoding histone H3, emerged as a particularly intriguing candidate. *S. pombe* possesses three histone H3-encoding genes (*hht1*, *hht2*, and *hht3*) with identical amino acid sequences. Among these, only the *hht1* expression shows dynamic fluctuation upon germination, while the *hht2* and *hht3* expressions remain relatively constant. Deletion analysis reveals that both *hht1*Δ and *hht2*Δ spores exhibit defects in germination progression, while *hht3*Δ spores germinate normally. Hht2 appears to function as a housekeeping histone because *hht2*Δ vegetative cells exhibit temperature sensitivity and growth defects, while *hht1*Δ and *hht3*Δ cells grow normally. These findings establish Hht1 as a germination-specific important histone, while Hht2 serves essential functions in both germination and vegetative growth. The reduction in Hht1 amount during the onset of germination may promote the relaxation of chromatin structure, facilitating a global transcriptional activation. For a comprehensive overview of other germination-related candidate genes and their expression patterns, readers are directed to the review by Tsuyuzaki et al. [49]. Determining the precise functions of these candidate genes during germination would be a valuable future direction. In addition, the established single-cell transcriptome methodology offers powerful opportunities when applied to mutant strains for characterizing gene expression dynamics throughout germination. A promising candidate for such analysis is Pka1, as discussed in Section 3.1.1, which could illuminate the molecular mechanisms governing germination in *S. pombe*.

#### 3.1.3. Nutrients Required for Germination in *S. pombe*

The progression of entire germination requires the appropriate nutrients beyond glucose. A simple glucose solution lacking other basic nutrients can trigger initial germination, which is measured by the characteristic bright-to-dark transition of spore cytoplasm observed under phase contrast microscopy [58]. Similar initial germination responses can be triggered by simple sucrose and mannose solutions [58]. However, subsequent morphological changes, such as spore swelling and germ tube formation, fail to occur in these minimal solutions, indicating that additional nutrients are necessary for later stages of germination. Copper and iron ions play crucial roles in spore outgrowth [59,60]. Under copper- or iron-insufficient conditions, germinating spores arrest their differentiation immediately after isotropic swelling [59,60]. Successful outgrowth requires the uptake systems for the specific ions and biosynthesis in these conditions, including copper transporters Ctr4 and Ctr5, the copper sensor Cuf1, ferrichrome biosynthesis enzymes Sib1 and Sib2, and the transporter Str1 [59,60]. In the future, a systematic investigation of nutrient requirements for germination would yield valuable insights into the molecular mechanisms that drive germination progression. Currently, glucose function is well established; it acts both as a primary carbon source and an activator of the cAMP-PKA signaling pathway. In contrast, our knowledge of other nutrients, particularly copper and iron ions, remains limited, and many potentially essential nutrients await discovery and characterization. Deciphering the complete nutritional landscape necessary for germination would likely reveal important regulatory mechanisms governing this fundamental biological process.

#### 3.1.4. Morphological Changes During Germination in *S. pombe*

The molecular mechanisms underlying morphological changes during spore germination have been extensively investigated. Germinating spores undergo two distinct morphological changes: isotropic swelling and germ tube formation (outgrowth). Inhibition of either protein translation or target of rapamycin (TOR) signaling, which regulates ribosome biogenesis, prevents these morphological changes [37]. These observations demonstrate that protein synthesis is required for proper germination progression. Consistent with this, protein content increases linearly during spore outgrowth [91]. Cytoskeletal organization, involving both actin and microtubule networks, plays essential roles in germ tube formation [37,48]. F-actin patches initially distribute randomly throughout the cortical region of germinated spores and subsequently become concentrated at a specific cortical site in swollen spores. After that, germ tube formation occurs at this actin-enriched site [47]. This actin localization precedes germ tube formation, indicating that cell polarity is established during the spore swelling phase. A recent study has elucidated the molecular mechanisms governing this cell polarization during spore germination [48]. The vegetative fission yeast cells establish polarity through the formation of a polar cap at the growing tip. This polar cap consists of various polarity proteins, including a Rho-type GTPase Cdc42, actin regulators, and cell-wall remodeling proteins. Germinating spores establish a single polar cap that exhibits dynamic behavior, undergoing stochastic movement across the cell surface through cycles of assembly and disassembly. The polar cap eventually stabilizes at a specific site when spores reach approximately twice their initial volume, making the location of future germ tube emergence. The polar cap includes GTP-bound Cdc42, actin nucleation-promoting factor Bud6, and (1,3)β-D-glucan synthase Bgs4 [48,54]. The site where the polar cap is stabilized has a singular rupture in the outer spore wall (OSW), releasing OSW constraints on polarized cell growth. Therefore, successful outgrowth requires both polar cap stability and appropriate spore wall mechanics. The spatial regulation of this process is further controlled by Cdc42 GTPase-activating proteins (GAP), which is necessary for establishing monopolar outgrowth during germination [55]. Although three Cdc42 GAPs (Rga3, Rga4, and Rga6) have been identified in *S. pombe* [97], only Rga6 has a key role in maintaining GTP-bound Cdc42 at the outgrowth zone [55].

### 3.2. *S. cerevisiae*

The spore germination process of *S. cerevisiae* can be divided into roughly four stages from the perspective of morphological changes: (1) spore uncoating, (2) polarized cell growth, (3) non-polarized cell growth, and (4) bud emergence [43–46]. Following these morphological changes, yeast spores re-enter the mitotic cell cycle and restart vegetative growth. Upon germination induction with a nutrient-rich medium, the spore cytoplasm exhibits a decreased electron density with swelling, followed by local degradation of the outer two spore layers: dityrosine and chitosan (spore uncoating) [46,62]. The inner two spore layers (mannan and β-1,3-glucan) remain intact and subsequently function as the vegetative cell wall [98]. During the next phase, the spore elongates unidirectionally along its long axis without any obvious growth along its short axis (polarized cell growth). Subsequently, the spore expands uniformly in both the long and short axes until bud emergence (non-polarized cell growth). *S. cerevisiae* spores complete their entire germination process in approximately 4 to 6 h, which is relatively fast compared to the 10 to 12 h required for *S. pombe* spores. The major events during spore germination are illustrated in Figure 3.

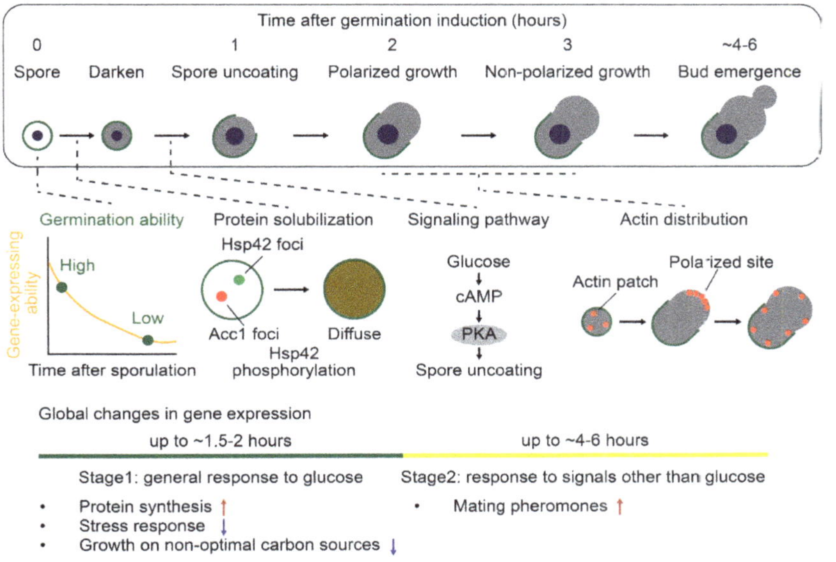


**Figure 3.** The events during spore germination in *S. cerevisiae*. After the completion of sporulation, the germination ability of budding yeast spores gradually decreases over time. This decline is due to the decreased gene-expressing ability, such as a decrease in the RNA polymerase II amount. In the presence of glucose, spores sense and respond to it through the cAMP-PKA signaling pathway, initiating germination. In the early stage of germination, small heat shock protein Hsp42 foci dissolved upon its phosphorylation, promoting the solubilization of other proteins, including the metabolic enzyme Acc1. Subsequently, actin patches (red dots) accumulate at the tip region to facilitate polarized growth and later disperse again, potentially promoting non-polarized cell growth. Furthermore, gene expression undergoes global changes following germination induction. The gene expression pattern can be categorized into two stages: the general response to glucose (up to ~1.5–2 h) and the response to signals other than glucose (up to ~4–6 h).

### 3.2.1. Characteristics for *S. cerevisiae* Germination: Intact Ascus and Interspore Bridge

One of the unique features of *S. cerevisiae* is that the ascus wall remains intact after sporulation is completed, whereas in *S. pombe*, the ascus wall is digested by glucanases, Agn2 and Eng2, immediately after the completion of meiosis [23–25]. In addition to the intact ascus wall, *S. cerevisiae* spores are interconnected by interspore bridges, which hold spores closely even after digestion of the ascus wall [31]. Unlike *S. pombe*, *S. cerevisiae* naturally exists as diploid cells. A potential role of the intact ascus wall and the interspore bridge is to keep sibling spores close, enabling efficient mating and producing diploids after germination begins. Indeed, germinating spores of *S. cerevisiae* can mate with spores of an opposite mating type before budding and the mitotic cell cycle [52,68]. Global gene expression analysis has revealed that genes associated with mating are upregulated in the early germination stage [43,44]. In naturally isolated strains, mating between sibling spores is the usual event that follows germination [68]. The mating process has been mainly investigated with haploid lab strains, which are heterothallic and incapable of self-mating; thus, further studies focused on homothallic wild yeasts would provide intriguing insights into mating behavior. Another unique feature of *S. cerevisiae* is that spores are inherently prepolarized to outgrow away from interspore bridges, which occurs before budding or mating [52]. As mentioned above, *S. pombe* spores undergo symmetry breaking and polarization at random sites on the cell surface during germination (refer to Section 3.1.4. Morphological Changes during Germination in *S. pombe*) [48]. In *S. cerevisiae* spores, prepolarized sites contain the septin Cdc10, the cortical landmark protein Bud8, and the GTPase Rsr1 GEF Bud5 [52,53].

#### 3.2.2. Methods to Monitor *S. cerevisiae* Germination

Germination progression can be monitored using various methods in *S. cerevisiae* spores. Numerous studies have determined the onset of germination by purifying spores before inducing germination and assessing the loss of stress resistance in the spores. For example, after germination onset, budding yeast spores are sensitive to heat shock and digestive enzymes (e.g., Zymolyase) [43,50,62,99]. The increased stress sensitivity can be attributed to spore uncoating, which leads to the rupture of stress-resistant spore wall structures. Additionally, optical density decreases concurrently with increased sensitivity or slightly later [56,62,100–102]. Similar to *S. pombe* spores, the spore cytoplasm exhibits a bright-to-dark transition under the phase contrast microscope around the same time [62,101]. A recent study investigated spore germination by monitoring the emergence of elongated spores from the ascus following its rupture [82]. This approach is unique to budding yeast, unlike fission yeast, in which the ascus ruptures immediately after meiosis completion. The ascus rupture is due to the polarized cell growth and/or bud emergence, and therefore, this is an indicator of the later germination stage compared to increased stress sensitivity.

#### 3.2.3. Molecular Mechanisms Regulating *S. cerevisiae* Germination

The molecular mechanisms regulating germination progression have been studied in *S. cerevisiae*. As in *S. pombe*, the cAMP-PKA glucose-sensing pathway is required for the germination initiation [50]. Ras2 (GTP-binding protein), Cdc25 (GEF for Ras proteins), and Cyr1 (adenylate cyclase) are central components of this pathway, and the absence of these genes results in germination defects. This indicates that the cAMP-PKA pathway has a crucial role in the early germination stage, including spore uncoating. In contrast, many key cell cycle regulators are dispensable for the early germination stage. The spores lacking Cdc28 (cyclin-dependent kinase) and Cdc37 (Hsp90 co-chaperone required to pass through the START) undergo spore uncoating and polarized cell growth after germination induction. Other cell cycle regulators, including Cdc4 and Cdc34 (G1/S transition), Cdc7 (S phase), and Cdc24 (GEF for Cdc42), are also not necessary for the early germination stage. Furthermore, actin distribution dynamically changes during the germination process [46]. Actin patches randomly distribute upon spore uncoating, then accumulate at the tip region to support polarized growth, and disperse again after a few hours, probably promoting the non-polarized cell growth. Then, they localize to the bud site upon bud emergence. The spores treated with an actin depolymerizer, latrunculin A, exhibit no polarized growth even after germination induction. The polarized site includes polarisome proteins (Spa2, Pea2, Bud6, and Bni1), and these proteins are considered to establish cell polarization during germination. A recent study based on single-cell level analysis has revealed that dormant spores gradually lose their gene expression ability at 30°C over days to months [82]. With this decreased ability, the amount of RNA polymerase II (RNAPII) also decreases over time. Interestingly, spores with higher levels of RNAPII are more likely to germinate than those with lower levels of RNAPII after approximately a month of incubation at 30°C. These results show that gene expression ability associated with RNAPII levels determines the germination ability. Other recent findings have shown that many proteins assemble to form cytoplasmic foci in dormant spores, which disperse upon germination induction [62,103]. The loss of Hsp42, a small heat shock protein, disrupts the dissolution of these foci, implying its involvement in controlling protein assembly during germination.

#### 3.2.4. Transcriptome, Proteome, and Phosphoproteome Profiles During *S. cerevisiae* Germination

One advantage of studying germination in budding yeast is the potential to achieve synchronization of germination timing. This approach allows us to investigate global changes in transcriptome, proteome, and phosphoproteome during germination. Transcriptomic analysis revealed that the germination process can be divided into two different stages after glucose addition [43,44]. The first stage (from germination start to ~1.5–2 h) is the gene expression pattern similar to the general response of budding yeast cells to glucose; glucose addition to stationary phase cells or glucose-starved cells. For example, genes associated with protein synthesis (e.g., ribosome biogenesis, rRNA processing) are upregulated, while genes related to stress response and growth on non-optimal carbon sources (e.g., gluconeogenesis, TCA cycle) are downregulated. However, there are some exceptions during germination (e.g., cell cycle-related genes) compared to the general glucose response, indicating that these differences make the germination-specific mechanisms. This gene expression in the first stage can also be induced when spores are incubated in glucose alone, not nutrient-rich medium. The second stage (from 2 h to the entering of the first mitotic cell cycle) makes spores capable of sensing and responding to environmental conditions other than glucose. For example, mating-related genes are upregulated, thereby allowing spores to respond to mating pheromones. Other transcriptomic analyses identified a complex transcription factor network regulating gene expression during germination [104]. Additionally, recent proteome analysis identified the changes in global protein solubility during germination [62]. In particular, metabolic enzymes involved in carbohydrate, lipid, and nitrogen metabolisms solubilize during germination, reflecting the activation of spore metabolism. In the study, phosphoproteome analysis determined the dynamic phosphorylation of S233 of Hsp42 during germination. The Hsp42 phosphorylation is correlated with its solubility and is crucial for the metabolic enzyme Acc1 solubilization. Therefore, these dynamic changes in transcriptome, proteome, and phosphoproteome provide insight into the comprehensive mechanisms to regulate germination progression. Investigating the relationship between enzyme solubilization and subsequent enzymatic activity during the initiation of germination represents a compelling avenue for future research.

## 4. The Biophysical Properties of the Cytoplasm During Dormancy

### *4.1. Overview*

The cytoplasm is a highly complex and crowded milieu, densely packed with diverse cellular components, such as proteins, nucleic acids, and organelles. The crowded conditions impact the biochemical reactions by slowing molecular diffusion and reducing the encounter rates, thereby impeding diffusion-limited reactions. Conversely, the crowded environment enhances molecular assembly through entropic effects (the depletion attraction force) and promotes reaction-limited reactions. Recent findings have revealed that the cytoplasm is not merely a crowded environment but exhibits a complex porous architecture [105–107], which is not reproduced by in vitro assays such as bovine serum albumin (BSA) solutions [108,109]. Additionally, the cytoplasm is dynamically agitated by metabolic activities that arise from the combined effects of molecular motor motion, cytoskeletal structures and dynamics, and enzymatic reactions. Indeed, ATP depletion causes the cytoplasm of bacterial and yeast cells to exhibit solid or glass-like properties [103,110–113], highlighting the crucial role of ATP-driven intracellular processes in maintaining cytoplasmic fluidity. These properties, along with cytoplasmic viscoelasticity and pH levels, together contribute significantly to cellular organization and macromolecular movement [103,114].

Dormant cells exhibit more viscous, rigid, and dense cytoplasm, which is believed to contribute to their stress resistance. For example, the spore cytoplasm of *Bacillus subtilis* immobilizes proteins rotationally, potentially serving as a mechanism to prevent protein denaturation [115]. In more extreme cases, plant seeds and several animal species achieve immobilization of the entire cytoplasm through vitrification by replacing water with disaccharides [116–118]. However, some dormant cells exist in intermediate states, where their cytoplasm has low fluidity but still maintains metabolic activities. The cytoplasm of such intermediate dormancy states, including that of yeast spores, remains an active area of research.

#### 4.1.1. What Cellular Processes Are Affected by Cytoplasmic Properties?

The biophysical properties of the cytoplasm are modulated in response to environmental perturbations, including osmotic stress, temperature fluctuations, nutrient availability, and mechanical forces. These alterations in the cytoplasmic properties impact myriad biochemical reactions and cellular organization. A key example is microtubule dynamics, which are influenced by osmotic stress in fission yeast, mammalian, and moss cells [119]. Hyperosmotic stress reduces cell volume with water efflux, increasing cytoplasmic concentration, while hypoosmotic stress expands cell volume, decreasing the concentration. Microtubule polymerization and depolymerization rates vary linearly and inversely with cytoplasmic concentration, likely due to changes in cytoplasmic viscosity. Moreover, increased cytoplasmic crowding upon severe hyperosmotic stress slows down diffusion rates and nuclear import of stress-response signaling molecules in budding yeast cells [120]. Protein production and growth rate also decrease with increased cytoplasmic crowding upon osmotic stress, indicating that these processes are diffusion-limited [121]. Similarly, when yeast cells are grown in confined spaces, growth-induced mechanical pressure increases cytoplasmic crowding, which hinders gene expression and impedes growth rate [121]. Another important example involves liquid-liquid phase separation (LLPS), which is controlled by cytoplasmic crowding in yeast and mammalian cells [122]. When cells are treated with rapamycin, a mTORC1 inhibitor, ribosome concentration drastically decreases and particle diffusion increases, showing that ribosomes function as a cytoplasmic crowding agent. Phase separation using a synthetic system proportionally increases with ribosome concentration in the cytoplasm [122]. Finally, kinase phosphorylation of the ERK MAP kinase pathway is controlled by molecular crowding in mammalian cells [123]. The activation loop of ERK is processively phosphorylated by MEK kinase in cells. Processive phosphorylation can be reproduced in vitro with crowding agents, but without them, the reaction shifts to distributive phosphorylation. This suggests that cytoplasmic crowding plays a key role in regulating the successive phosphorylation reactions.

#### 4.1.2. Homeostasis Mechanisms Regulating Cytoplasmic Properties

Cells possess homeostasis mechanisms that control the biophysical properties of the cytoplasm and adapt to environmental changes. Recent studies have identified WNK1 as a molecular crowding sensor in mammalian cells, which facilitates cell volume recovery and alleviates crowding under hyperosmotic stress [124]. As another example, when exposed to elevated temperatures, budding yeast cells produce trehalose and glycogen to enhance cytoplasmic viscosity, thereby counteracting the increase in protein diffusivity [125]. Given that the cytoplasmic properties affect cellular reactions and processes, extreme changes in cytoplasmic biophysics can lead to cellular dysfunctions. Indeed, excessive cell growth without

DNA replication causes cytoplasmic dilution, contributing to cellular senescence [126]. These homeostasis mechanisms are therefore critical for maintaining the proper cytoplasmic environment and cellular functions, though it is still unclear the causal relation between cytoplasmic properties and cellular functions.

#### 4.1.3. Methods to Evaluate the Cytoplasmic Properties

Various techniques allow us to measure the biophysical properties of the cytoplasm. Here, we focus on the cytoplasmic fluidity inside the cell, which can be measured by tracking the mobility of proteins, nucleic acids, and particles and computing their diffusion coefficients. These measurements enable us to estimate the viscoelasticity, molecular crowding, cellular structure, and organization inside the cell. The diffusion of molecules with an average protein size (~5 nm) can be assessed using several methods, such as fluorescence recovery after photobleaching (FRAP), fluorescence correlation spectroscopy (FCS), fluorescence loss in photobleaching (FLIP), and fluorescence decay after photostimulation (FDAP) [127]. Each technique offers unique benefits; for example, FRAP is well-suited for slow diffusion (<1.0 $\mu m^2$/sec), while FCS specializes in fast diffusion (>1.0 $\mu m^2$/sec). The mobility of large molecules (~10 to 1000 nm) can be assessed using particle-tracking microrheology [128]. Recently, genetically encoded nanoparticles have been developed and applied across various biological contexts [129]. These nanoparticles consist of a scaffolding domain and fluorescent protein, self-assembling into bright spherical particles inside the cell. Their size can be adjusted by substituting the scaffolding domain. Several types of nanoparticles are available, such as μNS (50–150 nm) [110] and genetically encoded multimeric nanoparticles (GEMs) with 20 nm (20 nm-GEMs), 40 nm (40 nm-GEMs), and 50 nm (50 nm-GEMs) in diameter [122,130]. In addition, GEMs can be localized to different organelles, including the nucleus (nucGEMs) [131] and endoplasmic reticulum (ER-AqLs) [132]. Genetically encoded nanoparticles overcome several technical challenges; for example, unlike endogenous structures, they are bio-orthogonal, and therefore they do not interact with intracellular components, allowing simplification of mobility analysis. Additionally, genetically encoded nanoparticles can be utilized in organisms with rigid cell walls, where microinjection of exogenous particles is not feasible. Genetically encoded nanoparticles have been successfully implemented in numerous biological systems, spanning bacteria, yeasts (*S. pombe, S. cerevisiae, C. albicans*), fungi (*Ashbya gossypii*), mammalian cells, and *Drosophila* [37,122,133–135]. This broad applicability highlights the versatility of this technique for studying cytoplasmic properties across evolutionarily distant organisms. Despite its utility, this methodology presents several important technical challenges. Notably, it exhibits high sensitivity to expression levels, potentially resulting in aggregate formation when overexpressed. Such aggregation may also be triggered under specific stress conditions, which can confound experimental interpretations and limit applicability in certain cellular states. Overall, genetically encoded nanoparticles have emerged as transformative tools for investigating cytoplasmic biophysics across numerous biological contexts. Their application over the past decade has catalyzed unprecedented advances in our understanding of intracellular biophysical properties, generating a wealth of evidence and insights that have fundamentally reshaped our conceptual framework in this field. A particularly exciting prospect for advancing this technique involves comprehensive genetic screening to identify key regulators of cytoplasmic properties. The methodology is ideally suited for organisms with rigid cell walls, such as bacteria and yeasts, which offer robust platforms for genetic approaches. Large-scale screens would likely offer unprecedented insights into the fundamental organization of the cytoplasm and potentially reveal novel mechanisms that orchestrate its biophysical regulation. This approach could bridge molecular genetics with biophysics, establishing new paradigms for understanding cellular dormancy and its reversal.

### 4.2. Dormancy and Dormancy Breaking

#### 4.2.1. The Biophysical Properties of the Cytoplasm in Dormant Yeast Spores

Two recent studies using microrheological measurements have demonstrated that the biophysical properties of the cytoplasm are drastically altered during dormancy in *S. pombe* and *S. cerevisiae* [37,62]. In *S. pombe*, the mobility of 40 nm-GEMs and μNS (50–150 nm) particles is strongly restricted in spores compared to vegetative cells [37]. Similarly, in *S. cerevisiae*, the mobility of μNS particles is reduced in spores [62], showing that spore cytoplasm hinders macromolecular diffusion during dormancy. In contrast to these particles, the fluorescent proteins, comparable to the average protein size (~5 nm), are relatively free to diffuse even in the spore cytoplasm in *S. pombe* [37]. This size-dependent limitation on cytoplasmic diffusion might explain how dormant cells, despite the diffusion and reaction barriers imposed by the crowded spore cytoplasm, can rapidly resume growth when environmental conditions improve. Specifically, it is proposed that small signaling molecules are relatively free to diffuse in the spore cytoplasm and transmit dormancy-breaking signals upon germination initiation, whereas the mobility of large protein complexes, such as ribosomes (~ 30 nm), is hindered. Recent in vitro studies with *E. coli* and *Xenopus* cytoplasms revealed that an overcrowded cytoplasm can inhibit ribosome mobility, leading to reduced translational activity [136,137]. Similarly, the crowded milieu of spore cytoplasm may suppress translational activity. Indeed, the protein expression rate slows down in spores compared to vegetative cells in *S. cerevisiae* [82]. The reduced protein production may help spores conserve energy sources (e.g., ATP) and substrates (e.g., amino acids, nucleic acids), thereby reducing essential metabolic activity and enabling their extended survival without nutrients.

#### 4.2.2. The Biophysical Properties of the Cytoplasm During Spore Germination

The biophysical properties of the cytoplasm during *S. pombe* germination are shown in Figure 4. Germination triggers dynamic changes in the spore cytoplasm. In *S. pombe*, 40 nm-GEMs mobility exhibits a rapid increase within an hour after germination induction [37]. This cytoplasmic fluidization may contribute to an increase in metabolic activity, driving the subsequent germination progression. Of note, the increased mobility of 40 nm-GEMs precedes a bright-to-dark transition under phase contrast microscopy, the traditionally recognized earliest hallmark of germination initiation [37,47,48]. In both *S. pombe* and *S. cerevisiae*, larger particles (μNS, 50–150 nm) behave slightly differently from 40 nm-GEMs, showing gradually increasing mobility during germination [37,62]. These results suggest that cellular structures are reorganized on a scale of 50 to 150 nm over time. *S. cerevisiae* spores contain cytoplasmic foci of specific proteins (e.g., Acc1, Ura7), which gradually disassemble and then diffuse throughout the cytoplasm as μNS mobility increases [62]. The overcrowded spore cytoplasm might entropically drive these cytoplasmic foci, while cytoplasmic fluidization may accelerate their disassembly. This proposed mechanism is supported by the finding that Acc1 foci formation is facilitated by crowding agents [138].

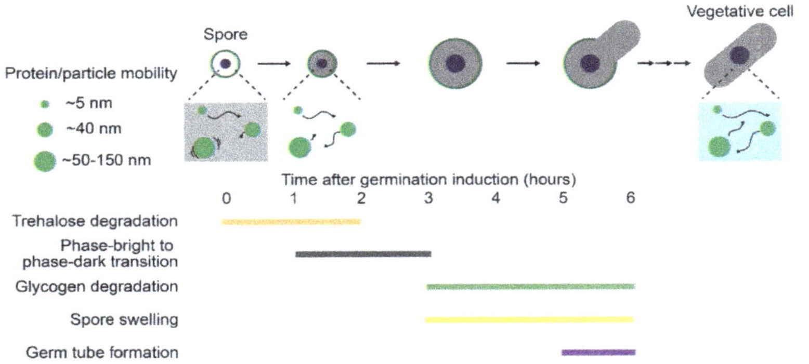


**Figure 4.** The biophysical properties of the cytoplasm during dormancy and germination in *S. pombe*. Protein diffusion and particle mobility are illustrated with the timing of the germination events. In dormant spores, the mobility of particles (~40 nm and ~50–150 nm in diameter) is restricted, whereas small proteins can diffuse relatively freely. Upon germination induction, particle mobility increases rapidly.

#### 4.2.3. The Mechanisms Regulating Cytoplasmic Properties During Dormancy and Germination

Previous studies have proposed several mechanisms for the molecular and physico-chemical regulation of the cytoplasmic properties in vegetative yeast cells and other organisms: (1) pH levels [103], (2) cellular volume changes [111,119], (3) cytoskeleton structure and dynamics [139], (4) viscogen concentration (e.g., trehalose, glycogen, sorbitol) [125,140], (5) ATP-driven metabolic activity [110], and (6) cytoplasmic crowder concentration (e.g., polysome, ribosome, and RNA) [112,122,141]. Some of these potential mechanisms have been investigated during dormancy and germination in *S. pombe* and *S. cerevisiae* spores. First, pH-driven cytoplasmic reorganization is known to trigger cytoplasmic solidification. In *S. cerevisiae*, the pH level in dormant spores (pH ~5.9) drops and then increases during germination, eventually approaching the value found in vegetative cells (pH ~7.4) [62,63]. However, the increase in μNS mobility is delayed compared to an increase in intracellular pH level. These findings suggest that while decreased pH levels may contribute to cytoplasmic properties during dormancy, other mechanisms must also be involved in the regulation of the cytoplasmic fluidity. In *S. pombe*, while spore germination involves cell swelling, the cytoplasmic fluidization based on 40 nm-GEMs mobility precedes the changes in cell sizes [37]. Thus, water uptake accompanied by cell swelling may not be required for cytoplasmic fluidization during germination. Furthermore, actin and microtubule dynamics are not involved in cytoplasmic fluidization in *S. pombe* germinating spores [37]. A recent study has proposed that trehalose, known to act as cytoplasmic viscogen [125], plays a key role in regulating cytoplasmic fluidity during dormancy and germination in *S. pombe* [37]. *S. pombe* spores contain large amounts of trehalose during dormancy [37,38]. Upon germination initiation, the trehalose is quickly degraded concurrently with increased 40 nm-GEM mobility. This trehalose degradation requires the glucose-sensing by cAMP-PKA pathway and natural trehalase Ntp1. Deletion of genes encoding components in cAMP-PKA-Ntp1 signaling impedes trehalose degradation and cytoplasmic fluidization, suggesting a direct link between trehalose and cytoplasmic fluidity. Similarly, in *S. cerevisiae*, trehalose accumulates in spores and rapidly degrades during germination [62], suggesting a conserved mechanism for controlling cytoplasmic properties. While metabolic activity is generally reduced during dormancy, ATP-driven processes could influence cytoplasmic properties [37]. Further investigation is also required to verify whether cytoplasmic crowder concentration regulates cytoplasmic properties during dormancy and germination in yeast cells.

## 5. Summary and Perspective

In this review, we have explored the molecular and biophysical aspects of dormancy and dormancy breaking in yeast models. Numerous studies using *S. cerevisiae* and *S. pombe* have elucidated the complex molecular machinery governing these processes, including signaling pathways, metabolic adaptations, and gene regulatory networks. Recent findings have particularly highlighted the crucial role of cytoplasmic biophysical properties in dormancy and its breaking. Several key questions remain to be addressed: (1) How do cytoplasmic properties regulate cellular functions during dormancy and their reactivation during germination? In particular, it remains unclear whether dormancy-specific proteins undergo phase separation, reminiscent of the reversible phase separation observed for the prion-like protein FLOE1 during seed germination [142]. (2) What is the mechanistic link between trehalose accumulation and cytoplasmic properties in dormant cells? (3) How does dormancy affect the diffusion of endogenous proteins? (4) What are the biophysical properties of other cellular components, such as the nucleus, during dormancy? In addition to the biophysical aspects, critical questions remain in other areas: (1) It is still technically challenging to identify genes required specifically for germination but not for meiosis and sporulation. Future improvements in genome-wide screening would offer new insights into the molecular mechanisms during germination. (2) There is limited knowledge regarding the advantages of sporulation in natural environments, particularly in *S. pombe*. *S. pombe* in its natural habitats might possess unique sporulation characteristics not observed in laboratory strains. In the case of *S. japonicus*, a species in the same genus as *S. pombe*, the naturally isolated strains from *Drosophila* sporulate more efficiently than wild strains isolated from fruits and plants [143].

The comparative study of dormancy mechanisms across different yeast species offers valuable evolutionary insights. Similarities and differences in how these distantly related yeasts regulate dormancy may reveal core conserved mechanisms essential for cellular survival as well as species-specific adaptations to particular ecological niches. Advanced techniques such as cryo-electron tomography, super-resolution microscopy, and single-molecule tracking will likely provide unprecedented views of the dormant cellular state at the molecular level in coming years.

## 6. Conclusions

In this review, we have provided a comprehensive overview of dormancy and germination in two model yeast species, *S. pombe* and *S. cerevisiae*, with particular emphasis on molecular mechanisms and biophysical properties. Using these simple model systems, researchers have elucidated comprehensive changes in gene and protein expression, metabolic profiles, as well as alterations in the biophysical properties of the cytoplasm and its control during dormancy and germination. The cytoplasm in dormant spores differs markedly from that in vegetative cells, characterized by an acidic, viscous, and overcrowded environment. The integration of molecular, biophysical, and ecological approaches in yeast dormancy research continues to reveal fundamental principles of cellular survival. These insights not only deepen our understanding of basic cell biology but also have important implications for biotechnology and medicine, where controlling cellular dormancy is of crucial importance. Future studies focusing on the mechanistic links between cytoplasmic properties and cellular functions during dormancy, as well as improved genome-wide screening methods to identify germination-specific genes, will further advance our understanding of these remarkable survival strategies that are conserved across diverse species.

**Funding:** K.A. was supported by JSPS KAKENHI grants (nos. 18H02444, 19H05798, and 22H02625), Nagase Science Foundation, Takeda Science Foundation, and Joint Research of the Exploratory Research Center on Life and Living Systems (ExCELLS) (ExCELLS program No. 25EXC603). Y.G. was supported by a JST ACT-X grant (no. JPMJAX22B8), JSPS KAKENHI grants (nos. 19K16050 and 22K15110), a Jigami Yoshifumi Memorial Research grant, and a Sumitomo Research grant. K.S. was supported by a JSPS KAKENHI grant (no. 22J10844) and The Graduate University for Advanced Studies, SOKENDAI (SOKENDAI Student Dispatch Program). Y.K. was supported by a JSPS KAKENHI grant (no. 19H05675).

**Acknowledgments:** We thank all members of the Aoki Laboratory for their helpful discussions and assistance. We also thank members of the Tokai Tor Conference (ToToCo) for the general discussion.

**Conflicts of Interest:** The authors declare no conflicts of interest.

## Abbreviations

The following abbreviations are used in this manuscript:

| | |
|---|---|
| cAMP-PKA | cyclic Adenosine MonoPhosphate-Protein Kinase A |
| FRAP | Fluorescence Recovery After Photobleaching |
| FCS | Fluorescence Correlation Spectroscopy |
| FLIP | Fluorescence Loss in Photobleaching |
| FDAP | Fluorescence Decay After Photostimulation |
| LLPS | Liquid-Liquid Phase Separation |
| GEMs | Genetically Encoded Multimeric nanoparticles |
| GAP | GTPase-Activating Proteins |
| TOR | Target Of Rapamycin |
| RNAPII | RNA Polymerase II |
| BSA | Bovine Serum Albumin |

## References

1. Lennon, J.T.; Jones, S.E. Microbial Seed Banks: The Ecological and Evolutionary Implications of Dormancy. *Nat. Rev. Microbiol.* **2011**, *9*, 119–130. [CrossRef] [PubMed]
2. Rittershaus, E.S.C.; Baek, S.-H.; Sassetti, C.M. The Normalcy of Dormancy: Common Themes in Microbial Quiescence. *Cell Host Microbe* **2013**, *13*, 643–651. [CrossRef]
3. Setlow, P. Germination of Spores of Bacillus Species: What We Know and Do Not Know. *J. Bacteriol.* **2014**, *196*, 1297–1305. [CrossRef]
4. Park, H.-S.; Yu, J.-H. Genetic Control of Asexual Sporulation in Filamentous Fungi. *Curr. Opin. Microbiol.* **2012**, *15*, 669–677. [CrossRef]
5. Zhao, Y.; Lin, J.; Fan, Y.; Lin, X. Life Cycle of Cryptococcus Neoformans. *Annu. Rev. Microbiol.* **2019**, *73*, 17–42. [CrossRef] [PubMed]
6. Née, G.; Xiang, Y.; Soppe, W.J. The Release of Dormancy, a Wake-up Call for Seeds to Germinate. *Curr. Opin. Plant Biol.* **2017**, *35*, 8–14. [CrossRef] [PubMed]
7. Keilin, D. The Leeuwenhoek Lecture—The Problem of Anabiosis or Latent Life: History and Current Concept. *Proc. R. Soc. Lond. Ser. B-Biol. Sci.* **1959**, *150*, 149–191. [CrossRef]
8. Clegg, J.S. Cryptobiosis—A Peculiar State of Biological Organization. *Comp. Biochem. Physiol. B Biochem. Mol. Biol.* **2001**, *128*, 613–624. [CrossRef]
9. Hibshman, J.D.; Clegg, J.S.; Goldstein, B. Mechanisms of Desiccation Tolerance: Themes and Variations in Brine Shrimp, Roundworms, and Tardigrades. *Front. Physiol.* **2020**, *11*, 592016. [CrossRef]
10. Sogame, Y.; Kikawada, T. Current Findings on the Molecular Mechanisms Underlying Anhydrobiosis in Polypedilum Vanderplanki. *Curr. Opin. Insect Sci.* **2017**, *19*, 16–21. [CrossRef]
11. Wharton, D.A. Anhydrobiosis. *Curr. Biol.* **2015**, *25*, R1114–R1116. [CrossRef] [PubMed]
12. Sipiczki, M. Where Does Fission Yeast Sit on the Tree of Life? *Genome Biol.* **2000**, *1*, 1–4. [CrossRef]

13. Wood, V.; Gwilliam, R.; Rajandream, M.-A.; Lyne, M.; Lyne, R.; Stewart, A.; Sgouros, J.; Peat, N.; Hayles, J.; Baker, S.; et al. The Genome Sequence of Schizosaccharomyces Pombe. *Nature* **2002**, *415*, 871–880. [CrossRef] [PubMed]
14. van Werven, F.J.; Amon, A. Regulation of Entry into Gametogenesis. *Philos. Trans. R. Soc. B Biol. Sci.* **2011**, *366*, 3521–3531. [CrossRef] [PubMed]
15. Neiman, A.M. Sporulation in the Budding Yeast Saccharomyces Cerevisiae. *Genetics* **2011**, *189*, 737–765. [CrossRef]
16. Ohtsuka, H.; Imada, K.; Shimasaki, T.; Aiba, H. Sporulation: A Response to Starvation in the Fission Yeast Schizosaccharomyces Pombe. *MicrobiologyOpen* **2022**, *11*, e1303. [CrossRef]
17. Sun, S.; Gresham, D. Cellular Quiescence in Budding Yeast. *Yeast Chichester Engl.* **2021**, *38*, 12–29. [CrossRef]
18. Yanagida, M. Cellular Quiescence: Are Controlling Genes Conserved? *Trends Cell Biol.* **2009**, *19*, 705–715. [CrossRef]
19. Walker, R.M.; Sanabria, V.C.; Youk, H. Microbial Life in Slow and Stopped Lanes. *Trends Microbiol.* **2024**, *32*, 650–662. [CrossRef]
20. Li, L.; Miles, S.; Melville, Z.; Prasad, A.; Bradley, G.; Breeden, L.L. Key Events during the Transition from Rapid Growth to Quiescence in Budding Yeast Require Posttranscriptional Regulators. *Mol. Biol. Cell* **2013**, *24*, 3697–3709. [CrossRef]
21. Peter, J.; De Chiara, M.; Friedrich, A.; Yue, J.-X.; Pflieger, D.; Bergström, A.; Sigwalt, A.; Barre, B.; Freel, K.; Llored, A.; et al. Genome Evolution across 1011 Saccharomyces Cerevisiae Isolates. *Nature* **2018**, *556*, 339–344. [CrossRef] [PubMed]
22. Freese, E.B.; Chu, M.I.; Freese, E. Initiation of Yeast Sporulation by Partial Carbon, Nitrogen, or Phosphate Deprivation. *J. Bacteriol.* **1982**, *149*, 840–851. [CrossRef]
23. Dekker, N.; van Rijssel, J.; Distel, B.; Hochstenbach, F. Role of the α-Glucanase Agn2p in Ascus-Wall Endolysis Following Sporulation in Fission Yeast. *Yeast* **2007**, *24*, 279–288. [CrossRef] [PubMed]
24. Encinar del Dedo, J.; Dueñas, E.; Arnáiz, Y.; del Rey, F.; Vázquez de Aldana, C.R. β-Glucanase Eng2 Is Required for Ascus Wall Endolysis after Sporulation in the Fission Yeast Schizosaccharomyces Pombe. *Eukaryot. Cell* **2009**, *8*, 1278–1286. [CrossRef] [PubMed]
25. Guo, H.; King, M.C. A Quality Control Mechanism Linking Meiotic Success to Release of Ascospores. *PLoS ONE* **2013**, *8*, e82758. [CrossRef]
26. Briza, P.; Winkler, G.; Kalchhauser, H.; Breitenbach, M. Dityrosine Is a Prominent Component of the Yeast Ascospore Wall. A Proof of Its Structure. *J. Biol. Chem.* **1986**, *261*, 4288–4294. [CrossRef]
27. Briza, P.; Ellinger, A.; Winkler, G.; Breitenbach, M. Chemical Composition of the Yeast Ascospore Wall. The Second Outer Layer Consists of Chitosan. *J. Biol. Chem.* **1988**, *263*, 11569–11574. [CrossRef]
28. Liu, J.; Tang, X.; Wang, H.; Balasubramanian, M. Bgs2p, a 1,3-β-Glucan Synthase Subunit, Is Essential for Maturation of Ascospore Wall in Schizosaccharomyces Pombe. *FEBS Lett.* **2000**, *478*, 105–108. [CrossRef]
29. Tougan, T.; Chiba, Y.; Kakihara, Y.; Hirata, A.; Nojima, H. Meu10 Is Required for Spore Wall Maturation in Schizosaccharomyces Pombe. *Genes Cells* **2002**, *7*, 217–231. [CrossRef]
30. Arellano, M.; Cartagena-Lirola, H.; Nasser Hajibagheri, M.A.; Durán, A.; Henar Valdivieso, M. Proper Ascospore Maturation Requires the Chs1+ Chitin Synthase Gene in Schizosaccharomyces Pombe. *Mol. Microbiol.* **2000**, *35*, 79–89. [CrossRef]
31. Coluccio, A.; Neiman, A.M. Interspore Bridges: A New Feature of the Saccharomyces Cerevisiae Spore Wall. *Microbiology* **2004**, *150*, 3189–3196. [CrossRef] [PubMed]
32. Yang, Y.; Yang, G.; Li, Z.-J.; Liu, Y.-S.; Gao, X.-D.; Nakanishi, H. Studies on the Proteinaceous Structure Present on the Surface of the Saccharomyces Cerevisiae Spore Wall. *J. Fungi* **2023**, *9*, 392. [CrossRef]
33. Tahara, Y.O.; Miyata, M.; Nakamura, T. Quick-Freeze, Deep-Etch Electron Microscopy Reveals the Characteristic Architecture of the Fission Yeast Spore. *J. Fungi* **2021**, *7*, 7. [CrossRef] [PubMed]
34. Zhang, B.; Teraguchi, E.; Imada, K.; Tahara, Y.O.; Nakamura, S.; Miyata, M.; Kagiwada, S.; Nakamura, T. The Fission Yeast RNA-Binding Protein Meu5 Is Involved in Outer Forespore Membrane Breakdown during Spore Formation. *J. Fungi* **2020**, *6*, 284. [CrossRef] [PubMed]
35. Walther, T.; Létisse, F.; Peyriga, L.; Alkim, C.; Liu, Y.; Lardenois, A.; Martin-Yken, H.; Portais, J.-C.; Primig, M.; François, J.M. Developmental Stage-Dependent Metabolic Regulation during Meiotic Differentiation in Budding Yeast. *BMC Biol.* **2014**, *12*, 60. [CrossRef]
36. Kane, S.M.; Roth, R. Carbohydrate Metabolism During Ascospore Development in Yeast. *J. Bacteriol.* **1974**, *118*, 8–14. [CrossRef]
37. Sakai, K.; Kondo, Y.; Goto, Y.; Aoki, K. Cytoplasmic Fluidization Contributes to Breaking Spore Dormancy in Fission Yeast. *Proc. Natl. Acad. Sci. USA* **2024**, *121*, e2405553121. [CrossRef]
38. Beltran, F.F.; Castillo, R.; Vicente-Soler, J.; Cansado, J.; Gacto, M. Role for Trehalase during Germination of Spores in the Fission Yeast Schizosaccharomyces Pombe. *FEMS Microbiol. Lett.* **2000**, *193*, 117–121. [CrossRef]
39. Inoue, H.; Shimoda, C. Changes in Trehalose Content and Trehalase Activity during Spore Germination in Fission Yeast, Schizosaccharomyces Pombe. *Arch. Microbiol.* **1981**, *129*, 19–22. [CrossRef]
40. Takaine, M.; Ueno, M.; Kitamura, K.; Imamura, H.; Yoshida, S. Reliable Imaging of ATP in Living Budding and Fission Yeast. *J. Cell Sci.* **2019**, *132*, jcs230649. [CrossRef]

41. Brengues, M.; Pintard, L.; Lapeyre, B. mRNA Decay Is Rapidly Induced after Spore Germination of Saccharomyces Cerevisiae. *J. Biol. Chem.* **2002**, *277*, 40505–40512. [CrossRef] [PubMed]
42. Hugener, J.; Xu, J.; Wettstein, R.; Ioannidi, L.; Velikov, D.; Wollweber, F.; Henggeler, A.; Matos, J.; Pilhofer, M. FilamentID Reveals the Composition and Function of Metabolic Enzyme Polymers during Gametogenesis. *Cell* **2024**, *187*, 3303–3318.e18. [CrossRef]
43. Joseph-Strauss, D.; Zenvirth, D.; Simchen, G.; Barkai, N. Spore Germination in Saccharomyces Cerevisiae: Global Gene Expression Patterns and Cell Cycle Landmarks. *Genome Biol.* **2007**, *8*, R241. [CrossRef]
44. Geijer, C.; Joseph-Strauss, D.; Simchen, G.; Barkai, N.; Hohmann, S. Saccharomyces Cerevisiae Spore Germination. In *Dormancy and Resistance in Harsh Environments*; Lubzens, E., Cerda, J., Clark, M., Eds.; Springer: Berlin/Heidelberg, Germany, 2010; pp. 29–41. ISBN 978-3-642-12422-8.
45. Sando, N.; Oguchi, T.; Nagano, M.; Osumi, M. Morphological Changes in Ascospores of Saccharomyces Cerevisiae During Aerobic and Anaerobic Germination. *J. Gen. Appl. Microbiol.* **1980**, *26*, 403–412. [CrossRef]
46. Kono, K.; Matsunaga, R.; Hirata, A.; Suzuki, G.; Abe, M.; Ohya, Y. Involvement of Actin and Polarisome in Morphological Change during Spore Germination of Saccharomyces Cerevisiae. *Yeast* **2005**, *22*, 129–139. [CrossRef] [PubMed]
47. Hatanaka, M.; Shimoda, C. The Cyclic AMP/PKA Signal Pathway Is Required for Initiation of Spore Germination in Schizosaccharomyces Pombe. *Yeast* **2001**, *18*, 207–217. [CrossRef] [PubMed]
48. Bonazzi, D.; Julien, J.-D.; Romao, M.; Seddiki, R.; Piel, M.; Boudaoud, A.; Minc, N. Symmetry Breaking in Spore Germination Relies on an Interplay between Polar Cap Stability and Spore Wall Mechanics. *Dev. Cell* **2014**, *28*, 534–546. [CrossRef]
49. Tsuyuzaki, H.; Ujiie, R.; Sato, M. Wake-up Alarm: Virtual Time-Lapse Gene Expression Landscape Illuminates Mechanisms Underlying Dormancy Breaking of Germinating Spores. *Curr. Genet.* **2021**, *67*, 519–534. [CrossRef]
50. Herman, P.K.; Rine, J. Yeast Spore Germination: A Requirement for Ras Protein Activity during Re-entry into the Cell Cycle. *EMBO J.* **1997**, *16*, 6171–6181. [CrossRef]
51. Maeda, T.; Watanabe, Y.; Kunitomo, H.; Yamamoto, M. Cloning of the Pka1 Gene Encoding the Catalytic Subunit of the cAMP-Dependent Protein Kinase in Schizosaccharomyces Pombe. *J. Biol. Chem.* **1994**, *269*, 9632–9637. [CrossRef]
52. Heasley, L.R.; Singer, E.; Cooperman, B.J.; McMurray, M.A. Saccharomyces Spores Are Born Prepolarized to Outgrow Away from Spore–Spore Connections and Penetrate the Ascus Wall. *Yeast* **2021**, *38*, 90–101. [CrossRef] [PubMed]
53. Cooperman, B.; McMurray, M. Roles for the Canonical Polarity Machinery in the de Novo Establishment of Polarity in Budding Yeast Spores. *Mol. Biol. Cell* **2025**, *36*(3), 1–17. [CrossRef]
54. Cortés, J.C.G.; Carnero, E.; Ishiguro, J.; Sánchez, Y.; Durán, A.; Ribas, J.C. The Novel Fission Yeast (1,3)β-D-Glucan Synthase Catalytic Subunit Bgs4p Is Essential during Both Cytokinesis and Polarized Growth. *J. Cell Sci.* **2005**, *118*, 157–174. [CrossRef]
55. Wei, W.; Zheng, B.; Zheng, S.; Wu, D.; Chu, Y.; Zhang, S.; Wang, D.; Ma, X.; Liu, X.; Yao, X.; et al. The Cdc42 GAP Rga6 Promotes Monopolar Outgrowth of Spores. *J. Cell Biol.* **2022**, *222*, e202202064. [CrossRef] [PubMed]
56. Donnini, C.; Artoni, N.; Marmiroli, N. Germination Conditions That Require Mitochondrial Function in Saccharomyces Cerevisiae: Utilization of Acetate and Galactose. *J. Bacteriol.* **1986**, *168*, 1250–1253. [CrossRef]
57. Xu, G.; West, T.P. Nutritional and Physiological Factors Affecting Germination of Heterothallic Saccharomyces Cerevisiae Ascospores. *Microbios* **1992**, *72*, 27–34.
58. Shimoda, C. Differential Effect of Glucose and Fructose on Spore Germination in the Fission Yeast, Schizosaccharomyces Pombe. *Can. J. Microbiol.* **1980**, *26*, 741–745. [CrossRef]
59. Plante, S.; Normant, V.; Ramos-Torres, K.M.; Labbé, S. Cell-Surface Copper Transporters and Superoxide Dismutase 1 Are Essential for Outgrowth during Fungal Spore Germination. *J. Biol. Chem.* **2017**, *292*, 11896–11914. [CrossRef]
60. Plante, S.; Labbé, S. Spore Germination Requires Ferrichrome Biosynthesis and the Siderophore Transporter Str1 in Schizosaccharomyces Pombe. *Genetics* **2019**, *211*, 893–911. [CrossRef]
61. Tsuyuzaki, H.; Hosokawa, M.; Arikawa, K.; Yoda, T.; Okada, N.; Takeyama, H.; Sato, M. Time-Lapse Single-Cell Transcriptomics Reveals Modulation of Histone H3 for Dormancy Breaking in Fission Yeast. *Nat. Commun.* **2020**, *11*, 1265. [CrossRef]
62. Plante, S.; Moon, K.-M.; Lemieux, P.; Foster, L.J.; Landry, C.R. Breaking Spore Dormancy in Budding Yeast Transforms the Cytoplasm and the Solubility of the Proteome. *PLOS Biol.* **2023**, *21*, e3002042. [CrossRef] [PubMed]
63. Barton, J.K.; den Hollander, J.A.; Lee, T.M.; MacLaughlin, A.; Shulman, R.G. Measurement of the Internal pH of Yeast Spores by 31P Nuclear Magnetic Resonance. *Proc. Natl. Acad. Sci. USA* **1980**, *77*, 2470–2473. [CrossRef] [PubMed]
64. Coluccio, A.E.; Rodriguez, R.K.; Kernan, M.J.; Neiman, A.M. The Yeast Spore Wall Enables Spores to Survive Passage through the Digestive Tract of Drosophila. *PLoS ONE* **2008**, *3*, e2873. [CrossRef] [PubMed]
65. Knight, S.J.; Goddard, M.R. Sporulation in Soil as an Overwinter Survival Strategy in Saccharomyces Cerevisiae. *FEMS Yeast Res.* **2016**, *16*, fov102. [CrossRef]
66. Stefanini, I.; Dapporto, L.; Legras, J.-L.; Calabretta, A.; Di Paola, M.; De Filippo, C.; Viola, R.; Capretti, P.; Polsinelli, M.; Turillazzi, S.; et al. Role of Social Wasps in Saccharomyces Cerevisiae Ecology and Evolution. *Proc. Natl. Acad. Sci. USA* **2012**, *109*, 13398–13403. [CrossRef]

67. Brysch-Herzberg, M.; Jia, G.-S.; Seidel, M.; Assali, I.; Du, L.-L. Insights into the Ecology of Schizosaccharomyces Species in Natural and Artificial Habitats. *Antonie Van Leeuwenhoek* **2022**, *115*, 661–695. [CrossRef]
68. McClure, A.W.; Jacobs, K.C.; Zyla, T.R.; Lew, D.J. Mating in Wild Yeast: Delayed Interest in Sex after Spore Germination. *Mol. Biol. Cell* **2018**, *29*, 3119–3127. [CrossRef]
69. Mata, J.; Lyne, R.; Burns, G.; Bähler, J. The Transcriptional Program of Meiosis and Sporulation in Fission Yeast. *Nat. Genet.* **2002**, *32*, 143–147. [CrossRef]
70. Krapp, A.; Hamelin, R.; Armand, F.; Chiappe, D.; Krapp, L.; Cano, E.; Moniatte, M.; Simanis, V. Analysis of the S. pombe Meiotic Proteome Reveals a Switch from Anabolic to Catabolic Processes and Extensive Post-Transcriptional Regulation. *Cell Rep.* **2019**, *26*, 1044–1058.e5. [CrossRef]
71. Sivakova, B.; Wagner, A.; Kretova, M.; Jakubikova, J.; Gregan, J.; Kratochwill, K.; Barath, P.; Cipak, L. Quantitative Proteomics and Phosphoproteomics Profiling of Meiotic Divisions in the Fission Yeast Schizosaccharomyces Pombe. *Sci. Rep.* **2024**, *14*, 23105. [CrossRef]
72. Billmyre, R.B.; Eickbush, M.T.; Craig, C.J.; Lange, J.J.; Wood, C.; Helston, R.M.; Zanders, S.E. Genome-Wide Quantification of Contributions to Sexual Fitness Identifies Genes Required for Spore Viability and Health in Fission Yeast. *PLOS Genet.* **2022**, *18*, e1010462. [CrossRef] [PubMed]
73. Ucisik-Akkaya, E.; Leatherwood, J.K.; Neiman, A.M. A Genome-Wide Screen for Sporulation-Defective Mutants in Schizosaccharomyces Pombe. *G3 GenesGenomesGenetics* **2014**, *4*, 1173–1182. [CrossRef] [PubMed]
74. Fukunishi, K.; Miyakubi, K.; Hatanaka, M.; Otsuru, N.; Hirata, A.; Shimoda, C.; Nakamura, T. The Fission Yeast Spore Is Coated by a Proteinaceous Surface Layer Comprising Mainly Isp3. *Mol. Biol. Cell* **2014**, *25*, 1549–1559. [CrossRef] [PubMed]
75. Yang, H.-J.; Osakada, H.; Kojidani, T.; Haraguchi, T.; Hiraoka, Y. Lipid Droplet Dynamics during Schizosaccharomyces Pombe Sporulation and Their Role in Spore Survival. *Biol. Open* **2017**, *6*, 217–222. [CrossRef]
76. Ünal, E.; Kinde, B.; Amon, A. Gametogenesis Eliminates Age-Induced Cellular Damage and Resets Life Span in Yeast. *Science* **2011**, *332*, 1554–1557. [CrossRef]
77. Dawes, I.W.; Hardie, I.D. Selective Killing of Vegetative Cells in Sporulated Yeast Cultures by Exposure to Diethyl Ether. *Mol. Gen. Genet. MGG* **1974**, *131*, 281–289. [CrossRef]
78. Ho, K.H.; Miller, J.J. Free Proline Content and Sensitivity to Desiccation and Heat during Yeast Sporulation and Spore Germination. *Can. J. Microbiol.* **1978**, *24*, 312–320. [CrossRef]
79. Briza, P.; Breitenbach, M.; Ellinger, A.; Segall, J. Isolation of Two Developmentally Regulated Genes Involved in Spore Wall Maturation in Saccharomyces Cerevisiae. *Genes Dev.* **1990**, *4*, 1775–1789. [CrossRef]
80. Suda, Y.; Rodriguez, R.K.; Coluccio, A.E.; Neiman, A.M. A Screen for Spore Wall Permeability Mutants Identifies a Secreted Protease Required for Proper Spore Wall Assembly. *PLoS ONE* **2009**, *4*, e7184. [CrossRef]
81. Pammer, M.; Briza, P.; Ellinger, A.; Schuster, T.; Stucka, R.; Feldmann, H.; Breitenbach, M. DIT101 (CSD2, CAL1), a cell cycle-regulated yeast gene required for synthesis of chitin in cell walls and chitosan in spore walls. *Yeast* **1992**, *8*, 1089–1099. [CrossRef]
82. Maire, T.; Allertz, T.; Betjes, M.A.; Youk, H. Dormancy-to-death Transition in Yeast Spores Occurs Due to Gradual Loss of Gene-expressing Ability. *Mol. Syst. Biol.* **2020**, *16*, e9245. [CrossRef] [PubMed]
83. Sawadogo, M.; Roeder, R.G. Energy Requirement for Specific Transcription Initiation by the Human RNA Polymerase II System. *J. Biol. Chem.* **1984**, *259*, 5321–5326. [CrossRef]
84. Takaine, M.; Imamura, H.; Yoshida, S. High and Stable ATP Levels Prevent Aberrant Intracellular Protein Aggregation in Yeast. *eLife* **2022**, *11*, e67659. [CrossRef]
85. Bell, W.; Klaassen, P.; Ohnacker, M.; Boller, T.; Herweijer, M.; Schoppink, P.; Vanderzee, P.; Wiemken, A. Characterization of the 56-kDa Subunit of Yeast Trehalose-6-Phosphate Synthase and Cloning of Its Gene Reveal Its Identity with the Product of CIF1, a Regulator of Carbon Catabolite Inactivation. *Eur. J. Biochem.* **1992**, *209*, 951–959. [CrossRef] [PubMed]
86. De Silva-Udawatta, M.N.; Cannon, J.F. Roles of Trehalose Phosphate Synthase in Yeast Glycogen Metabolism and Sporulation. *Mol. Microbiol.* **2001**, *40*, 1345–1356. [CrossRef] [PubMed]
87. Shi, L.; Sutter, B.M.; Ye, X.; Tu, B.P. Trehalose Is a Key Determinant of the Quiescent Metabolic State That Fuels Cell Cycle Progression upon Return to Growth. Mol. Biol. *Cell* **2010**, *21*, 1982–1990. [CrossRef]
88. Ewald, J.C.; Kuehne, A.; Zamboni, N.; Skotheim, J.M. The Yeast Cyclin-Dependent Kinase Routes Carbon Fluxes to Fuel Cell Cycle Progression. *Mol. Cell* **2016**, *62*, 532–545. [CrossRef]
89. Zhao, G.; Chen, Y.; Carey, L.; Futcher, B. Cyclin-Dependent Kinase Co-Ordinates Carbohydrate Metabolism and Cell Cycle in S. cerevisiae. *Mol. Cell* **2016**, *62*, 546–557. [CrossRef]
90. Reuter, M.; Bell, G.; Greig, D. Increased Outbreeding in Yeast in Response to Dispersal by an Insect Vector. *Curr. Biol.* **2007**, *17*, R81–R83. [CrossRef]
91. Johnke, R.; Padilla, G.M. Germination and Outgrowth of Schizosaccharomyces Pombe Spores Isolated by a Simple Batch Centrifugation Technique. *Microbiology* **1979**, *115*, 255–258. [CrossRef]

92. Sakai, K.; Aoki, K.; Goto, Y. Live-Cell Fluorescence Imaging and Optogenetic Control of PKA Kinase Activity in Fission Yeast Schizosaccharomyces Pombe. *Yeast* **2024**, *41*, 349–363. [CrossRef] [PubMed]
93. Yu, G.; Li, J.; Young, D. The Schizosaccharomyces Pombe Pka1 Gene, Encoding a Homolog of cAMP-Dependent Protein Kinase. *Gene* **1994**, *151*, 215–220. [CrossRef] [PubMed]
94. Hoffman, C.S. Glucose Sensing via the Protein Kinase A Pathway in Schizosaccharomyces Pombe. *Biochem. Soc. Trans.* **2005**, *33*, 257–260. [CrossRef]
95. Cansado, J.; Soto, T.; Fernandez, J.; Vicente-Soler, J.; Gacto, M. Characterization of Mutants Devoid of Neutral Trehalase Activity in the Fission Yeast Schizosaccharomyces Pombe: Partial Protection from Heat Shock and High-Salt Stress. *J. Bacteriol.* **1998**, *180*, 1342–1345. [CrossRef]
96. Blázquez, M.A.; Stucka, R.; Feldmann, H.; Gancedo, C. Trehalose-6-P Synthase Is Dispensable for Growth on Glucose but Not for Spore Germination in Schizosaccharomyces Pombe. *J. Bacteriol.* **1994**, *176*, 3895–3902. [CrossRef]
97. Gallo Castro, D.; Martin, S.G. Differential GAP Requirement for Cdc42-GTP Polarization during Proliferation and Sexual Reproduction. *J. Cell Biol.* **2018**, *217*, 4215–4229. [CrossRef] [PubMed]
98. Hashimoto, T.; Conti, S.F.; Naylor, H.B. Fine Structure of Microorganisms. *J. Bacteriol.* **1958**, *76*, 406–416. [CrossRef]
99. Plante, S.; Landry, C.R. Purification of Yeast Spores to Investigate Their Dynamics of Activation. *Curr. Protoc. Microbiol.* **2020**, *59*, e123. [CrossRef]
100. Choih, S.J.; Ferro, A.J.; Shapiro, S.K. Function of S-Adenosylmethionine in Germinating Yeast Ascospores. *J. Bacteriol.* **1977**, *131*, 63–68. [CrossRef]
101. Rousseau, P.; Halvorson, H.O.; Bulla, L.A.; Julian, G.S. Germination and Outgrowth of Single Spores of Saccharomyces Cerevisiae Viewed by Scanning Electron and Phase-Contrast Microscopy. *J. Bacteriol.* **1972**, *109*, 1232–1238. [CrossRef]
102. Tingle, M.A.; Küenzi, M.T.; Halvorson, H.O. Germination of Yeast Spores Lacking Mitochondrial Deoxyribonucleic Acid. *J. Bacteriol.* **1974**, *117*, 89–93. [CrossRef] [PubMed]
103. Munder, M.C.; Midtvedt, D.; Franzmann, T.; Nüske, E.; Otto, O.; Herbig, M.; Ulbricht, E.; Müller, P.; Taubenberger, A.; Maharana, S.; et al. A pH-Driven Transition of the Cytoplasm from a Fluid- to a Solid-like State Promotes Entry into Dormancy. *eLife* **2016**, *5*, e09347. [CrossRef]
104. Geijer, C.; Pirkov, I.; Vongsangnak, W.; Ericsson, A.; Nielsen, J.; Krantz, M.; Hohmann, S. Time Course Gene Expression Profiling of Yeast Spore Germination Reveals a Network of Transcription Factors Orchestrating the Global Response *BMC Genomics* **2012**, *13*, 554. [CrossRef] [PubMed]
105. Luby-Phelps, K.; Taylor, D.L.; Lanni, F. Probing the Structure of Cytoplasm. *J. Cell Biol.* **1986**, *102*, 2015–2022. [CrossRef]
106. Moeendarbary, E.; Valon, L.; Fritzsche, M.; Harris, A.R.; Moulding, D.A.; Thrasher, A.J.; Stride, E.; Mahadevan, L.; Charras, G.T. The Cytoplasm of Living Cells Behaves as a Poroelastic Material. *Nat. Mater.* **2013**, *12*, 253–261. [CrossRef]
107. Etoc, F.; Balloul, E.; Vicario, C.; Normanno, D.; Liße, D.; Sittner, A.; Piehler, J.; Dahan, M.; Coppey, M. Non-Specific Interactions Govern Cytosolic Diffusion of Nanosized Objects in Mammalian Cells. *Nat. Mater.* **2018**, *17*, 740–746. [CrossRef]
108. Di Rienzo, C.; Piazza, V.; Gratton, E.; Beltram, F.; Cardarelli, F. Probing Short-Range Protein Brownian Motion in the Cytoplasm of Living Cells. *Nat. Commun.* **2014**, *5*, 5891. [CrossRef] [PubMed]
109. Nishizawa, K.; Fujiwara, K.; Ikenaga, M.; Nakajo, N.; Yanagisawa, M.; Mizuno, D. Universal Glass-Forming Behavior of in Vitro and Living Cytoplasm. *Sci. Rep.* **2017**, *7*, 15143. [CrossRef]
110. Parry, B.R.; Surovtsev, I.V.; Cabeen, M.T.; O'Hern, C.S.; Dufresne, E.R.; Jacobs-Wagner, C. The Bacterial Cytoplasm Has Glass-like Properties and Is Fluidized by Metabolic Activity. *Cell* **2014**, *156*, 183–194. [CrossRef]
111. Joyner, R.P.; Tang, J.H.; Helenius, J.; Dultz, E.; Brune, C.; Holt, L.J.; Huet, S.; Müller, D.J.; Weis, K. A Glucose-Starvation Response Regulates the Diffusion of Macromolecules. *eLife* **2016**, *5*, e09376. [CrossRef]
112. Xie, Y.; Shu, T.; Liu, T.; Spindler, M.-C.; Mahamid, J.; Hocky, G.M.; Gresham, D.; Holt, L.J. Polysome Collapse and RNA Condensation Fluidize the Cytoplasm. *Mol. Cell* **2024**, *84*, 2698–2716.e9. [CrossRef] [PubMed]
113. Marini, G.; Nüske, E.; Leng, W.; Alberti, S.; Pigino, G. Reorganization of Budding Yeast Cytoplasm upon Energy Depletion. *Mol. Biol. Cell* **2020**, *31*, 1232–1245. [CrossRef]
114. Xie, J.; Najafi, J.; Le Borgne, R.; Verbavatz, J.-M.; Durieu, C.; Sallé, J.; Minc, N. Contribution of Cytoplasm Viscoelastic Properties to Mitotic Spindle Positioning. *Proc. Natl. Acad. Sci. USA* **2022**, *119*, e2115593119. [CrossRef]
115. Sunde, E.P.; Setlow, P.; Hederstedt, L.; Halle, B. The Physical State of Water in Bacterial Spores. *Proc Natl. Acad. Sci. USA* **2009**, *106*, 19334–19339. [CrossRef]
116. Buitink, J.; Leprince, O. Intracellular Glasses and Seed Survival in the Dry State. *C. R. Biol.* **2008**, *331*. 788–795. [CrossRef]
117. Sakurai, M.; Furuki, T.; Akao, K.; Tanaka, D.; Nakahara, Y.; Kikawada, T.; Watanabe, M.; Okuda, T. Vitrification Is Essential for Anhydrobiosis in an African Chironomid, Polypedilum Vanderplanki. *Proc. Natl. Acad. Sci. USA* **2008**, *105*, 5093–5098. [CrossRef] [PubMed]
118. Hengherr, S.; Heyer, A.G.; Köhler, H.-R.; Schill, R.O. Trehalose and Anhydrobiosis in Tardigrades—Evidence for Divergence in Responses to Dehydration. *FEBS J.* **2008**, *275*, 281–288. [CrossRef] [PubMed]

119. Molines, A.T.; Lemière, J.; Gazzola, M.; Steinmark, I.E.; Edrington, C.H.; Hsu, C.-T.; Real-Calderon, P.; Suhling, K.; Goshima, G.; Holt, L.J.; et al. Physical Properties of the Cytoplasm Modulate the Rates of Microtubule Polymerization and Depolymerization. *Dev. Cell* **2022**, *57*, 466–479.e6. [CrossRef]
120. Miermont, A.; Waharte, F.; Hu, S.; McClean, M.N.; Bottani, S.; Léon, S.; Hersen, P. Severe Osmotic Compression Triggers a Slowdown of Intracellular Signaling, Which Can Be Explained by Molecular Crowding. *Proc. Natl. Acad. Sci. USA* **2013**, *110*, 5725–5730. [CrossRef]
121. Alric, B.; Formosa-Dague, C.; Dague, E.; Holt, L.J.; Delarue, M. Macromolecular Crowding Limits Growth under Pressure. *Nat. Phys.* **2022**, *18*, 411–416. [CrossRef]
122. Delarue, M.; Brittingham, G.P.; Pfeffer, S.; Surovtsev, I.V.; Pinglay, S.; Kennedy, K.J.; Schaffer, M.; Gutierrez, J.I.; Sang, D.; Poterewicz, G.; et al. mTORC1 Controls Phase Separation and the Biophysical Properties of the Cytoplasm by Tuning Crowding. *Cell* **2018**, *174*, 338–349.e20. [CrossRef] [PubMed]
123. Aoki, K.; Yamada, M.; Kunida, K.; Yasuda, S.; Matsuda, M. Processive Phosphorylation of ERK MAP Kinase in Mammalian Cells. *Proc. Natl. Acad. Sci. USA* **2011**, *108*, 12675–12680. [CrossRef] [PubMed]
124. Boyd-Shiwarski, C.R.; Shiwarski, D.J.; Griffiths, S.E.; Beacham, R.T.; Norrell, L.; Morrison, D.E.; Wang, J.; Mann, J.; Tennant, W.; Anderson, E.N.; et al. WNK Kinases Sense Molecular Crowding and Rescue Cell Volume via Phase Separation. *Cell* **2022**, *185*, 4488–4506.e20. [CrossRef]
125. Persson, L.B.; Ambati, V.S.; Brandman, O. Cellular Control of Viscosity Counters Changes in Temperature and Energy Availability. *Cell* **2020**, *183*, 1572–1585.e16. [CrossRef]
126. Neurohr, G.E.; Terry, R.L.; Lengefeld, J.; Bonney, M.; Brittingham, G.P.; Moretto, F.; Miettinen, T.P.; Vaites, L.P.; Soares, L.M.; Paulo, J.A.; et al. Excessive Cell Growth Causes Cytoplasm Dilution And Contributes to Senescence. *Cell* **2019**, *176*, 1083–1097.e18. [CrossRef] [PubMed]
127. Matsuda, T.; Nagai, T. Quantitative Measurement of Intracellular Protein Dynamics Using Photobleaching or Photoactivation of Fluorescent Proteins. *Microscopy* **2014**, *63*, 403–408. [CrossRef]
128. Wirtz, D. Particle-Tracking Microrheology of Living Cells: Principles and Applications. *Annu. Rev. Biophys.* **2009**, *38*, 301–326. [CrossRef]
129. Bonucci, M.; Shu, T.; Holt, L.J. How It Feels in a Cell. *Trends Cell Biol.* **2023**, *33*, 924–938. [CrossRef]
130. Hernandez, C.M.; Duran-Chaparro, D.C.; van Eeuwen, T.; Rout, M.P.; Holt, L.J. Development and Characterization of 50 Nanometer Diameter Genetically Encoded Multimeric Nanoparticles. *bioRxiv* **2024**. bioRxiv:2024.07.05.602291.
131. Shu, T.; Szórádi, T.; Kidiyoor, G.R.; Xie, Y.; Herzog, N.L.; Bazley, A.; Bonucci, M.; Keegan, S.; Saxena, S.; Ettefa, F.; et al. nucGEMs Probe the Biophysical Properties of the Nucleoplasm. *bioRxiv* **2021**. bioRxiv:2021.11.18.469159.
132. Chambers, J.E.; Zubkov, N.; Kubánková, M.; Nixon-Abell, J.; Mela, I.; Abreu, S.; Schwiening, M.; Lavarda, G.; López-Duarte, I.; Dickens, J.A.; et al. Z-A1-Antitrypsin Polymers Impose Molecular Filtration in the Endoplasmic Reticulum after Undergoing Phase Transition to a Solid State. *Sci. Adv.* **2022**, *8*, eabm2094. [CrossRef] [PubMed]
133. Serrano, A.; Puerner, C.; Plumb, E.; Chevalier, L.; Elferich, J.; Sinn, L.R.; Grigorieff, N.; Ralser, M.; Delarue, M.; Bassilana, M.; et al. Dynamic Cytoplasmic Fluidity during Morphogenesis in a Human Fungal Pathogen. *bioRxiv* **2024**. bioRxiv:2024.11.16.623909.
134. McLaughlin, G.A.; Langdon, E.M.; Crutchley, J.M.; Holt, L.J.; Forest, M.G.; Newby, J.M.; Gladfelter, A.S. Spatial Heterogeneity of the Cytosol Revealed by Machine Learning-Based 3D Particle Tracking. *Mol. Biol. Cell* **2020**, *31*, 1498–1511. [CrossRef] [PubMed]
135. Lu, W.; Lakonishok, M.; Serpinskaya, A.S.; Gelfand, V.I. A Novel Mechanism of Bulk Cytoplasmic Transport by Cortical Dynein in Drosophila Ovary. *eLife* **2022**, *11*, e75538. [CrossRef]
136. Vibhute, M.A.; Schaap, M.H.; Maas, R.J.M.; Nelissen, F.H.T.; Spruijt, E.; Heus, H.A.; Hansen, M.M.K.; Huck, W.T.S. Transcription and Translation in Cytomimetic Protocells Perform Most Efficiently at Distinct Macromolecular Crowding Conditions. *ACS Synth. Biol.* **2020**, *9*, 2797–2807. [CrossRef]
137. Chen, Y.; Huang, J.-H.; Phong, C.; Ferrell, J.E. Viscosity-Dependent Control of Protein Synthesis and Degradation. *Nat. Commun.* **2024**, *15*, 2149. [CrossRef]
138. Petrovska, I.; Nüske, E.; Munder, M.C.; Kulasegaran, G.; Malinovska, L.; Kroschwald, S.; Richter, D.; Fahmy, K.; Gibson, K.; Verbavatz, J.-M.; et al. Filament Formation by Metabolic Enzymes Is a Specific Adaptation to an Advanced State of Cellular Starvation. *eLife* **2014**, *3*, e02409. [CrossRef]
139. Guo, M.; Ehrlicher, A.J.; Jensen, M.H.; Renz, M.; Moore, J.R.; Goldman, R.D.; Lippincott-Schwartz, J.; Mackintosh, F.C.; Weitz, D.A. Probing the Stochastic, Motor-Driven Properties of the Cytoplasm Using Force Spectrum Microscopy. *Cell* **2014**, *158*, 822–832. [CrossRef]
140. Torrino, S.; Oldham, W.M.; Tejedor, A.R.; Burgos, I.S.; Nasr, L.; Rachedi, N.; Fraissard, K.; Chauvet, C.; Sbai, C.; O'Hara, B.P.; et al. Mechano-Dependent Sorbitol Accumulation Supports Biomolecular Condensate. *Cell* **2024**, *188*, 447–464.e20. [CrossRef]
141. Gade, V.R.; Heinrich, S.; Paloni, M.; Gómez-García, P.A.; Dzanko, A.; Oswald, A.; Marchand, D.; Khawaja, S.; Barducci, A.; Weis, K. Polysomes and mRNA Control the Biophysical Properties of the Eukaryotic Cytoplasm. *bioRxiv* **2024**. bioRxiv:2024.11.14.623620.

142. Dorone, Y.; Boeynaems, S.; Flores, E.; Jin, B.; Hateley, S.; Bossi, F.; Lazarus, E.; Pennington, J.G.; Michiels, E.; Decker, M.D.; et al. A Prion-like Protein Regulator of Seed Germination Undergoes Hydration-Dependent Phase Separation. *Cell* **2021**, *184*, 4284–4298.e27. [CrossRef] [PubMed]
143. Seike, T.; Sakata, N.; Matsuda, F.; Furusawa, C. Elevated Sporulation Efficiency in Fission Yeast Schizosaccharomyces Japonicus Strains Isolated from Drosophila. *J. Fungi* **2021**, *7*, 350. [CrossRef] [PubMed]

biomolecules

MDPI

*Review*

# The Critical Balance Between Quiescence and Reactivation of Neural Stem Cells

**Adam M. Elkin †, Sarah Robbins †, Claudia S. Barros * and Torsten Bossing ***

Peninsula Medical School, Faculty of Health, University of Plymouth, Plymouth PL6 8BU, UK; adam.elkin@plymouth.ac.uk (A.M.E.); sarah.robbins@plymouth.ac.uk (S.R.)
* Correspondence: claudia.barros@plymouth.ac.uk (C.S.B.); torsten.bossing@plymouth.ac.uk (T.B.)
† These authors contributed equally to this work.

**Abstract:** Neural stem cells (NSC) are multipotent, self-renewing cells that give rise to all neural cell types within the central nervous system. During adulthood, most NSCs exist in a quiescent state which can be reactivated in response to metabolic and signalling changes, allowing for long-term continuous neurogenesis and response to injury. Ensuring a critical balance between quiescence and reactivation is required to maintain the limited NSC reservoir and neural replenishment throughout lifetime. The precise mechanisms and signalling pathways behind this balance are at the focus of current research. In this review, we highlight and discuss recent studies using *Drosophila*, mammalian and zebrafish models contributing to the understanding of molecular mechanisms underlying quiescence and reactivation of NSCs.

**Keywords:** neural stem cells; quiescence; reactivation; damage repair

## 1. Introduction

The central nervous system (CNS) is a highly complex network allowing for many functions, including movement, learning, and memory. The billions of neurons and glia required to form this intricate system are derived from a relatively small population of progenitor cells termed neural stem cells (NSCs). Tissue stem cells have a remarkable self-renewal and multilineage capacity, allowing them to asymmetrically divide to generate diverse cell types during development, tissue homeostasis in adulthood, and during regenerative or damage repair processes [1–3]. This capacity is spatially and temporally regulated, allowing for the differing progeny that contribute to drive complex behaviours [4]. However, most NSCs during adulthood are maintained in a quiescent state, allowing for the precious, limited pool of NSCs to be preserved and reactivated as needed including following disease or injury [4–6]. Organisms with a high regenerative capacity such as zebrafish (*Danio rerio*) and salamanders utilise quiescent NSCs (qNSCs) to replenish neurons and glia within newly regenerated limbs [3]. The balance between quiescence and reactivation is critical, too little reactivation leads to impaired repair and growth, while excess reactivation could lead to uncontrolled growth and exhaustion of the NSC pool [3,5,7–9]. The mechanisms behind stem cell quiescence and reactivation may also be utilised by cancer cells, allowing quiescent cells to resist treatment targeted at rapidly dividing cells and re-activate the cell cycle at a later time-point [1,4]. While much progress has been achieved in the last years, knowledge of the complex mechanisms and signalling pathways controlling quiescence and reactivation of NSCs is far from comprehensive, therefore their better understanding is critical for the development of NSC-based therapies directed at tissue regeneration, repair and restoring function.

 https://doi.org/10.3390/biom15050672

Different model systems have been utilised to study quiescence and reactivation of NSCs. A powerful system for investigating NSCs is the fruit fly *Drosophila* model. *Drosophila* NSCs share many properties with vertebrate NSCs, along with aspects of brain development, with all neurons and glia arising from a small set of progenitor cells [4]. *Drosophila* NSCs also have quiescent and reactivation states. In the embryo, NSCs form between 3.5h and 5.5h after fertilisation (a.f.) and generate progeny until the end of embryogenesis (17h a.f.) when they enter quiescence, a state where they remain until they are reactivated by dietary nutrients during early larval stages. An exception being four mushroom body NSCs per brain lobe, which continuously divide from embryonic stages to larval stages [4,10–13]. In the adult mammalian brain, the generation of new neurons and glia from NSCs is predominant in two major neurogenic niches, the ventricular-subventricular zone (V-SVZ) lining the lateral ventricles, and the subgranular zone (SGZ) of the dentate gyrus (DG) in the hippocampus [14–16]. NSCs in the V-SVZ, also known as radial glial-like, type B, cells, generate different interneurons of the mouse olfactory bulb and can give rise to glia, including oligodendrocytes and astrocytes, albeit in lower numbers in homeostasis. SGZ NSCs (also known as Type I cells) generate granule neurons, and at least a Nestin-expressing subpopulation has been reported to convert into astrocytes upon terminal differentiation. When cultured with growth factors, both V-SVZ and SGZ derived NSCs can generate the three neural lineages, highlighting their multipotency [5,7,17–24]. Another powerful model increasingly used to study NSCs is the zebrafish [25]. Compared to mammals, adult zebrafish have the incredible ability to heal injuries and repair damage within the CNS through the precise control of NSCs throughout the high abundance of stem cell niches [25,26]. In this review, we aim to bring together findings acquired using each model to highlight and discuss current perspectives on quiescence and reactivation of NSCs (Figure 1).

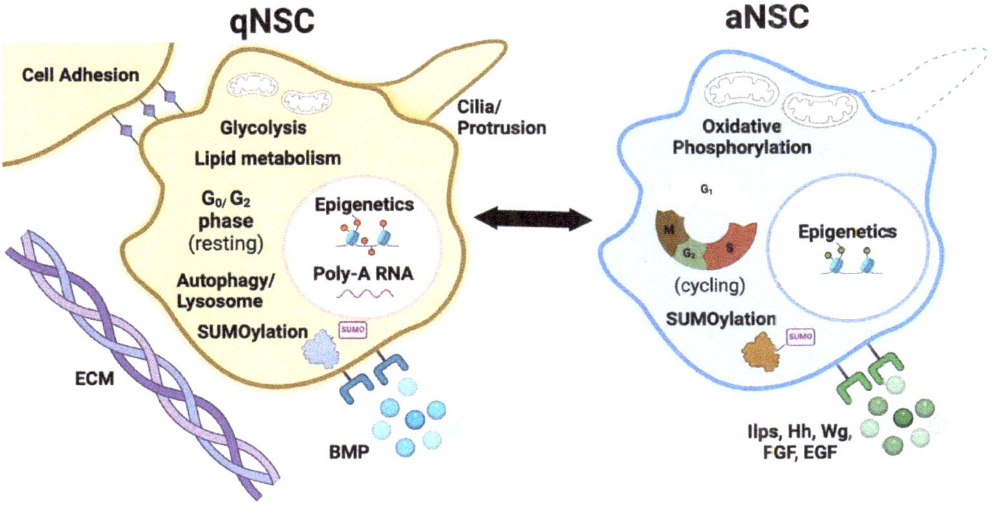


**Figure 1.** Several main features of quiescent and active NSCs. Alterations in metabolism, cell adhesion, Extracellular matrix (ECM), cilia/ protrusion, autophagy/ lysosomes, epigenetics regulation, SUMOylation, poly-A RNA nuclear accumulation and several ligand-receptors are depicted [6,12,17,27–32]. qNSC: quiescent NSC; aNSC: active NSC. Adapted from [33]. Created in BioRender [34].

## 2. NSC Quiescent States, Activation and the Niche Influence

Quiescence is defined as a reversible exit from the cell cycle, with quiescent cells observed in the canonical $G_0$ as well as in $G_2$ states [4,10,35–37]. It is a functionally important cellular state to protect stem cells from proliferative exhaustion and maintain their longevity and self-renewal capacity for tissue homeostasis and repair. In addition, quiescence protects cells from environmental stresses such as hypoxia or nutrient starvation and allows them to avoid accumulation of DNA, protein and mitochondrial damage, which occur in dividing cells [5,27,36,38,39]. While quiescence is associated with a low metabolic rate, reduced translation, transcription and ribosome biogenesis, a growing body of evidence indicates it is an actively maintained state, involving signalling pathways to prevent senescence, suppress terminal differentiation, ensuring reversibility and re-entry into the proliferative cell cycle. Extrinsic signalling, conveyed by the cells' niche, and intrinsic factors converge to regulate the transition between quiescence and proliferative active states [5,39,40].

Quiescence has been well defined in *Drosophila* NSCs, with temporal transcription factors and co-factors Pdm, Castor (Cas), Squeeze (Sqz) and Nab, involved in triggering quiescence, while the Hox proteins Antp and Adb-A spatially control quiescence entry [41]. *Drosophila* NSC (also known as neuroblast, NB) lineages are stereotyped, well described and named according to their position in the CNS [42]. Nab also functions in an alternate pathway in specifying temporal identity within the NSC 3-3T lineage (NB 3-3T, the NSC located in the third row and third column in each thoracic segment of the CNS ventral nerve cord), as NSCs retain their identity during quiescence [41]. Interestingly, it has been characterised that 75% of *Drosophila* qNSCs are arrested in $G_2$ and only 25% in $G_0$ [43]. In the CNS, the dorsal region primarily contains $G_0$ NSCs, while the ventral contains primarily $G_2$, which is facilitated by dorsal-ventral patterning cues [44]. These dorsal-ventral differences are induced by the dorsal patterning transcription factor Drop (Dr)/ Msx, which promotes p57/Dacapo (Dap) in the dorsal NSCs and thus inducing $G_0$ quiescence [3,44]. Both Dr/Msx and p57/Dap are evolutionarily conserved, suggesting similar patterning-mechanisms may induce differing states of quiescence in other systems [44]. In ventral NSCs, NKX/Vnd is expressed and induces $G_2$ quiescence. $G_2$ qNSCs reactivate faster than $G_0$ qNSCs and primarily express the evolutionarily conserved pseudokinase Tribbles (Trbl) [43]. Trbl inhibits insulin signalling by binding to Akt and preventing its phosphorylation, which along with the degradation of $Cdc25^{String}$, maintains quiescence in $G_2$ qNSCs [43]. Reversible $G_2$ arrest through nutrient restriction has also been shown in *Xenopus laevis* tadpole NSCs, suggesting that $G_2$ quiescence may be a cellular mechanism to control brain development in response to the availability of nutrients [45].

$G_2$ quiescence is yet to be described in mammalian models but, both spatial and temporal patterning mechanisms appear to be present, as qNSCs in the V-SVZ line the ventricular wall while aNSCs reside in the periventricular region and area-specific regulation of quiescence has been reported [46,47]. The majority of NSCs reside in a $G_0$ quiescent state in the adult mammalian brain reactivating to maintain homeostasis, responding to physiological cues and upon brain ischemia or injury [5,7,22]. Once reactivated, NSCs have the ability to self-renew and differentiate to generate new neural cells. Dysregulation of quiescence or reactivation of NSCs can result in depletion of their pool and decreased neurogenesis [5,7,22].

Comparable to $G_0$ and $G_2$ quiescent stages in *Drosophila*, quiescence in vertebrates is also not a single state but a continuum of varying degrees of depth [5,39,48]. Shallow qNSCs have a shorter reactivation time and are said to exist in a 'GAlert state', whereas deeper qNSCs require a longer period of exposure to physiological stimulus to reactivate [48]. In response to brain ischemia, a transition of deep qNSCs into a 'primed' quiescent state, in which NSCs are ready for rapid cell cycle re-entry, is promoted by interferon gamma signalling [27,48]. In mice, NSCs that have previously proliferated and return to a shallow quiescent state are said to be 'resting', compared to NSCs that have not yet proliferated which are termed 'dormant' [3,48]. NSCs in a shallow quiescence state are more likely to re-enter the cell cycle compared to NSCs which have not recently divided [48]. The proneural transcription factor Achaete-scute homolog 1 (Ascl1) is expressed in the V-SVZ and SGZ by active (aNSCs) and at a significantly lower level in qNSCs [5,48]. Despite low levels of Ascl1 expression in qNSCs, it is required to promote the transition of hippocampal NSCs from deep to shallow quiescence by modulating the activity of key signalling pathways [48]. Deletion of Ascl1 has been shown to move the population of qNSCs into a deeper quiescent state [48]. Through the transition from deep to shallow quiescence, Ascl1 induces the expression of Mycn via direct transcriptional activation, driving NSCs through shallow quiescence towards an activated state, via increased transcription and translation [48]. In Mycn KO mice, a greater percentage of NSCs are identified in a quiescent state, as qNSC cannot progress through shallow quiescence [48]. Zebrafish NSCs have also been shown to exhibit states of shallow and deep quiescence [49].

NSCs form an heterogenous population, coexisting in quiescent and proliferating states, showing region specific molecular properties in the mouse V-SVZ and giving rise to different progeny types [5,22]. This diversity is only beginning to be uncovered but increasing evidence highlights key signals that are received from, and exchanged with, the part of the neurogenic niche in which NSCs reside, heavily influencing their behaviour including their transition between quiescent and active states [5,22]. Different physiological stimuli impact NSCs and their niches. These cues include the presence of nutrients and feeding behaviour triggering hypothalamic-derived signals that control activation of a V-SVZ NSC subsets [50], and physical exercise such as running, which activates quiescent SGZ radial NSCs and has since long known to increase numbers of neurons in the hippocampal DG [51,52]. Pregnancy and early motherhood including the lactation stage have also been reported to activate V-SVZ NSC subpopulations, with specific olfactory neuron subtypes generated during the former [53,54]. Using live imaging in freely behaving mice, the day/ night cycle was shown to control V-SVZ NSC activation versus quiescence, a process dependent on darkness-induced circadian hormone melatonin signalling and favouring proliferation during day light [55], providing another example of physiological signals, which can be highly dynamic, influencing NSCs. In their niche compartments, adult NSCs respond to these and other stimuli as well as more local signals, interacting with the varied surrounding types of neural, vascular and immune cells, ECM and cerebrospinal fluid (CSF) in the case of the V-SVZ [5–7,22,56]. Neuronal activity, and in particular neurotransmitter and neuromodulators release by neuronal cells in the NSC niche, or located further away in the brain, can modulate NSC states, as exemplified by studies demonstrating GABAergic signalling limits proliferation [57,58]. Very recent findings show that a feedback mechanism also exists in which the proliferative state of V-SVZ NSCs is repressed by their rapidly dividing progeny of transient amplifying cells (TAPs), via Ephrin (Efn) signaling and $Ca^{2+}$ dynamics [59]. This is consistent with studies in other tissue stem cells, such as by Walker and colleagues showing activation of mesenchymal stem cells of the mouse incisor to be coordinated by the respective TAPs [60]. The vasculature is another prominent niche component, with both NSCs and TAPs contacting directly endothelial cells [61]. Interestingly, while vasculature secreted factors have been shown to promote adult V-SVZ NSC lineage progression [62], diffuse endothelial and pericyte signals from the cortex were reported to be more potent at eliciting proliferation than those from the V-SVZ niche [63]. Moreover, direct cell–cell interactions between NSCs and endothelial cells, via exposure to EphrinB2 and Jagged1 ligands, have been demonstrated to restrict NSC proliferation and differentiation towards preventing stem cell pool depletion [64]. The studies suggest differential modulation of the vascular niche on NSC behaviour.

During ageing, there is a decline in NSC function resulting in decreased production of neural cells, accompanied by reduced cognitive and olfactory discrimination [65,66]. This decline is thought to involve varied processes including an increase in the fraction of quiescent NSCs, which also enter deeper states of quiescence and have reduced ability to activate during homeostasis or upon injury [48,67]. Several age-related changes occur in NSCs affecting quiescence and activation [65]. Interestingly, the NSC functional decline is connected to an increase in glucose uptake, with transient glucose starvation or knockout of the GLUT4 glucose transporter Slc2a4, restoring old NSCs' ability to activate [68]. Aging not only affects NSCs but also their niche components. Notably, the transcriptome of young and old NSCs was reported not to differ greatly, and old NSCs to function similarly to young ones in that they can generate similar progeny numbers [67]. These findings favour argument that age-related changes in NSC quiescence are at least in part due to niche alternations and signalling exchanged with it. In support, changes in inflammatory signals from the niche in aging brains have been linked to quiescence, with inhibition of acute inflammation in the old V-SVZ niche demonstrated to activate NSCs [67]. Adult NSC interactions with their niche in homeostasis and upon aging have been extensively reviewed [5,22,65,66]. In the next sections, we focus on reviewing molecular changes and signalling pathways underlying intrinsic and extrinsic factors controlling NSC quiescence and activation.

## 3. Alterations in mRNA Processing During Quiescence

In *Drosophila* qNSCs, messenger RNA and polyadenylated [poly(A)] RNA accumulate in high levels in nuclei in comparison to active aNSCs [69]. These alterations in nucleocytoplasmic portioning of poly(A) RNA and mRNA likely regulate the reduction of protein synthesis in qNSCs mentioned in the above section [40], but may also allow reactivation under the appropriate cues [69,70]. This trait of qNSCs is also present in adult mouse hippocampal NSCs, suggesting nucleocytoplasmic portioning might be an evolutionarily conserved hallmark of qNSCs [69,71]. Both *Drosophila* and mouse NSCs can be induced into quiescence when Nup98-96, nucleoporins involved in regulating the transport of macromolecules between the nucleus and cytoplasm, is reduced [69]. Other nucleoporins, including Nup155 and Tpr along with 204 out of 228 detected splicing factors are also downregulated [69]. One nucleocytoplasmic transport factor that may be downregulated in qNSCs of *Drosophila* is RanGEF/Rcc1/Bj1, which is thought to exclude the homeodomain transcription factor Prospero from NSC nuclei, thus its downregulation might enable the induction of quiescence by nuclear Prospero accumulation [72,73]. In mammals, nuclear transport is also altered in quiescent stem cells. The expression of Exportin-5 has also been shown to be down-regulated at protein level by autophagy, and at mRNA level by miR-34a, which may be facilitated by Exportin-1 [71]. Unlike Exportin-5, Exportin-1 is required for the processing of miRNAs that are induced through quiescence, allowing for pri-miR-34a to enter the nucleus and downregulate mRNA levels [71]. These Exportin-1 dependent changes have also been characterized in zebrafish NSCs, suggesting conserved parallels in mRNA biogenesis and localisation in quiescence across organisms [69,74]. In mammal, some miRNAs have been shown to play a role in regulating quiescence. For example, the overexpression of miR-221 and miR-222, which target p27 and p57 mRNA, was shown to induce S-phase bypassing quiescence when overexpressed [75].

Interestingly, in the study by Rossi and colleagues, 12 of the 228 detected splicing factors were detected upregulated in qNSCs, which may ensure the regulation of genes required for quiescence and priming NSCs for reactivation [69]. The differential expression of splicing factors suggests that there is a complex regulation of mature mRNA and immature poly(A)-RNA, facilitated through the alterations of many splicing and nucleocytoplasmic transport factors. This appears also to be the case in zebrafish, with one example being the nuclear localisation of miR-9 and Argonaut (Ago) proteins specific within qNSCs of the adult telencephalon, potentiating Notch signalling and targeting the FGF signalling pathway [76,77]. The localisation of nuclear miR-9 and Ago proteins is also conserved in the adult mouse brain [77]. Thus, alterations in nucleocytoplasmic portioning of poly(A) RNA and mRNA are likely to be a conserved mechanism inducing quiescence features and reducing protein synthesis in NSCs, whilst also priming them for reactivation under certain conditions.

## 4. Promotion and Maintenance of Quiescence Through Notch Signalling

The first *Notch* mutant was identified in *Drosophila* in 1913, and sequencing of the *Notch* gene in the mid 80s revealed it encoded a large transmembrane receptor-like protein. Since then, Notch signalling has been intensively studied, found conserved among species and with key roles in cell-cell communication involving cell fate specification, proliferation, differentiation and death [78–81]. There is, however, limited knowledge surrounding the involvement of Notch in *Drosophila* qNSCs. Only recently it has been shown that Notch signalling regulates central brain NSC quiescence, through the activation of the cell cycle exit gene *dap*, promoting $G_0$ arrest [40]. Notch signalling plays an important role in maintaining adult mammalian NSC quiescence through cell-cell contact [5,7,82]. In response to dietary nutrients, low Notch levels are required to allow for the reactivation of qNSCs [40]. Notch ligands Delta (Dll1 and Dll4) and Jagged (Jag1) are expressed in the V-SVZ and SGZ by aNSCs and bind Notch receptors, activating the release of Notch intracellular domain (NICD) from the cell membrane to the cytoplasm [7,83]. NCID is then translocated to the nucleus where it forms a complex with DNA-binding CSL protein (Rbpj) and induces the expression of transcriptional repressor genes, such as hairy and enhancer of split 1 and 5 (*Hes1* and *Hes5*), which inhibit proneural genes such as Ascl1, thereby suppressing cell cycle re-entry and inhibiting neuronal differentiation [7,84]. Loss of Notch effector genes *Hes1* and *Hes5* or deletion of downstream mediator *Rbpj* results in activation and premature differentiation of NSCs and depletion of the NSC pool in the CNS of adult mice [5,82–84]. Overall, Notch is required for the maintenance of the NSC pool, by inducing NSCs in a quiescent state, whereas inactivation of Notch signalling leads to NSC proliferation and differentiation, depleting the NSC population [83]. Yet, expression of *Notch1*, *Notch2*, and *Notch3* varies between cellular states [7]. *Notch1* is highly expressed by proliferating NSCs and has been shown to promote proliferation of aNSCs without affecting qNSCs [5,46,82]. On the other hand, *Notch2* and *Notch3* are expressed to a higher extent in qNSCs, and deletion of *Notch2* and *Notch3* results in the activation of qNSCs and eventual depletion of the NSC pool [5,7,82].

In zebrafish, Notch plays a very similar role to the mammalian system, with a majority of functionally conserved orthologue ligands and receptors [25,28]. The pallial radial glial cells present in the telencephalon are similar both molecularly and cellularly to mouse adult NSCs. Here, Notch signalling is also a master regulator of quiescence and reactivation, with Notch3 controlling quiescence through *hey1* expression while Notch1/1b signalling is involved in injury response [28,85–87] Notch signalling controls the regenerative response to injury in the telencephalon [88,89]. Injury induces the proliferation of *olig2*$^+$ radial glia cells, likely through the inhibition of Notch, allowing for the regeneration of oligodendrocytes [88]. On the other hand, injury also activates Notch signalling in proliferating cells in the ventricular zone of the telencephalon, producing neuronal precursor cells that migrate to the injury site to differentiate into mature neurons, a process which may be partially conserved in mammals [89].

## 5. Bone Morphogenetic Proteins in Quiescence and Regeneration

Bone morphogenic proteins (BMPs) belong to the superfamily of ligands that take part in the Transforming growth factor-β (TGF-β) pathways, which have key roles during CNS development and adult function [90]. Interestingly, BMP signalling appears to diverge in promoting proliferation of NSCs in late larval stages of *Drosophila* versus contributing to NSC quiescence in the brains of adult mice [91,92]. If BMP is required for *Drosophila* NSC reactivation is yet to be determined. In adult mammalian NSCs, BMP induces NSC quiescence by transducing signals through Smad and non-Smad signalling pathways [93,94]. BMP4 in combination with FGF2 has been shown to induce a shallow quiescent state whereas BMP4 alone induces a deeper quiescent state [95]. In the mouse postnatal dorsal V-SVZ, single cell transcriptomics shows that subpallial NSCs remain primed for activation, whereas pallial NSCs enter a state of deep quiescence characterized by high BMP signalling, reduced transcriptional activity and Hopx expression. These findings highlight the heterogeneity of NSCs in the adult mammalian brain [96]. The two distinct levels of quiescence in those regions have been associated with the rapid closure of glutamatergic neurogenesis postnatally whereas gabaergic neurogenesis continues throughout life [96]. In the SGZ niche, BMPs produced by NSCs and granule neurons inhibit proliferation and maintain NSC quiescence by promoting the expression of inhibitory transcription factors such as Inhibitor of DNA Binding 1 (*Id1*), *Id2* and *Id3* [7,93,94]. BMP type IA receptor (BMPR1A) is a key mediator of quiescence in adult NSCs as genetic deletion of *Bmpr1a* and *Smad4* results in short-term increased proliferation in the SGZ [94]. Also, the BMP antagonist Noggin inhibits BMP signalling by preventing the activation of BMPRs, allowing NSCs to re-activate, proliferate and differentiate [97]. However, long term deletion of *Bmpr1a* and *Smad4* and exposure to Noggin results in loss of NSCs regenerative capacity and depletion of newborn neurons [5,94].

In the zebrafish adult brain, BMP is primarily expressed in the neurons of the telencephalon, which communicates with radial glial cells that express the components of the canonical BMP signalling pathway, inducing *Id1* expression and promoting quiescence [25,98]. BMP signalling is induced in response to injury, indicating that BMP/Id1 has an essential role in balancing quiescence and reactivation in order to maintain the regenerative capacity of NSCs [98,99]. There may be links between the BMP and Notch signalling pathways, as both pathways have a positive effect on *her4.1* in expression within qNSCs, reinforcing the notion that the regulation of quiescence is a complex network of different signalling pathways acting together [98].

## 6. Hippo Signalling Maintains NSC Quiescence

Originally discovered in *Drosophila*, the Hippo pathway is one of the major signalling cascades that control organ growth and proliferation and is highly conserved throughout the animal kingdom [25,26,100]. The major kinase cascade within this pathway is composed of Hippo [MST1/2 in mammals], Salvador [SAV1], Wats [LATS1/2], Mats [MOB1] and Yorkie [YAP, TAZ] [2,8,100–106] and is regulated by numerous upstream inputs. The Hippo/Salvador kinase complex phosphorylates and activates the Wts-Mats kinase complex, leading to the phosphorylation of Yorkie, resulting in its cytoplasmic retention and degradation and thus maintenance of quiescence in both *Drosophila* and mammalian NSCs [8,32,102–107].

Several mechanisms regulate the Hippo pathway in larval qNSCs in *Drosophila*. For example, it can be activated by niche glial cells expressing Crumbs (Crb) and Echinoid

(Ed) and the secretion of Anachromism (Ana) [107]. In response to nutrition, Crb and Ed are downregulated which inactivates the Hippo pathway, allowing Yki to relocate into the nucleus and initiate reactivation through activation of numerous genes in the NSC including the microRNA *bantam* [32,107]. Members of the conserved Striating-interacting phosphatase and kinases (STRIPAK) complex, which include Monopolar spindle-one-binder family member (Mob4), Connector of kinase to AP-1 (Cka) and the catalytic subunit of Protein phosphatase 2A (PP2A [Mts]), have been shown to orchestrate the Hippo and Insulin Receptor (InR/PI3K/AKT) signalling pathways [108–110]. PP2A/Mts has been shown to play different roles in quiescence and reactivation of early postembryonic NSCs [111]. During reactivation, Mob4 and Cka cooperate to associate PP2A/Mts with Hippo, dephosphorylating the Hippo kinase and inactivating the pathway allowing cells to reactivate [110–112]. PP2A/Mts and its regulatory subunit Wdb inactivate Akt, independent of Cka and Mob4, allowing Hippo signalling to remain active and thus maintain NSC quiescence [111]. Along with the regulation of the Hippo pathway, it has been reported that PP2A also dephosphorylates Cubitus interruptus (Ci) to increase Hedgehog (Hh) signalling, another pathway implicated in quiescence [113].

In the adult mammalian brain, the Hippo pathway is evolutionarily conserved and also active during quiescence [102,114,115]. Inhibition of the Hippo pathway via a variety of intrinsic and extrinsic factors allows NSCs to transition from a quiescent to active state [102,103]. Reduced Hippo pathway activity results in dephosphorylation and subsequent nuclear localisation of YAP/TAZ, which then bind with TEA domain transcription factors (TEADs) and promote the expression of genes that drive cell cycle entry and proliferation [8,104–106]. The Hippo pathway is tightly controlled, as overexpression of YAP does not result in NSC activation in the adult hippocampal SGZ, however overexpression of a mutant form of YAP (YAP-5SA; a protein insensitive to phosphorylation-dependent inhibition by LATs), promotes qNSCs in the hippocampus to re-enter the cell cycle in vivo and in vitro [102]. Analysis of single cell RNA-sequencing data reveals YAP highly enriched in aNSCs in the adult hippocampus, specifically in Sox2-positive cells of the SGZ [102]. YAP immunoreactivity co-localises with nuclear Sox2 expression in a subpopulation of these NSCs, suggesting NSC activation is accompanied by YAP nuclear translocation [102]. Furthermore, genetic deletion of YAP results in a significant reduction in the ratio of active versus qNSCs [102]. As mentioned in the above section, BMP signalling is one of the major factors involved in mammalian NSC quiescence. Specifically, BMP2 inhibits the proliferation of mouse NSCs through reduction of YAP nuclear translocation and YAP/TEAD interaction. The BMP2 effector Smad1/4 compete with YAP for the interaction with TEAD1 [105,106], linking Hippo and BMP pathways in maintaining qNSCs.

The Hippo pathway is also well characterised and conserved in zebrafish [116]. In this model Yap/Taz also promote regeneration [117]. *Yap1* and *wwtr1* [Taz] are both upregulated following spinal cord injury and localise to bridging glia [118]. To induce glial bridging, the ventral ependymoglia, CSF-facing cells with a radial morphology typical of the zebrafish CNS, undergo an epithelial-to-mesenchymal transition which has been linked to Yap1-Ctgfa signalling [117,118]. A better understanding of the role of the Hippo pathway in regeneration would be vital for the translation into the mammals and potentially provide therapeutical targets towards enhancing neural tissues regeneration. Mob4 has been shown to be highly expressed in the zebrafish central nervous system and implicated in neurodevelopment and cell proliferation [119,120]. PP2A has also been shown, along with p38 isoforms, to be expressed in quiescent Müller glia cells and in response to injury [121]. It was proposed that PP2A, along with Notch signalling, act downstream of the transforming growth factor TGFb3 to regulate Müller glia quiescence [121]; however little has been shown of its role in NSCs or radial glia cells.

## 7. Nutrient-Dependent Reactivation Through the InR/PI3K/Akt Cascade

The Insulin receptor InR/PI3K/Akt cascade is a highly conserved pathway that is regulated by insulin and involved in controlling growth during development and response to nutrient availability [122]. This pathway promotes reactivation of NSCs, with intake of nutrients, mainly amino acids, being its primary activator. In *Drosophila*, amino acids are sensed by the fat body, an organ serving roles of adipose tissue, liver and immune system, via detected by the amino acid transporter Slimfast (Slif, SLC7A in mammals), which activates the TOR pathway and generates a hormonal signal that stimulates PI3K/TOR signalling in the larval blood-brain barrier glia, inducing the release of Insulin like peptides (Ilps) [123]. Trbl, which is involved in $G_2$ quiescence, is overridden by the nutrition-dependent secretion of dilp, enabling the reactivation of NSCs [43].

The secretion of Ilps by glial cells in *Drosophila*, or insulin-like growth factor 1 (IGF-1) from astrocytes, neurons and NSCs in the mammalian V-SVZ and SGZ recruits PI3K to the cell membrane and converts phosphoinositol(4,5)P2 (PIP2) to phosphoinositol(3,4,5)P3 (PIP3) [6,124]. PIP3 recruits several proteins including Akt, of which activation leads to reactivation of NSCs through numerous means, including the phosphorylation of FOXO (FOXO3 in mammals) excluding it from the nucleus and translocating it to the cytoplasm for degradation by ubiquitination [29,124–126]. FOXO3 mediates transcription of genes involved in $G_1/_2$ cell cycle arrest [8,126]. FOXO3 shares target genes with Ascl1 and suppresses Ascl1 target genes associated with cell cycle progression [5,127]. Therefore, FOXO3 inhibits NSC proliferation and maintains qNSC in the SGZ and V-SVZ by inhibiting Ascl1-mediated cell cycle entry [5,8,9,127]. Mice deficient in FOXO3 show an initial increase in NSC proliferation followed by premature exhaustion of the stem cell pool [9,29,126,128]. The PI3K/Akt phosphorylation cascade also leads to the activation of the mammalian target of rapamycin complex 1 (mTORC1) [29].

Milk-fat globule-epidermal growth factor (Mfge8) is secreted by SGZ NSCs, and astrocytes and promotes NSC quiescence via suppression of the PI3K-Akt-mTOR pathway. Mfge8 expression is enriched in qNSCs where it binds to Integrin$\beta$1 and activates Phosphatase and tensin homolog (PTEN). PTEN negatively regulates PI3K activation, which inhibits Akt activation, therefore suppressing Akt-mediated activation of mTOR and NSC proliferation [5,7,92,128–130]. Loss of PTEN results in the activation of adult SGZ NSCs resulting in depletion of the NSC pool [5,9,92]. FOXO signalling has been well characterised between mammals and *Drosophila* [131,132]. In zebrafish, it is expressed in developing embryos and has also been implicated in promoting neurogenesis in the adult brain, with two isoforms of InR (*insra* and *insrb*) being highly conserved [133–135]. However, very little is known about TOR signalling in adult zebrafish NSCs.

Activation of the InR pathway in *Drosophila* NSCs also requires numerous other factors, including the physical association of the Heat shock protein 83 (Hsp83) and its co-chaperone Cdc37 in the presence of dietary amino acids [136]. Lack of amino acids leads to a large downregulation of *hsp83* mRNA, preventing the activation of the InR pathway in qNSCs. The Gap junction proteins Inx1 and Inx2 are also required in blood-brain barrier glia to ensure the translation of metabolic signals into synchronised calcium pulses in order to secrete Insulin [137].

One downstream target of the InR/PI3K/Akt pathway in *Drosophila* is Chromator (Chro), a spindle matrix protein critical in the reactivation of NSCs [138]. Chro is suggested to reactivate NSCs through activation of Grainyhead (Grh), and both inactivate the transcriptional repressor Pros [72,139]. Pros is an important differentiation factor involved in the balance between self-renewal and differentiation in NSCs, repressing cell-fate and cell-cycle genes, but has also been implicated in inducing NSC quiescence [72,139]. While high levels of Prospero trigger differentiation, low levels of nuclear Pros can promote quiescence by inhibiting the expression of Miranda, Asense and Cyclin E, arresting the cell cycle without triggering differentiation. Another gene, Fragile X protein (FMRP) has also been shown to regulate Cyclin E and phosphorylate Akt, which leads to decreased FOXO expression and premature exit from quiescence of NSCs [140,141].

## 8. SUMOylation and NSC Reactivation

SUMOylation, a type of post-translational modification that involves the conjugation of a small-ubiquitin-related modifier (SUMO), has been shown to have a role in the reactivation of NSCs [32]. *Drosophila* have a single SUMO gene *smt3*, which along with the SUMO conjugating enzyme 9 (Ubc9) promotes the SUMOylation of Hippo pathway Wts kinase at Lys766, inhibiting its activity and function and thus leading to reactivation [32]. During reactivation, Akt promotes the increase of Smt3 protein, suggesting that the InR/PI3K/Akt signalling cascade may promote the SUMOylation of Wts to inactivate the Hippo pathway [32]. In mammals there are five SUMO isoforms, but only SUMO1-3 are expressed in the brain [142]. Overexpression of Ubc9 in NSCs [12,32] elevates SUMO1-3 levels and increases the proportion of cells in the S-phase of the cell cycle, suggesting reactivation of qNSCs [142,143]. SUMO1 is also required for the activation of peroxisome proliferator-activated receptor $\gamma$ (PPAR$\gamma$), which regulates NSC proliferation [142]. On the other hand, loss of SUMO1 or Ubc9 increases Nanog, a transcription factor which promotes NSC self-renewal via the Hedgehog pathway, indicating SUMOylation has a negative effect on Nanog induced NSC self-renewal [142,144]. In addition, Protein inhibitor of activated STAT3 (Pias3)-induced SUMOylation of Sox2 further inhibits Nanog expression, impairing NSC self-renewal capacity [142,144]. Additional Sox transcription factors are modified by SUMOylation. Sox6 increases NSC self-renewal by upregulating B-cell lymphoma 2 (Bcl-2) and *Hes1* expression and enhancing Akt phosphorylation, but SUMOylation reduces its transcriptional activity [142,145]. SUMOylation has also been characterised in zebrafish NSCs. SUMOylating lysine 247 of Sox2 represses its expression, along with numerous targets involved in cell cycle control [146]. Ubc9 has also been implicated in zebrafish development; it is ubiquitously expressed during early development and restricted to proliferative zones at later stages, likely in stem cell niches [147]. However, SUMOylation has yet to be shown to be involved in NSC quiescence or reactivation in zebrafish.

## 9. Lysosomal Activity and Autophagy in Quiescence and Reactivation

Lysosomal activity plays an important role in maintaining NSC quiescence in the V-SVZ of the adult mouse brain [148]. Enlarged lysosomes storing protein aggregates, higher expression of lysosomal genes, such as the lysosomal master regulator transcriptional factor EB (TFEB), and higher lysosomal activity have all been identified in qNSCs compared to aNSCs [129,148,149]. Nutrient starvation, used to induce quiescence in vitro, results in the dephosphorylation and activation of TFEB. Translocation of TFEB to the nucleus upregulates genes involved in lysosomal activity and degradation of activated EGF receptors (EGFR) [129,148]. Inhibition of lysosomal activity in vitro via deletion of TFEB, delays the reduction in EGF as well as Notch signalling, resulting in delayed NSC quiescence [129,148]. On the other hand, TFEB activation by Rapamycin inhibition of mTORC1, decreases aNSC proliferation [148]. Loss of TFEB in vivo prevents degradation of EGFR and Notch1, results in exit from quiescence and an increase in the number of aNSCs in the SGZ of young adult mice [148]. Conversely, elevation of lysosomal activity by ectopic expression of TFEB in aNSCs leads to quiescence induction [148]. qNSCs in the V-SVZ of aged mice express similar levels of lysosomal protein Lamp-1 to aNSCs [149]. TFEB activation in the V-SVZ of aged mice promotes reactivation of qNSCs, interestingly opposing the role of lysosomes on NSCs in the SGZ of young mice [129,148,149]. In zebrafish, increased expression of the lysosomal hydrolase activator Prosaposin (Psap) has been implicated in deep quiescence, likely enhancing lysosomal activity and autophagy [49].

Similarly, autophagy has been identified to contribute to shift the population of NSCs from a proliferative to quiescence state in the SGZ [150]. An increase in expression of autophagy genes (LC3-II and SQSTM/p62) as well as Aggrephagy receptor (Tax1bp1), highlights the upregulation of autophagy in qNSCs, potentially in response to the protein aggregates accumulated in enlarged lysosomes [129,148,150]. Furthermore, phosphorylation and activation of the AMP-activated protein kinase/unc-51 like autophagy-activating kinase 1 (AMPK/ULK1) pathway due to low availability sensed by AMPK enhances autophagy induced NSCs quiescence [150]. Loss of a critical gene for autophagosome formation, *Atg7*, in NSCs impedes the induction of quiescence during the development of the hippocampal DG NSC niche [150].

## 10. Metabolic Shifts Between NSC Quiescence and Reactivation

When NSCs transition from quiescent to active and vice versa, they switch metabolic pathways used to obtain energy. Lipid metabolism modulates Hedgehog signalling, another conserved signalling pathway, and is involved in reactivation in *Drosophila* [113,151,152], through the Lipid storage regulator droplet-2 (Lsd-2) and Fatty acid synthase 1 (Fasn1) [153]. In zebrafish, Hedgehog signalling has been linked to redox control, generating a positive feedback loop with $H_2O_2$ that contributes to regeneration [154]. These reports indicate that along with dietary amino acids leading to activation of the TOR and InR pathway, lipid intake and redox levels also play role in the reactivation of NSCs [12].

For mammalian hippocampal NSCs to reactivate from quiescence, lipid and fatty acid metabolism are downregulated and a transition to oxidative metabolism in the mitochondria must occur [29,37,155,156]. Increasing evidence, including via single-cell transcriptome analysis, points to different metabolic pathways being used to obtain energy in quiescent, 'primed' and aNSCs [27,29,155]. Fatty acid metabolism is enriched in qNSCs in the V-SVZ and downregulated upon NSC reactivation [155]. Fatty acid oxidation (FAO), the breakdown of fatty acids into acetyl-coenzyme A (CoA) in the mitochondria, is regulated by its rate limiting step, Carnitine palmitoyltransferase 1a (Cpt1a), which mediates the transport of fatty acids to the mitochondria [157]. Cpt1a is highly expressed in quiescent compared to aNSCs and blockage of FAO by a Cpt1 inhibitor in qNSCs resulted in cell death, highlighting the critical role of FAO in maintaining adult hippocampal NSC quiescence [157]. In addition, Thyroid hormone-inducible hepatic protein (THRSP/SPOT14) regulates lipid metabolism and is highly expressed in qNSCs [29,155–157]. SPOT14 downregulates the expression of Fatty acid synthase (Fasn), a key enzyme involved in de novo lipogenesis, by reducing the availability of malonyl-CoA, a fundamental building block in fatty acid synthesis [29,156,157]. Conversely, Fasn expression is elevated in proliferating NSCs as lipid synthesis is required for NSC division [29,157].

Glycolytic metabolism has also been highlighted as a characteristic of quiescent adult NSCs, with glycolysis genes such as *aldolase A*, *aldolase C* and *Lactate Dehydrogenase B (Ldhb)* shown to be enriched in qNSCs but significantly decreased upon reactivation [27,155]. FOXO3 plays a role in inducing quiescence-enriched metabolic genes, resulting in glycolysis [5,8,9]. FOXO3 is not only proposed to coordinate a program of genes in NSCs involved in quiescence and glucose metabolism but also related to aging and hypoxia, the latter including Hypoxia-inducible factor 1 (HIF1) targets [9]. Interestingly, conditional *Hif1a* gene deletion in adult mouse SVZ NSCs leads to depletion of the NSC pool, preceded by regression of SVZ vasculature, suggesting HIF1 as an essential NSC maintenance factor possibly via promoting vasculature niche integrity [158]. Downregulation of glycolytic metabolism is also observed in mammalian NSCs in a 'primed' quiescent state, prior to reactivation, suggesting glycolytic metabolism might be enriched in deep quiescence [27]. Interestingly, birth-associated downregulation of glycolysis and upregulation of glutamine metabolism linked to increased expression of *Glutamine synthetase (Glul)*, was recently reported to be required for embryonic mouse NSCs (radial glia, RG) induction into quiescence via suppression of mTORC1 signalling [159]. However, insufficient upregulation of Glul in preterm birth leads to transient hyperactivation of mTORC1 signalling in RG, and impairs transition from active to quiescent states, resulting in depletion of the NSC pool [159]. Depletion of the NSC pool can be prevented by suppressing the hyperactivation of RG in preterm birth mice with rapamycin treatment [159]. During the first postnatal week of DG development, NSCs are actively proliferating to expand the number of long-lived NSCs, prior to entering a quiescent state [160]. Suppressor of Fused (Sufu) regulates Sonic Hedgehog (Shh) signalling activity to control the production and maintenance of long-lived SGZ NSCs and the timely establishment of the qNSC pool [160]. Accordingly, deletion of Sufu or Shh receptor Smoothened (Smo) in NSCs or removal of Shh ligands during early DG development decrease Shh signalling, compromising postnatal expansion of long-lived NSCs, resulting in premature NSC quiescence and a diminished NSC pool in adult mice [160].

Shh signalling plays also critical role in the adult mammalian V-SVZ, balancing quiescent and activated NSCs by regulating quiescence. Deletion of the Shh receptor Patched (PTC) results in activation of the Shh signalling and an initial increase in proliferation of both qNSCs and aNSCs in the V-SVZ [161]. On the other hand, long-term activation of the Shh signalling pathway results in an accumulation of qNSCs and a progressive depletion of the aNSC pool [161,162]. In the SGZ, normal levels of Shh signalling by the mossy cells, a subgroup of neurons, acts to reactivate qNSCs. Under excitotoxic conditions, Shh signalling is upregulated, resulting in decreased NSC activation and an accumulation of qNSCs [162]. The authors suggest this is a protective mechanism to preserve the NSC pool, as qNSCs are more resistant to these toxic conditions due to their reduced rate of proliferation [162]. In contrast, activation of the Shh signalling pathway in vitro causes an increase in aNSC proliferation by shortening of both G1 and S-G2/M phases of the cell cycle. This leads to an increased number of cells generated and the potential of aNSCs to form colonies [161]. Treatment with an antagonist of the receptor Smo reduced the number and size of colonies formed by aNSCs in vitro [161].

## 11. Cellular Protrusions and Adhesion in Quiescence and Reactivation

One hallmark of *Drosophila* qNSCs is the extension of a primary, microtubule-enriched protrusion from the cell body towards the neuropile [124,132,138,163–165]. These protrusions have the ability to regenerate upon injury, requiring the Golgi apparatus to act as a major acentrosomal microtubule-organising centre [30,166]. Patronin, a microtubule minus-end binding protein, is an important factor within these protrusions, required for promoting acentrosomal microtubule growth, the regeneration of the qNSC primary protrusions upon injury and the reactivation of NSCs [30]. Both the promotion of acentrosomal growth and the reactivation of qNSCs are governed by Patronin, a Golgi-resident GTPase Arf1, and the microtubule polymerase Mini spindles (Msps) [30,166]. Patronin is distributed in the cytoplasm of qNSCs and can physically associate to the GDP-bound form of Arf1, which is regulated by its guanidine nucleotide exchange factor, Sec71 [30]. Arf1 and Sec71 are critical in microtubule assembly and orientation in qNSCs, and localise towards the protrusion initial segment (PIS), an indispensable region required for qNSC protrusions [166]. Msps also physically associates with Arf1, both in GTP- and GDP- bound forms, likely acting as an effector of Arf1 during reactivation [166]. Msps functions upstream of Kinesin-2 which transports and allows E-cadherin (E-cad) to localise to NSC-neuropil contact sites and leads to NSC reactivation [164].

In mammals qNSC protrusions are also highly enriched in actin. The polymerisation of fine filamentous actine (F-actin) structures promotes the movement of Myocardin-related transcription factor (Mrtf) into the nucleus, leading to the reactivation of NSCs [167]. F-actin polymerisation is mediated by the activation of the G protein-coupled receptor Smog on astrocyte-like glia in the NSC niche, signalling down a cascade of G protein $\alpha_q$ subunit, Rho1 and Diaphanous (Dia)/Formin. This results in the flow of F-actin patches, promoting polymerisation of F-actin and thus translocation of Mrtf into nucleus to promote the transcription of *actin* and other genes required for cell proliferation [167]. Cellular protrusions are also highly enriched in clusters of mitochondria, potentially linking the role of protrusions and metabolic shift in regulating quiescence and reactivation [168].

Primary cilia have been shown to serve as hubs for a multitude of signalling pathways, such as those regulated by Hedgehog (HH), receptor-tyrosine kinases, G-protein-coupled receptors, Wnt and TGFβ/BMP [169]. As discussed above, BMPs plays an important role in inducing quiescence in adult mammalian NSCs [93,94]. BMP4-induced qNSCs show an upregulation of genes associated with the development, organisation and movement of primary cilia, compared to BMP4 plus FGF2-induced qNSCs and aNSCs [95]. Cilia are key microtubule-enriched cellular protrusions acting as hubs transducing signals from the extracellular space [170]. Cilium extension is known to occur in $G_1/G_0$ phase of the cell cycle, and most long term qNSCs display a primary cilium in adult mice V-SVZ [95,171]. A key regulator of motile ciliogenesis, Forkhead Box J1 (Foxj1), is significantly upregulated in BMP4-only induced qNSCs compared to BMP4 plus FGF2-induced qNSCs, and its expression increases with the duration of BMP4 treatment [95]. Morphological differences in cilia have also been observed with primary cilia being longer in BMP4-induced qNSCs than BMP4 plus FGF2-induced qNSCs and aNSCs [95]. The longer cilium in BMP4-induced qNSCs are more developed and corresponds with a longer duration in a deeper quiescent state [95]. When cells re-enter the cell cycle, extended cilia are dismantled, therefore, the length of cilia is negatively correlated with cell cycle progression [68,171]. However, non-ciliated qNSCs have also been identified [171]. Ablation of primary cilia in a ventral domain of the V-SVZ and the SGZ disrupts the Shh signalling pathway, leading to decreased postnatal NSC proliferation and loss of the qNSC pool [172–174]. Moreover, knockout of *Rsph3a*, a mouse orthologue of human *RSPH3* for which mutations lead to primary dyskinesia—a rare condition resulting in structural or functional defects in motile cilia and sperm flagella [175] - causes increased activation of NSCs in aged mice [68]. In zebrafish, it has been shown that the conserved orthologues of Patronin, CAMSAP1-3, are present within active radial glial cells and involved with controlling microtubule growth [176,177]. The radial glial cells do have protrusions named cilium, which connect to the ventricular cavity and cerebral spinal fluid, which provides growth factors and a niche that is necessary for retaining their stemness [178]. Loss of these cilium through laser ablation leads to a delayed cell-cycle re-entry and disrupts asymmetric division, implicating their potential role in quiescence and reactivation [179]. Collectively, these findings suggest that different microenvironment-controlled effects can be mediated by cilia triggering signalling pathways on qNSCs and their activation. The dynamic microenvironment provided by extracellular matrix (ECM) molecules influences virtually all aspects of development and function of the CNS and its NSCs [180]. In agreement, a plethora of literature highlights the importance of stem cell adhesion and position within the niche as determinants of quiescent and active NSCs [47]. Transcriptome profiles of qNSCs have been shown to be enriched in gene ontology categories associated with cell adhesion [181–183]. Several genes exhibit preferential transcription in qNSC compared to aNSCs in the V-SVZ and SGZ, such as Neural cell adhesion molecules 1 and 2 (NCAM1, NCAM2), Vascular cell adhesion molecule (VCAM-1) and N-cadherin [3,6,155,182–184]. Loss of N-cadherin in the

V-SVZ results in increased activation and proliferation of qNSCs, and depletion of the NSC pool [185]. Membrane-type metalloproteinase MT5-MPP regulates N-cadherin-mediated adhesion, promoting qNSCs activation [185]. Loss of VCAM1 induces qNSC reactivation and eventual NSC pool depletion [47,182,184,186]. VCAM1 is down regulated upon NSC exit from quiescence [182,184] and its deletion disrupts niche structure as well as leading to impaired hippocampal learning and memory [182,184,186]. WNT signalling is critical for the maintenance and differentiation of different types of stem cells [187]. Non-canonical Wnt signalling has been proposed as a key pathway in maintaining NSC quiescence in the SVZ, by activation of Rho-GTPase Cdc42, which maintains NSC adhesion to the apical niche [47]. Other studies have implicated canonical Wnt/β-catenin signalling. For example, NSC lineage tracing using an inducible Cre Axin2 mouse strain [188] showed that a subset of adult SVZ NSCs, derived from embryonically labeled $Axin2^{CreERT2+}$ precursors, are regionally and functionally restricted and retain Wnt/β-catenin responsiveness [189]. However, recent findings show that while both qNSCs and aNSCs express and can respond to components of the Wnt/β-catenin signalling pathway, it is dispensable for the maintenance of adult NSCs in a proliferative or quiescent state, and also for the transition of NSCs between quiescent and active states [190]. The authors comment that the apparent disparate results may be likely due to differing techniques used to modulate Wnt/β-catenin signalling in previous literature, as well as the use of in vivo and in vitro models, which may lead to different responses of NSCs. To understand if and when Wnt/β-catenin signalling has a role in NSCs, novel in vivo tools are needed allowing manipulation of Wnt gradients in vivo. Interestingly, as part of the canonical Wnt pathway, Wnt ligands bind to the class F (Frizzled) family of G-protein-coupled receptors, which localise in the primary cilia, however the functional relevance of this localisation is still unclear [169].

## 12. Mechanisms Behind qNSC Reactivation upon Injury

During injury, it is important to recruit and reactivate qNSCs to mount a regenerative response. In the *Drosophila* larval CNS, a coordinated neuro-glial response occurs upon injury [191]. Swim, a lipocalin-like transporter, is secreted from the glia in a HIF1- $\alpha$-dependent manner, either through direct activation or indirectly through other hypoxia-responsive factors, which promotes the distribution of Wingless (Wg)/Wnt in the local tissue environment [191]. Increased levels of Wg/Wnt signalling promotes qNSCs reactivation, thus mounting a regenerative response. Interestingly, the mouse Swim orthologue, Lcn7, has been shown to be induced by glial cells in the hippocampus upon brain injury, suggesting a conserved function [191].

NSCs and astrocytes of the adult mammalian neurogenic niches have been reported to secrete Wnts, such as Wnt3a and Wnt5a [5,17]. The response of NSCs to Wnt/β-catenin signalling is a current topic of investigation. Wnt activation seems to be dose-dependent and state-specific [190]. As mentioned above, non-canonical Wnt signalling has been proposed to maintain NSC quiescence in the SVZ by activation of Rho-GTPase Cdc42, which regulates Notch signalling activity [47]. However, in response to demyelination injury, non-canonical Wnt signalling is downregulated, and activation of canonical Wnt/β-catenin signalling is reported to occur to re-activate qNSCs, which divide symmetrical to increase the stem cell pool for tissue repair [7,29,47,190]. Wnt3a induces the expression of a canonical Wnt/β-catenin target, Axin2, whereas Wnt5a antagonises the Wnta3a-induced Axin2 levels, suggesting non-canonical Wnt signalling overrides canonical Wnt/β-catenin signalling in homeostatic conditions [47,189,190]. Therefore, a transient shift from non-canonical to canonical Wnt signalling may be required for the activation of qNSCs for brain repair [5,47].

T cell factors (TCFs) are main transcriptional mediators of Wnt signalling, which have also Wnt independent activities [192]. In contrast to TCF proteins in *Drosophila* and *C. elegans* that are encoded by a single *Tcf* gene, vertebrates have different TCFs that likely increase context-specific responses to Wnt signaling. Studies of vertebrate embryonic development reveal that TCF3 acts mainly as transcriptional repressor whereas TCF1, lymphoid enhancer factor 1 (LEF1) and TCF4 activate transcription [192]. Interestingly, a TCF switch has been reported in which TCF3 repressor is replaced by TCF1 activator to regulate the homeobox *Vent2* gene expression in developing *Xenopus* embryos. It remains to be demonstrated if a similar model acts in mammals [192,193]. How TCFs act in NSC quiescence and activation is also not fully understood. Using a TCF/LEF-reporter expressing an H2B-GFP fusion under the control of TCF/LEF binding sites [194], Kalamakis and colleagues argue that canonical WNT activity is predominant in presumably qNSCs (Ki67-negative) of the dorsal domain, as compared to the latero-ventral domain, of the V-SVZ [67]. They hypothesize that a switch to non-canonical WNT is required for NSC activation throughout life, in contrast to studies proposing non-canonical WNT signalling to maintain NSC quiescence (e.g., [47]). As mentioned previously, the authors show that qNSCs increase in the old brain and are more resistant to leave quiescence even upon injury. Interestingly, block of sFRP5, a known WNT antagonist found uniquely expressed in old qNSCs, is shown to increase active NSCs in the old brain in both homeostasis and following temozolamide-induced injury, which may contribute to explain age-related overall decline in active NSC function and regenerative capacity [67]. LEF1 and its partner β-catenin have also been found accumulating in quiescent hippocampal NSC cultures, concomitant with enhanced TCF/LEF1-driven transcription and preponderance of a full length LEF1 isoform contributing to its stability [195]. Interestingly, LEF1 interacts with the transcription factor NFIX, however, if these interact at specific enhancers in quiescent NSCs to regulate the cells' state requires further study [195].

Unlike *Drosophila* and mammalian NSCs, zebrafish are highly adept at healing injuries and repairing damage to their CNS, but the Wnt/ β-catenin signalling seems not required to reactivate qNSCs for regeneration [196]. As described above, Notch signalling has been shown to regulate regenerative response in the telencephalon [88,89], however BMP/Id1, Hedgehog and YAP/TAZ signalling have also been shown to be regulated following injury [31,117,118]. Leukotriene-C4 (LTC4) is reported to be sufficient to activate the regenerative response by inducing gata3 which promotes proliferation, along with the chemokine receptor Cxcr5, which shows increased expression after injury [87]. Another factor induced by injury in zebrafish is Islet Antigen-2 (Ia-2), which is highly conserved and involved in dense core vesicles to release Insulin, thus activating the InR signalling cascade [197]. Thus, the regenerative capabilities of zebrafish are regulated by numerous factors and pathways, many of which are present in mammals.

One potential reason the mammalian system lacks the regenerative capacity of zebrafish may be the presence of the tumour suppressor gene *ADP-ribosylation factor (ARF)*. ARF is important in suppressing cancer, but zebrafish, along with other organisms with the ability to regenerate, lack this gene [198]. Expression of human ARF in zebrafish was shown to localise into the nucleus along with Mdm2, stabilising Tp53 and inducing cell cycle arrest/apoptosis, suppressing fin regeneration [198]. This suggests that organisms with the ability to regenerate use widely conserved signalling pathways, but that mammals may express factors suppressing these signals. To potentially unlock the regenerative capacity in mammals without promoting cancer would be a major step towards tackling the rising burden of CNS injury and neurodegenerative diseases.

## 13. Conclusions

The research into the regulation of quiescence and reactivation of NSCs has been extensive across different model organisms. Together with supporting technology advance, it allowed better understanding of the complex network of signalling pathways and molecular mechanisms that regulate the critical balance between quiescence and reactivation. In this review, we have brought together findings using *Drosophila*, mammals and zebrafish models (Figure 2, Table 1). While much has been uncovered, there is still a long way to go including exploring the role of epigenetics in the control of NSC quiescence, which only recently has been entered into [199].

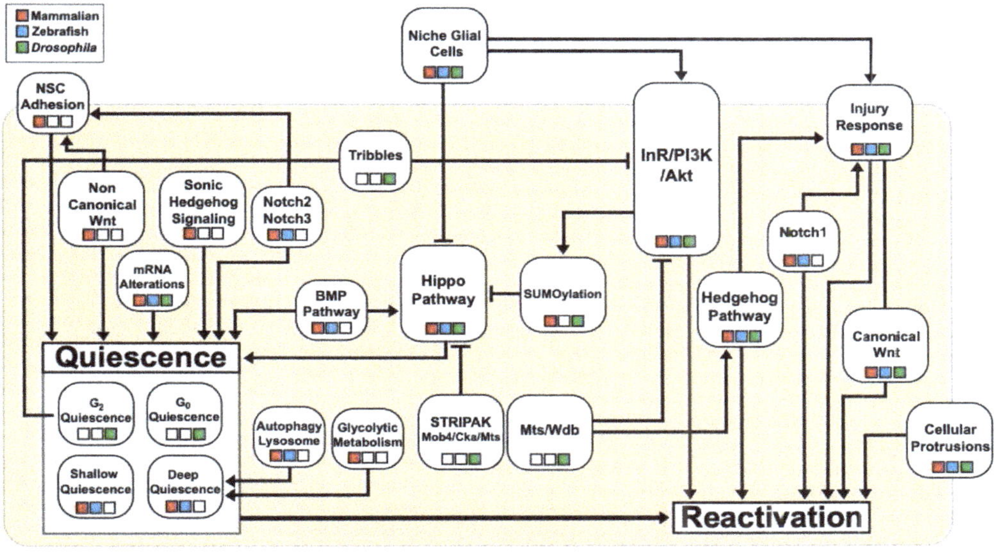


**Figure 2.** Key signalling pathways and processes involved in quiescence and reactivation of NSCs discussed in this review, and their interactions identified in *Drosophila*, mammalian or zebrafish models. Colour-coded boxes indicate of which model the data has been derived from. Arrow: activation; Bar: repression.

**Table 1.** Summary of findings and publication references discussed in this review using *Drosophila*, mammalian and zebrafish NSCs as model systems.

| | *Drosophila* | Mammals | Zebrafish |
|---|---|---|---|
| Quiescent states, activation and niche influence | G0 and G2 Quiescence: [3,41,43,44] | Shallow and Deep Quiescence: [3,5,7,25,27,35–37,39,40] [46–48,160]; Niche Influence: [22,51–64,158]; Activation and Aging: [65–68] | Shallow and Deep Quiescence: [49] |
| mRNA Alterations | Nucleocytoplasmic portioning of poly(A) RNA and mRNA regulates qNSC and prime for reactivation: [69,72,73] | Nucleocytoplasmic portioning of poly(A) RNA and mRNA regulates qNSC and prime for reactivation: [69–71,75] | Nucleocytoplasmic portioning of poly(A) RNA and mRNA regulates qNSC and prime for reactivation: [74,76,77] |
| Notch Signalling | Regulates quiescence: [40] | Varied Notch1, Notch2 and Notch3 in quiescence and reactivation: [5,7,46,82–84] | Varied Notch1, Notch2 and Notch3 in quiescence and reactivation: [25,28,85–89] |

**Table 1.** *Cont.*

| | *Drosophila* | Mammals | Zebrafish |
|---|---|---|---|
| BMP/Id1 | Not known in NSC quiescence/activation regulation. | Induces quiescence including through expression of Id1-3: [92–97] | Induces quiescence through expression of Id1; linked to Notch: [25,98,99] |
| Hippo Pathway | Maintains quiescence through Hippo/Salvador/Wats/Mats/Yorkie cascade: [2,32,100,107] | Maintains quiescence through MST1-2/SAV1/LATS1-2/MOB1/YAP/TAZ cascade: [2,8,100,102–106,114,115] | Maintains quiescence through Mst2/Sav1/Lats1-2/MOB1/Yap/Taz cascade: [25,26,116–118] |
| STRIPAK | Mob4 and Cka associate PP2A/Mts to inactivate Hippo Pathway PP2A/Mts and Wdb inactivate InR/PI3K/Akt pathway: [108–113] | Not implicated in quiescence or reactivation of NSCs. | Present within quiescent Müller glia cells and in response to injury, but limited knowledge in NSCs: [119–121] |
| Insulin Pathway | InR/PI3K/Akt/TOR pathway leads to reactivation of NSCs, including through inhibition of FOXO: [6,43,122–124,136–138,140,141] | InR/PI3K/Akt/mTOR pathway leads to reactivation of NSCs, including through inhibition of FOXO3: [5,7,9,29,92,125–132] | Limited knowledge. insra and insrb highly conserved: [133–135] |
| SUMOylation | smt3 and Ubc9 SUMOylate Wts leading to reactivation: [32] | Implicated in self-renewal and differentiation of NSCs: [142–145] | Ubc9 is present in early development and proliferative zones at later stages. Little known in NSCs: [146,147] |
| Lysosomal and Autophagy | Limited knowledge. | Lysosomal activity and autophagy are involved in maintaining NSC quiescence: [129,148–150] | Prosaposin (Psap) has been implicated in deep quiescence in zebrafish: [49] |
| Metabolism | Early lipid intake correlates to reactivation, modulated by Hedgehog: [113,151–153] | Fatty acid metabolism enriched in qNSCs; switch to oxidative metabolism required for reactivation; birth is associated with metabolic changes: [5,8,9,27,37,160] [29,155–157,159] | Limited knowledge; positive feedback loop of H2O2 contributes to regeneration through Hedgehog: [154] |
| Cellular Protusions and Adhesions | Cellular protrusions are hallmark of qNSCs; promotes reactivation through Golgi apparatus and Actin: [30,124,132,138,163–169] | Primary cilia are a hallmark of deep qNSCs; adhesions are highly important in determining quiescence or reactivation: [3,9,46,47,68,95,170–175,181,183–186,190] | Limited knowledge. Patronin is conserved in active radial glial cells; cilia are present and involved in retaining NSC stemness: [176–179] |
| Injury/Regeneration | Neuro-glial response coordinated by injury. Promotes Wg/Wnt distribution: [30,191] | NSCs responsive to canonical WNT signalling; Non-canonical Wnt signaling implicated in quiescence; TCFs; Canonical Wnt signaling required for brain repair: [5,7,29,47,67,189,190,194,195] | Highly adept at regeneration. Notch, BMP, Hedgehog, YAP/TAZ and InR pathways implicated in repair; ARF likely suppresses regeneration within mammalian system: [31,87–89,117,118,196–198] |

We only started to unravel the complex network of signalling pathways, epigenetics, genetic, and metabolic alterations; all contributing to the delicate balance that sustains the precious amount of NSCs required for development and repair. A detailed knowledge of how these cells maintain their states and depths of quiescence, how reactivation under certain conditions occurs and how to unlock the hidden regenerative potential that many regenerative organisms utilise are all critical milestones for the development of future regenerative therapies and combat the ever-rising burden of neurodegeneration.

**Author Contributions:** Conceptualisation: A.M.E., S.R., C.S.B. and T.B.; analysis of published work: A.M.E., S.R., C.S.B. and T.B.; writing—original draft preparation: A.M.E., S.R., C.S.B. and T.B.; writing—review and editing: C.S.B. and T.B.; figure preparation: A.M.E., S.R., C.S.B. and T.B.; supervision: C.S.B. and T.B.; funding acquisition: C.S.B. and T.B. All authors have read and agreed to the published version of the manuscript.

**Funding:** This research was funded by Peninsula Medical School, University of Plymouth, U.K. and Peninsula Medical Foundation.

**Institutional Review Board Statement:** Not applicable.

**Informed Consent Statement:** Not applicable.

**Data Availability Statement:** No new data were created or analyzed in this study.

**Conflicts of Interest:** The authors declare no conflicts of interest.

## References

1. Knoblich, J.A. Asymmetric cell division: Recent developments and their implications for tumour biology. *Nat. Rev. Mol. Cell Biol.* **2010**, *11*, 849–860. [CrossRef]
2. Pan, D. The hippo signaling pathway in development and cancer. *Dev. Cell* **2010**, *19*, 491–505. [CrossRef]
3. Otsuki, L.; Brand, A.H. Quiescent Neural Stem Cells for Brain Repair and Regeneration: Lessons from Model Systems. *Trends Neurosci.* **2020**, *43*, 213–226. [CrossRef]
4. Homem, C.C.; Knoblich, J.A. Drosophila neuroblasts: A model for stem cell biology. *Development* **2012**, *139*, 4297–4310. [CrossRef]
5. Urbán, N.; Blomfield, I.M.; Guillemot, F. Quiescence of Adult Mammalian Neural Stem Cells: A Highly Regulated Rest. *Neuron* **2019**, *104*, 834–848. [CrossRef]
6. Ding, W.Y.; Huang, J.; Wang, H. Waking up quiescent neural stem cells: Molecular mechanisms and implications in neurodevelopmental disorders. *PLoS Genet.* **2020**, *16*, e1008653. [CrossRef]
7. Matsubara, S.; Matsuda, T.; Nakashima, K. Regulation of Adult Mammalian Neural Stem Cells and Neurogenesis by Cell Extrinsic and Intrinsic Factors. *Cells* **2021**, *10*, 1145. [CrossRef]
8. Paik, J.H.; Ding, Z.; Narurkar, R.; Ramkissoon, S.; Muller, F.; Kamoun, W.S.; Chae, S.S.; Zheng, H.; Ying, H.; Mahoney, J.; et al. FoxOs cooperatively regulate diverse pathways governing neural stem cell homeostasis. *Cell Stem Cell* **2009**, *5*, 540–553. [CrossRef]
9. Renault, V.M.; Rafalski, V.A.; Morgan, A.A.; Salih, D.A.; Brett, J.O.; Webb, A.E.; Villeda, S.A.; Thekkat, P.U.; Guillerey, C.; Denko, N.C.; et al. FoxO3 regulates neural stem cell homeostasis. *Cell Stem Cell* **2009**, *5*, 527–539. [CrossRef]
10. Spéder, P.; Brand, A.H. Systemic and local cues drive neural stem cell niche remodelling during neurogenesis in. *Elife* **2018**, *7*. [CrossRef]
11. Britton, J.S.; Edgar, B.A. Environmental control of the cell cycle in Drosophila: Nutrition activates mitotic and endoreplicative cells by distinct mechanisms. *Development* **1998**, *125*, 2149–2158. [CrossRef]
12. Sousa-Nunes, R.; Yee, L.L.; Gould, A.P. Fat cells reactivate quiescent neuroblasts via TOR and glial insulin relays in Drosophila. *Nature* **2011**, *471*, 508–512. [CrossRef]
13. Ito, K.; Hotta, Y. Proliferation pattern of postembryonic neuroblasts in the brain of Drosophila melanogaster. *Dev. Biol.* **1992**, *149*, 134–148. [CrossRef]
14. Gage, F.H. Adult neurogenesis in mammals. *Science* **2019**, *364*, 827–828. [CrossRef]
15. Grégoire, C.A.; Goldenstein, B.L.; Floriddia, E.M.; Barnabé-Heider, F.; Fernandes, K.J. Endogenous neural stem cell responses to stroke and spinal cord injury. *Glia* **2015**, *63*, 1469–1482. [CrossRef]
16. Lin, R.; Iacovitti, L. Classic and novel stem cell niches in brain homeostasis and repair. *Brain Res.* **2015**, *1628*, 327–342. [CrossRef]
17. Lie, D.C.; Colamarino, S.A.; Song, H.J.; Désiré, L.; Mira, H.; Consiglio, A.; Lein, E.S.; Jessberger, S.; Lansford, H.; Dearie, A.R.; et al. Wnt signalling regulates adult hippocampal neurogenesis. *Nature* **2005**, *437*, 1370–1375. [CrossRef]

18. Ahn, S.; Joyner, A.L. In vivo analysis of quiescent adult neural stem cells responding to Sonic hedgehog. *Nature* **2005**, *437*, 894–897. [CrossRef]
19. Ortega, F.; Gascon, S.; Masserdotti, G.; Deshpande, A.; Simon, C.; Fischer, J.; Dimou, L.; Chichung Lie, D.; Schroeder, T.; Berninger, B. Oligodendrogliogenic and neurogenic adult subependymal zone neural stem cells constitute distinct lineages and exhibit differential responsiveness to Wnt signalling. *Nat. Cell Biol.* **2013**, *15*, 602–613. [CrossRef]
20. Sohn, J.; Orosco, L.; Guo, F.; Chung, S.H.; Bannerman, P.; Mills Ko, E.; Zarbalis, K.; Deng, W.; Pleasure, D. The subventricular zone continues to generate corpus callosum and rostral migratory stream astroglia in normal adult mice. *J. Neurosci.* **2015**, *35*, 3756–3763. [CrossRef]
21. Tong, C.K.; Fuentealba, L.C.; Shah, J.K.; Lindquist, R.A.; Ihrie, R.A.; Guinto, C.D.; Rodas-Rodriguez, J.L.; Alvarez-Buylla, A. A Dorsal SHH-Dependent Domain in the V-SVZ Produces Large Numbers of Oligodendroglial Lineage Cells in the Postnatal Brain. *Stem Cell Rep.* **2015**, *5*, 461–470. [CrossRef] [PubMed]
22. Chaker, Z.; Makarouni, E.; Doetsch, F. The Organism as the Niche: Physiological States Crack the Code of Adult Neural Stem Cell Heterogeneity. *Annu. Rev. Cell Dev. Biol.* **2024**, *40*, 381–406. [CrossRef] [PubMed]
23. Encinas, J.M.; Michurina, T.V.; Peunova, N.; Park, J.H.; Tordo, J.; Peterson, D.A.; Fishell, G.; Koulakov, A.; Enikolopov, G. Division-coupled astrocytic differentiation and age-related depletion of neural stem cells in the adult hippocampus. *Cell Stem Cell* **2011**, *8*, 566–579. [CrossRef]
24. Bond, A.M.; Ming, G.L.; Song, H. Adult Mammalian Neural Stem Cells and Neurogenesis: Five Decades Later. *Cell Stem Cell* **2015**, *17*, 385–395. [CrossRef]
25. Labusch, M.; Mancini, L.; Morizet, D.; Bally-Cuif, L. Conserved and Divergent Features of Adult Neurogenesis in Zebrafish. *Front. Cell Dev. Biol.* **2020**, *8*, 525. [CrossRef] [PubMed]
26. Lübke, L.; Zhang, G.; Strähle, U.; Rastegar, S. Expression Is Associated with Quiescent Neural Stem Cells during Constitutive and Reactive Neurogenesis in the Adult Zebrafish Telencephalon. *Brain Sci.* **2022**, *12*, 284. [CrossRef]
27. Llorens-Bobadilla, E.; Zhao, S.; Baser, A.; Saiz-Castro, G.; Zwadlo, K.; Martin-Villalba, A. Single-Cell Transcriptomics Reveals a Population of Dormant Neural Stem Cells that Become Activated upon Brain Injury. *Cell Stem Cell* **2015**, *17*, 329–340. [CrossRef]
28. Foley, T.; Thetiot, M.; Bally-Cuif, L. Neural Stem Cell Regulation in Zebrafish. *Annu. Rev. Genet.* **2024**, *58*, 249–272. [CrossRef]
29. Cavallucci, V.; Fidaleo, M.; Pani, G. Neural Stem Cells and Nutrients: Poised Between Quiescence and Exhaustion. *Trends Endocrinol. Metab.* **2016**, *27*, 756–769. [CrossRef]
30. Gujar, M.R.; Gao, Y.; Teng, X.; Ding, W.Y.; Lin, J.; Tan, Y.S.; Chew, L.Y.; Toyama, Y.; Wang, H. Patronin/CAMSAP promotes reactivation and regeneration of Drosophila quiescent neural stem cells. *EMBO Rep.* **2023**, *24*, e56624. [CrossRef]
31. Caron, A.; Trzuskot, L.; Lindsey, B.W. Uncovering the spectrum of adult zebrafish neural stem cell cycle regulators. *Front. Cell Dev. Biol.* **2022**, *10*, 941893. [CrossRef]
32. Gao, Y.; Tan, Y.S.; Lin, J.; Chew, L.Y.; Aung, H.Y.; Palliyana, B.; Gujar, M.R.; Lin, K.Y.; Kondo, S.; Wang, H. SUMOylation of Warts kinase promotes neural stem cell reactivation. *Nat. Commun.* **2024**, *15*, 8557. [CrossRef] [PubMed]
33. Blasco-Chamarro, L.; Fariñas, I. Fine-tuned Rest: Unveiling the Regulatory Landscape of Adult Quiescent Neural Stem Cells. *Neuroscience* **2023**, *525*, 26–37. [CrossRef]
34. Barros, C. Created in BioRender. 2025. Available online: https://BioRender.com/gzahgln (accessed on 4 March 2025).
35. Grandel, H.; Kaslin, J.; Ganz, J.; Wenzel, I.; Brand, M. Neural stem cells and neurogenesis in the adult zebrafish brain: Origin, proliferation dynamics, migration and cell fate. *Dev. Biol.* **2006**, *295*, 263–277. [CrossRef]
36. Martynoga, B.; Mateo, J.L.; Zhou, B.; Andersen, J.; Achimastou, A.; Urbán, N.; van den Berg, D.; Georgopoulou, D.; Hadjur, S.; Wittbrodt, J.; et al. Epigenomic enhancer annotation reveals a key role for NFIX in neural stem cell quiescence. *Genes Dev.* **2013**, *27*, 1769–1786. [CrossRef] [PubMed]
37. Alessio, N.; Aprile, D.; Cappabianca, S.; Peluso, G.; Di Bernardo, G.; Galderisi, U. Different Stages of Quiescence, Senescence, and Cell Stress Identified by Molecular Algorithm Based on the Expression of Ki67, RPS6, and Beta-Galactosidase Activity. *Int. J. Mol. Sci.* **2021**, *22*, 3102. [CrossRef] [PubMed]
38. Marqués-Torrejón, M.; Williams, C.A.C.; Southgate, B.; Alfazema, N.; Clements, M.P.; Garcia-Diaz, C.; Blin, C.; Arranz-Emparan, N.; Fraser, J.; Gammoh, N.; et al. LRIG1 is a gatekeeper to exit from quiescence in adult neural stem cells. *Nat. Commun.* **2021**, *12*, 2594. [CrossRef]
39. Cheung, T.H.; Rando, T.A. Molecular regulation of stem cell quiescence. *Nat. Rev. Mol. Cell Biol.* **2013**, *14*, 329–340. [CrossRef]
40. Sood, C.; Justis, V.T.; Doyle, S.E.; Siegrist, S.E. Notch signaling regulates neural stem cell quiescence entry and exit in Drosophila. *Development* **2022**, *149*. [CrossRef]
41. Tsuji, T.; Hasegawa, E.; Isshiki, T. Neuroblast entry into quiescence is regulated intrinsically by the combined action of spatial Hox proteins and temporal identity factors. *Development* **2008**, *135*, 3859–3869. [CrossRef]
42. Bossing, T.; Udolph, G.; Doe, C.Q.; Technau, G.M. The embryonic central nervous system lineages of Drosophila melanogaster. I. Neuroblast lineages derived from the ventral half of the neuroectoderm. *Dev. Biol.* **1996**, *179*, 41–64. [CrossRef] [PubMed]

43. Otsuki, L.; Brand, A.H. Cell cycle heterogeneity directs the timing of neural stem cell activation from quiescence. *Science* **2018**, *360*, 99–102. [CrossRef] [PubMed]
44. Otsuki, L.; Brand, A.H. Dorsal-Ventral Differences in Neural Stem Cell Quiescence Are Induced by p57. *Dev. Cell* **2019**, *49*, 293–300.e293. [CrossRef]
45. McKeown, C.R.; Cline, H.T. Nutrient restriction causes reversible G2 arrest in. *Development* **2019**, *146*. [CrossRef]
46. Luo, Y.; Coskun, V.; Liang, A.; Yu, J.; Cheng, L.; Ge, W.; Shi, Z.; Zhang, K.; Li, C.; Cui, Y.; et al. Single-cell transcriptome analyses reveal signals to activate dormant neural stem cells. *Cell* **2015**, *161*, 1175–1186. [CrossRef]
47. Chavali, M.; Klingener, M.; Kokkosis, A.G.; Garkun, Y.; Felong, S.; Maffei, A.; Aguirre, A. Non-canonical Wnt signaling regulates neural stem cell quiescence during homeostasis and after demyelination. *Nat. Commun.* **2018**, *9*, 36. [CrossRef] [PubMed]
48. Rigo, P.; Ahmed-de-Prado, S.; Johnston, R.L.; Choudhury, C.; Guillemot, F.; Harris, L. Sequential transcriptional programs underpin activation of quiescent hippocampal stem cells. *bioRxiv* **2024**. [CrossRef]
49. Labusch, M.; Thetiot, M.; Than-Trong, E.; Morizet, D.; Coolen, M.; Varet, H.; Legendre, R.; Ortica, S.; Mancini, L.; Bally-Cuif, L. Prosaposin maintains adult neural stem cells in a state associated with deep quiescence. *Stem Cell Rep.* **2024**, *19*, 515–528. [CrossRef]
50. Paul, A.; Chaker, Z.; Doetsch, F. Hypothalamic regulation of regionally distinct adult neural stem cells and neurogenesis. *Science* **2017**, *356*, 1383–1386. [CrossRef]
51. Lugert, S.; Basak, O.; Knuckles, P.; Haussler, U.; Fabel, K.; Gotz, M.; Haas, C.A.; Kempermann, G.; Taylor, V.; Giachino, C. Quiescent and active hippocampal neural stem cells with distinct morphologies respond selectively to physiological and pathological stimuli and aging. *Cell Stem Cell* **2010**, *6*, 445–456. [CrossRef]
52. van Praag, H.; Kempermann, G.; Gage, F.H. Running increases cell proliferation and neurogenesis in the adult mouse dentate gyrus. *Nat. Neurosci.* **1999**, *2*, 266–270. [CrossRef] [PubMed]
53. Shingo, T.; Gregg, C.; Enwere, E.; Fujikawa, H.; Hassam, R.; Geary, C.; Cross, J.C.; Weiss, S. Pregnancy-stimulated neurogenesis in the adult female forebrain mediated by prolactin. *Science* **2003**, *299*, 117–120. [CrossRef] [PubMed]
54. Chaker, Z.; Segalada, C.; Kretz, J.A.; Acar, I.E.; Delgado, A.C.; Crotet, V.; Moor, A.E.; Doetsch, F. Pregnancy-responsive pools of adult neural stem cells for transient neurogenesis in mothers. *Science* **2023**, *382*, 958–963. [CrossRef] [PubMed]
55. Gengatharan, A.; Malvaut, S.; Marymonchyk, A.; Ghareghani, M.; Snapyan, M.; Fischer-Sternjak, J.; Ninkovic, J.; Gotz, M.; Saghatelyan, A. Adult neural stem cell activation in mice is regulated by the day/night cycle and intracellular calcium dynamics. *Cell* **2021**, *184*, 709–722.e13. [CrossRef]
56. Alonso, M.I.; Gato, A. Cerebrospinal fluid and neural stem cell niche control. *Neural Regen. Res.* **2018**, *13*, 1546–1547. [CrossRef]
57. Liu, X.; Wang, Q.; Haydar, T.F.; Bordey, A. Nonsynaptic GABA signaling in postnatal subventricular zone controls proliferation of GFAP-expressing progenitors. *Nat. Neurosci.* **2005**, *8*, 1179–1187. [CrossRef]
58. Young, S.Z.; Taylor, M.M.; Bordey, A. Neurotransmitters couple brain activity to subventricular zone neurogenesis. *Eur. J. Neurosci.* **2011**, *33*, 1123–1132. [CrossRef]
59. Marymonchyk, A.; Rodriguez-Aller, R.; Willis, A.; Beaupre, F.; Warsi, S.; Snapyan, M.; Claver-Fournier, V.; Lavoie-Cardinal, F.; Kaplan, D.R.; Miller, F.D.; et al. Neural stem cell quiescence and activation dynamics are regulated by feedback input from their progeny under homeostatic and regenerative conditions. *Cell Stem Cell* **2025**, *32*, 445–462.e9. [CrossRef]
60. Walker, J.V.; Zhuang, H.; Singer, D.; Illsley, C.S.; Kok, W.L.; Sivaraj, K.K.; Gao, Y.; Bolton, C.; Liu, Y.; Zhao, M.; et al. Transit amplifying cells coordinate mouse incisor mesenchymal stem cell activation. *Nat. Commun.* **2019**, *10*, 3596. [CrossRef]
61. Tavazoie, M.; Van der Veken, L.; Silva-Vargas, V.; Louissaint, M.; Colonna, L.; Zaidi, B.; Garcia-Verdugo, J.M.; Doetsch, F. A specialized vascular niche for adult neural stem cells. *Cell Stem Cell* **2008**, *3*, 279–288. [CrossRef]
62. Calvo, C.F.; Fontaine, R.H.; Soueid, J.; Tammela, T.; Makinen, T.; Alfaro-Cervello, C.; Bonnaud, F.; Miguez, A.; Benhaim, L.; Xu, Y.; et al. Vascular endothelial growth factor receptor 3 directly regulates murine neurogenesis. *Genes Dev.* **2011**, *25*, 831–844. [CrossRef]
63. Crouch, E.E.; Liu, C.; Silva-Vargas, V.; Doetsch, F. Regional and stage-specific effects of prospectively purified vascular cells on the adult V-SVZ neural stem cell lineage. *J. Neurosci.* **2015**, *35*, 4528–4539. [CrossRef] [PubMed]
64. Ottone, C.; Krusche, B.; Whitby, A.; Clements, M.; Quadrato, G.; Pitulescu, M.E.; Adams, R.H.; Parrinello, S. Direct cell-cell contact with the vascular niche maintains quiescent neural stem cells. *Nat. Cell Biol.* **2014**, *16*, 1045–1056. [CrossRef]
65. Navarro Negredo, P.; Yeo, R.W.; Brunet, A. Aging and Rejuvenation of Neural Stem Cells and Their Niches. *Cell Stem Cell* **2020**, *27*, 202–223. [CrossRef] [PubMed]
66. Gage, F.H.; Temple, S. Neural stem cells: Generating and regenerating the brain. *Neuron* **2013**, *80*, 588–601. [CrossRef] [PubMed]
67. Kalamakis, G.; Brüne, D.; Ravichandran, S.; Bolz, J.; Fan, W.; Ziebell, F.; Stiehl, T.; Catalá-Martinez, F.; Kupke, J.; Zhao, S.; et al. Quiescence Modulates Stem Cell Maintenance and Regenerative Capacity in the Aging Brain. *Cell* **2019**, *176*, 1407–1419.e1414. [CrossRef]
68. Ruetz, T.J.; Pogson, A.N.; Kashiwagi, C.M.; Gagnon, S.D.; Morton, B.; Sun, E.D.; Na, J.; Yeo, R.W.; Leeman, D.S.; Morgens, D.W.; et al. CRISPR-Cas9 screens reveal regulators of ageing in neural stem cells. *Nature* **2024**, *634*, 1150–1159. [CrossRef]

69. Rossi, A.; Coum, A.; Madelenat, M.; Harris, L.; Miedzik, A.; Strohbuecker, S.; Chai, A.; Fiaz, H.; Chaouni, R.; Faull, P.; et al. Neural stem cells alter nucleocytoplasmic partitioning and accumulate nuclear polyadenylated transcripts during quiescence. *bioRxiv* **2021**. [CrossRef]
70. Suh, E.J.; Remillard, M.Y.; Legesse-Miller, A.; Johnson, E.L.; Lemons, J.M.; Chapman, T.R.; Forman, J.J.; Kojima, M.; Silberman, E.S.; Coller, H.A. A microRNA network regulates proliferative timing and extracellular matrix synthesis during cellular quiescence in fibroblasts. *Genome Biol.* **2012**, *13*, R121. [CrossRef]
71. Martinez, I.; Hayes, K.E.; Barr, J.A.; Harold, A.D.; Xie, M.; Bukhari, S.I.A.; Vasudevan, S.; Steitz, J.A.; DiMaio, D. An Exportin-1-dependent microRNA biogenesis pathway during human cell quiescence. *Proc. Natl. Acad. Sci. USA* **2017**, *114*, E4961–E4970. [CrossRef]
72. Lai, S.L.; Doe, C.Q. Transient nuclear Prospero induces neural progenitor quiescence. *Elife* **2014**, *3*. [CrossRef] [PubMed]
73. Joy, T.; Hirono, K.; Doe, C.Q. The RanGEF Bj1 promotes prospero nuclear export and neuroblast self-renewal. *Dev. Neurobiol.* **2015**, *75*, 485–493. [CrossRef] [PubMed]
74. Carmody, S.R.; Wente, S.R. mRNA nuclear export at a glance. *J. Cell Sci.* **2009**, *122*, 1933–1937. [CrossRef] [PubMed]
75. Medina, R.; Zaidi, S.K.; Liu, C.G.; Stein, J.L.; van Wijnen, A.J.; Croce, C.M.; Stein, G.S. MicroRNAs 221 and 222 bypass quiescence and compromise cell survival. *Cancer Res.* **2008**, *68*, 2773–2780. [CrossRef]
76. Leucht, C.; Stigloher, C.; Wizenmann, A.; Klafke, R.; Folchert, A.; Bally-Cuif, L. MicroRNA-9 directs late organizer activity of the midbrain-hindbrain boundary. *Nat. Neurosci.* **2008**, *11*, 641–648. [CrossRef]
77. Katz, S.; Cussigh, D.; Urbán, N.; Blomfield, I.; Guillemot, F.; Bally-Cuif, L.; Coolen, M. A Nuclear Role for miR-9 and Argonaute Proteins in Balancing Quiescent and Activated Neural Stem Cell States. *Cell Rep.* **2016**, *17*, 1383–1398. [CrossRef]
78. Artavanis-Tsakonas, S.; Muskavitch, M.A. Notch: The past, the present, and the future. *Curr. Top. Dev. Biol.* **2010**, *92*, 1–29. [CrossRef]
79. Wharton, K.A.; Johansen, K.M.; Xu, T.; Artavanis-Tsakonas, S. Nucleotide sequence from the neurogenic locus notch implies a gene product that shares homology with proteins containing EGF-like repeats. *Cell* **1985**, *43*, 567–581. [CrossRef]
80. Kopczynski, C.C.; Alton, A.K.; Fechtel, K.; Kooh, P.J.; Muskavitch, M.A. Delta, a Drosophila neurogenic gene, is transcriptionally complex and encodes a protein related to blood coagulation factors and epidermal growth factor of vertebrates. *Genes Dev.* **1988**, *2*, 1723–1735. [CrossRef]
81. Salazar, J.L.; Yamamoto, S. Integration of Drosophila and Human Genetics to Understand Notch Signaling Related Diseases. *Adv. Exp. Med. Biol.* **2018**, *1066*, 141–185. [CrossRef]
82. Kawai, H.; Kawaguchi, D.; Kuebrich, B.D.; Kitamoto, T.; Yamaguchi, M.; Gotoh, Y.; Furutachi, S. Area-Specific Regulation of Quiescent Neural Stem Cells by Notch3 in the Adult Mouse Subependymal Zone. *J. Neurosci.* **2017**, *37*, 11867–11880. [CrossRef] [PubMed]
83. Imayoshi, I.; Sakamoto, M.; Yamaguchi, M.; Mori, K.; Kageyama, R. Essential roles of Notch signaling in maintenance of neural stem cells in developing and adult brains. *J. Neurosci.* **2010**, *30*, 3489–3498. [CrossRef]
84. Hu, N.; Zou, L. Multiple functions of Hes genes in the proliferation and differentiation of neural stem cells. *Ann. Anat.* **2022**, *239*, 151848. [CrossRef]
85. Than-Trong, E.; Ortica-Gatti, S.; Mella, S.; Nepal, C.; Alunni, A.; Bally-Cuif, L. Neural stem cell quiescence and stemness are molecularly distinct outputs of the Notch3 signalling cascade in the vertebrate adult brain. *Development* **2018**, *145*. [CrossRef] [PubMed]
86. Chapouton, P.; Skupien, P.; Hesl, B.; Coolen, M.; Moore, J.C.; Madelaine, R.; Kremmer, E.; Faus-Kessler, T.; Blader, P.; Lawson, N.D.; et al. Notch activity levels control the balance between quiescence and recruitment of adult neural stem cells. *J. Neurosci.* **2010**, *30*, 7961–7974. [CrossRef] [PubMed]
87. Kizil, C. Mechanisms of Pathology-Induced Neural Stem Cell Plasticity and Neural Regeneration in Adult Zebrafish Brain. *Curr. Pathobiol. Rep.* **2018**, *6*, 71–77. [CrossRef]
88. März, M.; Schmidt, R.; Rastegar, S.; Strähle, U. Regenerative response following stab injury in the adult zebrafish telencephalon. *Dev. Dyn.* **2011**, *240*, 2221–2231. [CrossRef]
89. Kim, H.K.; Lee, D.W.; Kim, E.; Jeong, I.; Kim, S.; Kim, B.J.; Park, H.C. Notch Signaling Controls Oligodendrocyte Regeneration in the Injured Telencephalon of Adult Zebrafish. *Exp. Neurobiol.* **2020**, *29*, 417–424. [CrossRef]
90. Meyers, E.A.; Kessler, J.A. TGF-β Family Signaling in Neural and Neuronal Differentiation, Development, and Function. *Cold Spring Harb. Perspect. Biol.* **2017**, *9*. [CrossRef]
91. Kanai, M.I.; Kim, M.J.; Akiyama, T.; Takemura, M.; Wharton, K.; O'Connor, M.B.; Nakato, H. Regulation of neuroblast proliferation by surface glia in the Drosophila larval brain. *Sci. Rep.* **2018**, *8*, 3730. [CrossRef]
92. Bonaguidi, M.A.; Wheeler, M.A.; Shapiro, J.S.; Stadel, R.P.; Sun, G.J.; Ming, G.L.; Song, H. In vivo clonal analysis reveals self-renewing and multipotent adult neural stem cell characteristics. *Cell* **2011**, *145*, 1142–1155. [CrossRef]
93. Miyazono, K.; Kamiya, Y.; Morikawa, M. Bone morphogenetic protein receptors and signal transduction. *J. Biochem.* **2010**, *147*, 35–51. [CrossRef] [PubMed]

94. Mira, H.; Andreu, Z.; Suh, H.; Lie, D.C.; Jessberger, S.; Consiglio, A.; San Emeterio, J.; Hortigüela, R.; Marqués-Torrejón, M.A.; Nakashima, K.; et al. Signaling through BMPR-IA regulates quiescence and long-term activity of neural stem cells in the adult hippocampus. *Cell Stem Cell* **2010**, *7*, 78–89. [CrossRef] [PubMed]
95. Xu, S.; Zhang, X.; Li, Z.; Liu, C.; Liu, Q.; Chai, H.; Yao, H.; Luo, Y.; Li, S.; Li, C. Characteristics of quiescent adult neural stem cells induced by the bFGF/BMP4 combination or BMP4 alone. *Front. Cell Neurosci.* **2024**, *18*, 1391556. [CrossRef] [PubMed]
96. Marcy, G.; Foucault, L.; Babina, E.; Capeliez, T.; Texeraud, E.; Zweifel, S.; Heinrich, C.; Hernandez-Vargas, H.; Parras, C.; Jabaudon, D.; et al. Single-cell analysis of the postnatal dorsal V-SVZ reveals a role for Bmpr1a signaling in silencing pallial germinal activity. *Sci. Adv.* **2023**, *9*, eabq7553. [CrossRef]
97. Lim, D.A.; Tramontin, A.D.; Trevejo, J.M.; Herrera, D.G.; García-Verdugo, J.M.; Alvarez-Buylla, A. Noggin antagonizes BMP signaling to create a niche for adult neurogenesis. *Neuron* **2000**, *28*, 713–726. [CrossRef]
98. Zhang, G.; Lübke, L.; Chen, F.; Beil, T.; Takamiya, M.; Diotel, N.; Strähle, U.; Rastegar, S. Neuron-Radial Glial Cell Communication via BMP/Id1 Signaling Is Key to Long-Term Maintenance of the Regenerative Capacity of the Adult Zebrafish Telencephalon. *Cells* **2021**, *10*, 2794. [CrossRef]
99. Zhang, G.; Ferg, M.; Lübke, L.; Takamiya, M.; Beil, T.; Gourain, V.; Diotel, N.; Strähle, U.; Rastegar, S. Bone morphogenetic protein signaling regulates Id1-mediated neural stem cell quiescence in the adult zebrafish brain via a phylogenetically conserved enhancer module. *Stem Cells* **2020**, *38*, 875–889. [CrossRef]
100. Halder, G.; Johnson, R.L. Hippo signaling: Growth control and beyond. *Development* **2011**, *138*, 9–22. [CrossRef]
101. Harvey, K.; Tapon, N. The Salvador-Warts-Hippo pathway - an emerging tumour-suppressor network. *Nat. Rev. Cancer* **2007**, *7*, 182–191. [CrossRef]
102. Fan, W.; Jurado-Arjona, J.; Alanis-Lobato, G.; Péron, S.; Berger, C.; Andrade-Navarro, M.A.; Falk, S.; Berninger, B. The transcriptional co-activator Yap1 promotes adult hippocampal neural stem cell activation. *EMBO J.* **2023**, *42*, e110384. [CrossRef] [PubMed]
103. Ding, R.; Berger, C. Hippo pathway regulates neural stem cell quiescence. *Cell Cycle* **2016**, *15*, 1525–1526. [CrossRef] [PubMed]
104. Lavado, A.; Park, J.Y.; Paré, J.; Finkelstein, D.; Pan, H.; Xu, B.; Fan, Y.; Kumar, R.P.; Neale, G.; Kwak, Y.D.; et al. The Hippo Pathway Prevents YAP/TAZ-Driven Hypertranscription and Controls Neural Progenitor Number. *Dev. Cell* **2018**, *47*, 576–591.e578. [CrossRef]
105. Luo, J.; Li, P. Context-dependent transcriptional regulations of YAP/TAZ in stem cell and differentiation *Stem Cell Res. Ther.* **2022**, *13*, 10. [CrossRef] [PubMed]
106. Yao, M.; Wang, Y.; Zhang, P.; Chen, H.; Xu, Z.; Jiao, J.; Yuan, Z. BMP2-SMAD signaling represses the proliferation of embryonic neural stem cells through YAP. *J. Neurosci.* **2014**, *34*, 12039–12048. [CrossRef]
107. Ding, R.; Weynans, K.; Bossing, T.; Barros, C.S.; Berger, C. The Hippo signalling pathway maintains quiescence in Drosophila neural stem cells. *Nat. Commun.* **2016**, *7*, 10510. [CrossRef]
108. Shi, Z.; Jiao, S.; Zhou, Z. STRIPAK complexes in cell signaling and cancer. *Oncogene* **2016**, *35*, 4549–4557. [CrossRef]
109. Schulte, J.; Sepp, K.J.; Jorquera, R.A.; Wu, C.; Song, Y.; Hong, P.; Littleton, J.T. DMob4/Phocein regulates synapse formation, axonal transport, and microtubule organization. *J. Neurosci.* **2010**, *30*, 5189–5203. [CrossRef]
110. Couzens, A.L.; Knight, J.D.; Kean, M.J.; Teo, G.; Weiss, A.; Dunham, W.H.; Lin, Z.Y.; Bagshaw, R.D.; Sicheri, F.; Pawson, T.; et al. Protein interaction network of the mammalian Hippo pathway reveals mechanisms of kinase-phosphatase interactions. *Sci. Signal* **2013**, *6*, rs15. [CrossRef]
111. Gil-Ranedo, J.; Gonzaga, E.; Jaworek, K.J.; Berger, C.; Bossing, T.; Barros, C.S. STRIPAK Members Orchestrate Hippo and Insulin Receptor Signaling to Promote Neural Stem Cell Reactivation. *Cell Rep.* **2019**, *27*, 2921–2933.e2925. [CrossRef]
112. Ribeiro, P.S.; Josué, F.; Wepf, A.; Wehr, M.C.; Rinner, O.; Kelly, G.; Tapon, N.; Gstaiger, M. Combined functional genomic and proteomic approaches identify a PP2A complex as a negative regulator of Hippo signaling. *Mol. Cell* **2010**, *39*, 521–534. [CrossRef] [PubMed]
113. Chai, P.C.; Liu, Z.; Chia, W.; Cai, Y. Hedgehog signaling acts with the temporal cascade to promote neuroblast cell cycle exit. *PLoS Biol.* **2013**, *11*, e1001494. [CrossRef]
114. Yu, F.X.; Guan, K.L. The Hippo pathway: Regulators and regulations. *Genes Dev.* **2013**, *27*, 355–371. [CrossRef]
115. Moya, I.M.; Halder, G. Hippo-YAP/TAZ signalling in organ regeneration and regenerative medicine. *Nat. Rev. Mol. Cell Biol.* **2019**, *20*, 211–226. [CrossRef]
116. Zhao, X.; Le, T.P.; Erhardt, S.; Findley, T.O.; Wang, J. Hippo-Yap Pathway Orchestrates Neural Crest Ontogenesis. *Front. Cell Dev. Biol.* **2021**, *9*, 706623. [CrossRef] [PubMed]
117. Riley, S.E.; Feng, Y.; Hansen, C.G. Hippo-Yap/Taz signalling in zebrafish regeneration. *NPJ Regen. Med.* **2022**, *7*, 9. [CrossRef]
118. Klatt Shaw, D.; Saraswathy, V.M.; Zhou, L.; McAdow, A.R.; Burris, B.; Butka, E.; Morris, S.A.; Dietmann, S.; Mokalled, M.H. Localized EMT reprograms glial progenitors to promote spinal cord repair. *Dev. Cell* **2021**, *56*, 613–626.e617. [CrossRef]
119. Florindo, C.; Mimoso, J.M.; Palma, S.L.; Gonçalves, C.; Silvestre, D.; Campinho, M.; Tavares, Á. Mob4 is required for neurodevelopment in zebrafish. *MicroPubl. Biol.* **2023**, *2023*. [CrossRef]

120. Santos, I.B.; Garrido-Maraver, J.; Gonçalves, C.; Oliveira, B.I.; Tavares, Á.A. Role of MOB4 in Cell Proliferation and Neurogenesis. *BioChem* **2023**, *3*, 182–196. [CrossRef]
121. Lee, M.S.; Wan, J.; Goldman, D. Tgfb3 collaborates with PP2A and notch signaling pathways to inhibit retina regeneration. *Elife* **2020**, *9*. [CrossRef]
122. Vinayagam, A.; Kulkarni, M.M.; Sopko, R.; Sun, X.; Hu, Y.; Nand, A.; Villalta, C.; Moghimi, A.; Yang, X.; Mohr, S.E.; et al. An Integrative Analysis of the InR/PI3K/Akt Network Identifies the Dynamic Response to Insulin Signaling. *Cell Rep.* **2016**, *16*, 3062–3074. [CrossRef]
123. Homem, C.C.; Repic, M.; Knoblich, J.A. Proliferation control in neural stem and progenitor cells. *Nat. Rev. Neurosci.* **2015**, *16*, 647–659. [CrossRef]
124. Chell, J.M.; Brand, A.H. Nutrition-responsive glia control exit of neural stem cells from quiescence. *Cell* **2010**, *143*, 1161–1173. [CrossRef] [PubMed]
125. Zhao, Y.; Wang, Y.; Zhu, W.G. Applications of post-translational modifications of FoxO family proteins in biological functions. *J. Mol. Cell Biol.* **2011**, *3*, 276–282. [CrossRef] [PubMed]
126. Yu, J.; Chen, G.; Zhu, H.; Zhong, Y.; Yang, Z.; Jian, Z.; Xiong, X. Metabolic and proteostatic differences in quiescent and active neural stem cells. *Neural Regen. Res.* **2024**, *19*, 43–48. [CrossRef] [PubMed]
127. Webb, A.E.; Pollina, E.A.; Vierbuchen, T.; Urbán, N.; Ucar, D.; Leeman, D.S.; Martynoga, B.; Sewak, M.; Rando, T.A.; Guillemot, F.; et al. FOXO3 shares common targets with ASCL1 genome-wide and inhibits ASCL1-dependent neurogenesis. *Cell Rep.* **2013**, *4*, 477–491. [CrossRef]
128. Kobayashi, T.; Kageyama, R. Lysosomes and signaling pathways for maintenance of quiescence in adult neural stem cells. *FEBS J.* **2021**, *288*, 3082–3093. [CrossRef]
129. Zhou, Y.; Bond, A.M.; Shade, J.E.; Zhu, Y.; Davis, C.O.; Wang, X.; Su, Y.; Yoon, K.J.; Phan, A.T.; Chen, W.J.; et al. Autocrine Mfge8 Signaling Prevents Developmental Exhaustion of the Adult Neural Stem Cell Pool. *Cell Stem Cell* **2018**, *23*, 444–452.e444. [CrossRef]
130. Khalifeh-Soltani, A.; Ha, A.; Podolsky, M.J.; McCarthy, D.A.; McKleroy, W.; Azary, S.; Sakuma, S.; Tharp, K.M.; Wu, N.; Yokosaki, Y.; et al. α8β1 integrin regulates nutrient absorption through an Mfge8-PTEN dependent mechanism. *Elife* **2016**, *5*. [CrossRef]
131. Klotz, L.O.; Sánchez-Ramos, C.; Prieto-Arroyo, I.; Urbánek, P.; Steinbrenner, H.; Monsalve, M. Redox regulation of FoxO transcription factors. *Redox Biol.* **2015**, *6*, 51–72. [CrossRef]
132. Lehtinen, M.K.; Yuan, Z.; Boag, P.R.; Yang, Y.; Villén, J.; Becker, E.B.; DiBacco, S.; de la Iglesia, N.; Gygi, S.; Blackwell, T.K.; et al. A conserved MST-FOXO signaling pathway mediates oxidative-stress responses and extends life span. *Cell* **2006**, *125*, 987–1001. [CrossRef] [PubMed]
133. Schlueter, P.J.; Peng, G.; Westerfield, M.; Duan, C. Insulin-like growth factor signaling regulates zebrafish embryonic growth and development by promoting cell survival and cell cycle progression. *Cell Death Differ.* **2007**, *14*, 1095–1105. [CrossRef] [PubMed]
134. Toyoshima, Y.; Monson, C.; Duan, C.; Wu, Y.; Gao, C.; Yakar, S.; Sadler, K.C.; LeRoith, D. The role of insulin receptor signaling in zebrafish embryogenesis. *Endocrinology* **2008**, *149*, 5996–6005. [CrossRef]
135. Gence, L.; Fernezelian, D.; Meilhac, O.; Rastegar, S.; Bascands, J.L.; Diotel, N. Insulin signaling promotes neurogenesis in the brain of adult zebrafish. *J. Comp. Neurol.* **2023**, *531*, 1812–1827. [CrossRef]
136. Huang, J.; Wang, H. Hsp83/Hsp90 Physically Associates with Insulin Receptor to Promote Neural Stem Cell Reactivation. *Stem Cell Rep.* **2018**, *11*, 883–896. [CrossRef] [PubMed]
137. Spéder, P.; Brand, A.H. Gap junction proteins in the blood-brain barrier control nutrient-dependent reactivation of Drosophila neural stem cells. *Dev. Cell* **2014**, *30*, 309–321. [CrossRef]
138. Li, S.; Koe, C.T.; Tay, S.T.; Tan, A.L.K.; Zhang, S.; Zhang, Y.; Tan, P.; Sung, W.K.; Wang, H. Erratum: An intrinsic mechanism controls reactivation of neural stem cells by spindle matrix proteins. *Nat. Commun.* **2017**, *8*, 1298. [CrossRef]
139. Choksi, S.P.; Southall, T.D.; Bossing, T.; Edoff, K.; de Wit, E.; Fischer, B.E.; van Steensel, B.; Micklem, G.; Brand, A.H. Prospero acts as a binary switch between self-renewal and differentiation in Drosophila neural stem cells. *Dev. Cell* **2006**, *11*, 775–789. [CrossRef]
140. Callan, M.A.; Clements, N.; Ahrendt, N.; Zarnescu, D.C. Fragile X Protein is required for inhibition of insulin signaling and regulates glial-dependent neuroblast reactivation in the developing brain. *Brain Res.* **2012**, *1462*, 151–161. [CrossRef]
141. Callan, M.A.; Cabernard, C.; Heck, J.; Luois, S.; Doe, C.Q.; Zarnescu, D.C. Fragile X protein controls neural stem cell proliferation in the Drosophila brain. *Hum. Mol. Genet.* **2010**, *19*, 3068–3079. [CrossRef]
142. Queiroz, L.Y.; Kageyama, R.; Cimarosti, H.I. SUMOylation effects on neural stem cells self-renewal, differentiation, and survival. *Neurosci. Res.* **2024**, *199*, 1–11. [CrossRef]
143. Bernstock, J.D.; Peruzzotti-Jametti, L.; Leonardi, T.; Vicario, N.; Ye, D.; Lee, Y.J.; Maric, D.; Johnson, K.R.; Mou, Y.; Van Den Bosch, A.; et al. SUMOylation promotes survival and integration of neural stem cell grafts in ischemic stroke. *EBioMedicine* **2019**, *42*, 214–224. [CrossRef] [PubMed]

144. Wu, Y.; Guo, Z.; Wu, H.; Wang, X.; Yang, L.; Shi, X.; Du, J.; Tang, B.; Li, W.; Zhang, Y. SUMOylation represses Nanog expression via modulating transcription factors Oct4 and Sox2. *PLoS ONE* **2012**, *7*, e39606. [CrossRef]
145. Fernández-Lloris, R.; Osses, N.; Jaffray, E.; Shen, L.N.; Vaughan, O.A.; Girwood, D.; Bartrons, R.; Rosa, J.L.; Hay, R.T.; Ventura, F. Repression of SOX6 transcriptional activity by SUMO modification. *FEBS Lett.* **2006**, *580*, 1215–1221. [CrossRef]
146. Marelli, E.; Hughes, J.; Scotting, P.J. SUMO-dependent transcriptional repression by Sox2 inhibits the proliferation of neural stem cells. *PLoS ONE* **2024**, *19*, e0298818. [CrossRef]
147. Nowak, M.; Hammerschmidt, M. Ubc9 regulates mitosis and cell survival during zebrafish development. *Mol. Biol. Cell* **2006**, *17*, 5324–5336. [CrossRef] [PubMed]
148. Kobayashi, T.; Piao, W.; Takamura, T.; Kori, H.; Miyachi, H.; Kitano, S.; Iwamoto, Y.; Yamada, M.; Imayoshi, I.; Shioda, S.; et al. Enhanced lysosomal degradation maintains the quiescent state of neural stem cells. *Nat. Commun.* **2019**, *10*, 5446. [CrossRef]
149. Leeman, D.S.; Hebestreit, K.; Ruetz, T.; Webb, A.E.; McKay, A.; Pollina, E.A.; Dulken, B.W.; Zhao, X.; Yeo, R.W.; Ho, T.T.; et al. Lysosome activation clears aggregates and enhances quiescent neural stem cell activation during aging. *Science* **2018**, *359*, 1277–1283. [CrossRef]
150. Calatayud-Baselga, I.; Casares-Crespo, L.; Franch-Ibáñez, C.; Guijarro-Nuez, J.; Sanz, P.; Mira, H. Autophagy drives the conversion of developmental neural stem cells to the adult quiescent state. *Nat. Commun.* **2023**, *14*, 7541. [CrossRef] [PubMed]
151. Park, Y.; Rangel, C.; Reynolds, M.M.; Caldwell, M.C.; Johns, M.; Nayak, M.; Welsh, C.J.; McDermott, S.; Datta, S. Drosophila perlecan modulates FGF and hedgehog signals to activate neural stem cell division. *Dev. Biol.* **2003**, *253*, 247–257. [CrossRef]
152. Barrett, A.L.; Krueger, S.; Datta, S. Branchless and Hedgehog operate in a positive feedback loop to regulate the initiation of neuroblast division in the Drosophila larval brain. *Dev. Biol.* **2008**, *317*, 234–245. [CrossRef] [PubMed]
153. Dong, Q.; Zavortink, M.; Froldi, F.; Golenkina, S.; Lam, T.; Cheng, L.Y. Glial Hedgehog signalling and lipid metabolism regulate neural stem cell proliferation in Drosophila. *EMBO Rep.* **2021**, *22*, e52130. [CrossRef]
154. Meda, F.; Gauron, C.; Rampon, C.; Teillon, J.; Volovitch, M.; Vriz, S. Nerves Control Redox Levels in Mature Tissues Through Schwann Cells and Hedgehog Signaling. *Antioxid. Redox Signal.* **2016**, *24*, 299–311. [CrossRef] [PubMed]
155. Shin, J.; Berg, D.A.; Zhu, Y.; Shin, J.Y.; Song, J.; Bonaguidi, M.A.; Enikolopov, G.; Nauen, D.W.; Christian, K.M.; Ming, G.L.; et al. Single-Cell RNA-Seq with Waterfall Reveals Molecular Cascades underlying Adult Neurogenesis. *Cell Stem Cell* **2015**, *17*, 360–372. [CrossRef]
156. Knobloch, M.; Pilz, G.A.; Ghesquière, B.; Kovacs, W.J.; Wegleiter, T.; Moore, D.L.; Hruzova, M.; Zamboni, N.; Carmeliet, P.; Jessberger, S. A Fatty Acid Oxidation-Dependent Metabolic Shift Regulates Adult Neural Stem Cell Activity. *Cell Rep.* **2017**, *20*, 2144–2155. [CrossRef]
157. Knobloch, M.; Braun, S.M.; Zurkirchen, L.; von Schoultz, C.; Zamboni, N.; Araúzo-Bravo, M.J.; Kovacs, W.J.; Karalay, O.; Suter, U.; Machado, R.A.; et al. Metabolic control of adult neural stem cell activity by Fasn-dependent lipogenesis. *Nature* **2013**, *493*, 226–230. [CrossRef] [PubMed]
158. Li, L.; Candelario, K.M.; Thomas, K.; Wang, R.; Wright, K.; Messier, A.; Cunningham, L.A. Hypoxia inducible factor-1alpha (HIF-1alpha) is required for neural stem cell maintenance and vascular stability in the adult mouse SVZ. *J. Neurosci.* **2014**, *34*, 16713–16719. [CrossRef]
159. Kawase, K.; Nakamura, Y.; Wolbeck, L.; Takemura, S.; Zaitsu, K.; Ando, T.; Jinnou, H.; Sawada, M.; Nakajima, C.; Rydbirk, R.; et al. Significance of birth in the maintenance of quiescent neural stem cells. *Sci. Adv.* **2025**, *11*, eadn6377. [CrossRef]
160. Noguchi, H.; Castillo, J.G.; Nakashima, K.; Pleasure, S.J. Suppressor of fused controls perinatal expansion and quiescence of future dentate adult neural stem cells. *Elife* **2019**, *8*. [CrossRef]
161. Daynac, M.; Tirou, L.; Faure, H.; Mouthon, M.A.; Gauthier, L.R.; Hahn, H.; Boussin, F.D.; Ruat, M. Hedgehog Controls Quiescence and Activation of Neural Stem Cells in the Adult Ventricular-Subventricular Zone. *Stem Cell Rep.* **2016**, *7*, 735–748. [CrossRef]
162. Gonzalez-Reyes, L.E.; Chiang, C.C.; Zhang, M.; Johnson, J.; Arrillaga-Tamez, M.; Couturier, N.H.; Reddy, N.; Starikov, L.; Capadona, J.R.; Kottmann, A.H.; et al. Sonic Hedgehog is expressed by hilar mossy cells and regulates cellular survival and neurogenesis in the adult hippocampus. *Sci. Rep.* **2019**, *9*, 17402. [CrossRef]
163. Gherghina, L.-Y.; Tang, J.L.Y.; Brand, A.H. Quiescent neural stem cells transiently become 'neurons' to coordinate reactivation. *bioRxiv* **2024**. [CrossRef]
164. Deng, Q.; Tan, Y.S.; Chew, L.Y.; Wang, H. Msps governs acentrosomal microtubule assembly and reactivation of quiescent neural stem cells. *EMBO J.* **2021**, *40*, e104549. [CrossRef] [PubMed]
165. Truman, J.W.; Bate, M. Spatial and temporal patterns of neurogenesis in the central nervous system of Drosophila melanogaster. *Dev. Biol.* **1988**, *125*, 145–157. [CrossRef]
166. Gujar, M.R.; Gao, Y.; Teng, X.; Deng, Q.; Lin, K.Y.; Tan, Y.S.; Toyama, Y.; Wang, H. Golgi-dependent reactivation and regeneration of Drosophila quiescent neural stem cells. *Dev. Cell* **2023**, *58*, 1933–1949.e1935. [CrossRef]
167. Lin, K.Y.; Gujar, M.R.; Lin, J.; Ding, W.Y.; Huang, J.; Gao, Y.; Tan, Y.S.; Teng, X.; Christine, L.S.L.; Kanchanawong, P.; et al. Astrocytes control quiescent NSC reactivation via GPCR signaling-mediated F-actin remodeling. *Sci. Adv.* **2024**, *10*, eadl4694. [CrossRef]

168. Tiwari, S.K.; Mandal, S. Mitochondrial Control of Stem Cell State and Fate: Lessons From. *Front. Cell Dev. Biol.* **2021**, *9*, 606639. [CrossRef]
169. Anvarian, Z.; Mykytyn, K.; Mukhopadhyay, S.; Pedersen, L.B.; Christensen, S.T. Cellular signalling by primary cilia in development, organ function and disease. *Nat. Rev. Nephrol.* **2019**, *15*, 199–219. [CrossRef]
170. Singla, V.; Reiter, J.F. The primary cilium as the cell's antenna: Signaling at a sensory organelle. *Science* **2006**, *313*, 629–633. [CrossRef]
171. Khatri, P.; Obernier, K.; Simeonova, I.K.; Hellwig, A.; Hölzl-Wenig, G.; Mandl, C.; Scholl, C.; Wölfl, S.; Winkler, J.; Gaspar, J.A.; et al. Proliferation and cilia dynamics in neural stem cells prospectively isolated from the SEZ. *Sci. Rep.* **2014**, *4*, 3803. [CrossRef]
172. Han, Y.G.; Spassky, N.; Romaguera-Ros, M.; Garcia-Verdugo, J.M.; Aguilar, A.; Schneider-Maunoury, S.; Alvarez-Buylla, A. Hedgehog signaling and primary cilia are required for the formation of adult neural stem cells. *Nat. Neurosci.* **2008**, *11*, 277–284. [CrossRef] [PubMed]
173. Tong, C.K.; Han, Y.G.; Shah, J.K.; Obernier, K.; Guinto, C.D.; Alvarez-Buylla, A. Primary cilia are required in a unique subpopulation of neural progenitors. *Proc. Natl. Acad. Sci. USA* **2014**, *111*, 12438–12443. [CrossRef] [PubMed]
174. Breunig, J.J.; Sarkisian, M.R.; Arellano, J.I.; Morozov, Y.M.; Ayoub, A.E.; Sojitra, S.; Wang, B.; Flavell, R.A.; Rakic, P.; Town, T. Primary cilia regulate hippocampal neurogenesis by mediating sonic hedgehog signaling. *Proc. Natl. Acad. Sci. USA* **2008**, *105*, 13127–13132. [CrossRef]
175. Jeanson, L.; Copin, B.; Papon, J.F.; Dastot-Le Moal, F.; Duquesnoy, P.; Montantin, G.; Cadranel, J.; Corvol, H.; Coste, A.; Désir, J.; et al. RSPH3 Mutations Cause Primary Ciliary Dyskinesia with Central-Complex Defects and a Near Absence of Radial Spokes. *Am. J. Hum. Genet.* **2015**, *97*, 153–162. [CrossRef]
176. Coquand, L.; Victoria, G.S.; Tata, A.; Carpentieri, J.A.; Brault, J.B.; Guimiot, F.; Fraisier, V.; Baffet, A.D. CAMSAPs organize an acentrosomal microtubule network from basal varicosities in radial glial cells. *J. Cell Biol.* **2021**, *220*. [CrossRef] [PubMed]
177. Richard, C.A.; Seum, C.; Gonzalez-Gaitan, M. Microtubule polarity determines the lineage of embryonic neural precursor in zebrafish spinal cord. *Commun. Biol.* **2024**, *7*, 439. [CrossRef]
178. Jurisch-Yaksi, N.; Yaksi, E.; Kizil, C. Radial glia in the zebrafish brain: Functional, structural, and physiological comparison with the mammalian glia. *Glia* **2020**, *68*, 2451–2470. [CrossRef]
179. Alexandre, P.; Reugels, A.M.; Barker, D.; Blanc, E.; Clarke, J.D. Neurons derive from the more apical daughter in asymmetric divisions in the zebrafish neural tube. *Nat. Neurosci.* **2010**, *13*, 673–679. [CrossRef]
180. Barros, C.S.; Franco, S.J.; Müller, U. Extracellular matrix: Functions in the nervous system. *Cold Spring Harb. Perspect. Biol.* **2011**, *3*, a005108. [CrossRef]
181. Yeo, R.W.; Zhou, O.Y.; Zhong, B.L.; Sun, E.D.; Navarro Negredo, P.; Nair, S.; Sharmin, M.; Ruetz, T.J.; Wilson, M.; Kundaje, A.; et al. Chromatin accessibility dynamics of neurogenic niche cells reveal defects in neural stem cell adhesion and migration during aging. *Nat. Aging* **2023**, *3*, 866–893. [CrossRef]
182. Morizur, L.; Chicheportiche, A.; Gauthier, L.R.; Daynac, M.; Boussin, F.D.; Mouthon, M.A. Distinct Molecular Signatures of Quiescent and Activated Adult Neural Stem Cells Reveal Specific Interactions with Their Microenvironment. *Stem Cell Rep.* **2018**, *11*, 565–577. [CrossRef]
183. Codega, P.; Silva-Vargas, V.; Paul, A.; Maldonado-Soto, A.R.; Deleo, A.M.; Pastrana, E.; Doetsch, F. Prospective identification and purification of quiescent adult neural stem cells from their in vivo niche. *Neuron* **2014**, *82*, 545–559. [CrossRef]
184. Wang, D.Y.; Luo, A.F.; Bai, Q.R.; Gong, X.L.; Zheng, Y.; Shen, Q.; Hu, X.L.; Wang, X.M. VCAM1 Labels a Subpopulation of Neural Stem Cells in the Adult Hippocampus and Contributes to Spatial Memory. *Stem Cell Rep.* **2020**, *14*, 1093–1106. [CrossRef] [PubMed]
185. Porlan, E.; Martí-Prado, B.; Morante-Redolat, J.M.; Consiglio, A.; Delgado, A.C.; Kypta, R.; López-Otín, C.; Kirstein, M.; Fariñas, I. MT5-MMP regulates adult neural stem cell functional quiescence through the cleavage of N-cadherin. *Nat. Cell Biol.* **2014**, *16*, 629–638. [CrossRef] [PubMed]
186. Kokovay, E.; Wang, Y.; Kusek, G.; Wurster, R.; Lederman, P.; Lowry, N.; Shen, Q.; Temple, S. VCAM1 is essential to maintain the structure of the SVZ niche and acts as an environmental sensor to regulate SVZ lineage progression. *Cell Stem Cell* **2012**, *11*, 220–230. [CrossRef]
187. Van Camp, J.K.; Beckers, S.; Zegers, D.; Van Hul, W. Wnt signaling and the control of human stem cell fate. *Stem Cell Rev. Rep.* **2014**, *10*, 207–229. [CrossRef]
188. van Amerongen, R.; Bowman, A.N.; Nusse, R. Developmental stage and time dictate the fate of Wnt/beta-catenin-responsive stem cells in the mammary gland. *Cell Stem Cell* **2012**, *11*, 387–400. [CrossRef]
189. Bowman, A.N.; van Amerongen, R.; Palmer, T.D.; Nusse, R. Lineage tracing with Axin2 reveals distinct developmental and adult populations of Wnt/β-catenin-responsive neural stem cells. *Proc. Natl. Acad. Sci. USA* **2013**, *110*, 7324–7329. [CrossRef]
190. Austin, S.H.L.; Gabarró-Solanas, R.; Rigo, P.; Paun, O.; Harris, L.; Guillemot, F.; Urbán, N. Wnt/β-catenin signalling is dispensable for adult neural stem cell homeostasis and activation. *Development* **2021**, *148*. [CrossRef]

191. Simões, A.R.; Neto, M.; Alves, C.S.; Santos, M.B.; Fernández-Hernández, I.; Veiga-Fernandes, H.; Brea, D.; Durá, I.; Encinas, J.M.; Rhiner, C. Damage-responsive neuro-glial clusters coordinate the recruitment of dormant neural stem cells in Drosophila. *Dev. Cell* **2022**, *57*, 1661–1675.e1667. [CrossRef]
192. Sokol, S.Y. Maintaining embryonic stem cell pluripotency with Wnt signaling. *Development* **2011**, *138*, 4341–4350. [CrossRef]
193. Hikasa, H.; Sokol, S.Y. Phosphorylation of TCF proteins by homeodomain-interacting protein kinase 2. *J. Biol. Chem.* **2011**, *286*, 12093–12100. [CrossRef]
194. Ferrer-Vaquer, A.; Piliszek, A.; Tian, G.; Aho, R.J.; Dufort, D.; Hadjantonakis, A.K. A sensitive and bright single-cell resolution live imaging reporter of Wnt/ss-catenin signaling in the mouse. *BMC Dev. Biol.* **2010**, *10*, 121. [CrossRef]
195. Garcia-Corzo, L.; Calatayud-Baselga, I.; Casares-Crespo, L.; Mora-Martinez, C.; Julian Escribano-Saiz, J.; Hortiguela, R.; Asenjo-Martinez, A.; Jordan-Pla, A.; Ercoli, S.; Flames, N.; et al. The transcription factor LEF1 interacts with NFIX and switches isoforms during adult hippocampal neural stem cell quiescence. *Front. Cell Dev. Biol.* **2022**, *10*, 912319. [CrossRef] [PubMed]
196. Lindsey, B.W.; Aitken, G.E.; Tang, J.K.; Khabooshan, M.; Douek, A.M.; Vandestadt, C.; Kaslin, J. Midbrain tectal stem cells display diverse regenerative capacities in zebrafish. *Sci. Rep.* **2019**, *9*, 4420. [CrossRef]
197. Harrison, N.J.; Connolly, E.; Gascón Gubieda, A.; Yang, Z.; Altenhein, B.; Losada Perez, M.; Moreira, M.; Sun, J.; Hidalgo, A. Regenerative neurogenic response from glia requires insulin-driven neuron-glia communication. *Elife* **2021**, *10*. [CrossRef] [PubMed]
198. Hesse, R.G.; Kouklis, G.K.; Ahituv, N.; Pomerantz, J.H. The human ARF tumor suppressor senses blastema activity and suppresses epimorphic tissue regeneration. *Elife* **2015**, *4*. [CrossRef]
199. Radak, M.; Fallahi, H. The Epigenetic Regulation of Quiescent in Stem Cells. *Glob. Med. Genet.* **2023**, *10*, 339–344. [CrossRef]

MDPI AG
Grosspeteranlage 5
4052 Basel
Switzerland
Tel.: +41 61 683 77 34

*Biomolecules* Editorial Office
E-mail: biomolecules@mdpi.com
www.mdpi.com/journal/biomolecules

www.ingramcontent.com/pod-product-compliance
Lightning Source LLC
Chambersburg PA
CBHW061507210926
51850CB00008B/1846
*9783725879397*